INTRODUCTION TO SEDIMENTOLOGY

Introduction to Sedimentology

SUPRIYA SENGUPTA

Indian Institute of Technology
Kharagpur-721 302
India

CRC Press
Taylor & Francis Group
Boca Raton London New York

CRC Press is an imprint of the
Taylor & Francis Group, an **informa** business
A BALKEMA BOOK

Published by:
CRC Press/Balkema
P.O. Box 447, 2300 AK Leiden, The Netherlands
e-mail: Pub.NL@taylorandfrancis.com
www.crcpress.com – www.taylorandfrancis.com

ISBN 13: 978-90-5410-230-4 (hbk)

Visit the Taylor & Francis Web site at
http://www.taylorandfrancis.com

and the CRC Press Web site at
http://www.crcpress.com

Distributed in USA and Canada by: A.A. Balkema Publishers

To
The Memory of
My Mother

PREFACE

Sedimentology covers only a small part of the geology curriculum in most universities. Students of geology naturally do not find the time necessary to master comprehensive texts in sedimentology. While gathering piecemeal information from specialised texts they also often fail to appreciate the connections between the various aspects of the discipline.

The purpose of this book is to present a concise account of all the major branches of sedimentology (except chemical and biogenic structures) and to highlight the connecting links between them. The presentation is based on my conviction that in spite of the sophistications introduced in the current sedimentological literature, much of the subject can still be learnt by mastering a few basic principles. Once these have been grasped, it should not be difficult, following the references cited, to obtain specialised knowledge on any selected topic.

This book is expected to cater to the needs of students at various levels. The first four chapters are written in a style suitable for the undergraduate students. The first few sections of Chapters 7 (Tectonics and Sedimentation) and 8 (Stratigraphy and Sedimentation) should also be preferably taught at this level. The rest of the book is meant for those at the graduate level.

A book of this type could not have been prepared without borrowing heavily from published sources. I am grateful to the authors and publishers who have freely permitted me to reproduce information from their publications. Their names appear in appropriate places in the text. Dr. Indranil Banerjee of the Institute of Sedimentary and Petroleum Geology, Calgary, Canada, critically read an earlier version of the manuscript, provided valuable suggestions, and also made available to me a large volume of current sedimentological literature. I am indebted to him for all his help. I have also profited through discussions with Professor K. Naha of the Indian Institute of Technology (IIT), Kharagpur who reviewed some of the chapters.

Others who provided information during preparation of the manuscript and read earlier versions of the topics mentioned against their names include: Dr. Sibdas Bandyopadhyay (Indian Statistical Institute, Calcutta; Statistical Analysis), Dr. Abhijit Basu (Indiana University, Bloomington, USA; Provenance), Prof. Ajit K. Bhattacharyya (Jadavpur University, Calcutta; Depositional Environments), Prof. Amitabha Chakrabarti (IIT, Kharagpur; Sedimentary Structures, Tidal Flats), Prof. R.L. Folk (University of Texas, Austin, USA; Carbonate Sediments), Prof. G.M. Friedman (Northeastern Science Foundation, Inc., USA; Sedimentary Textures), Dr. Sibdas Ghosh (Geological Survey of India, Calcutta), Prof. Mandakini Majumdar (IIT, Kharagpur; Hydraulics), Prof. G.V. Middleton (McMaster University, Hamilton, Canada; History of Sedimentology), Prof. Asoke Mookherjee (IIT, Kharagpur), Prof. H. Okada (Kyushu University, Japan; Siliciclastic Sedimentary Rocks), Prof. H.G. Reading (University of Oxford, Oxford, U.K.), Prof. K.K. Roy (IIT, Kharagpur; Wireline Logging) Prof. S.K. Sen (IIT, Kharagpur), Dr. R.N. Singh and Dr. R. Srinivasan (National Geophysical Research Institute, Hyderabad; Tectonics and Sedimentation) Prof. Å Sundborg (Uppsala University, Uppsala, Sweden; Hydraulics and Sediment Transportation). Profs. A. Chakrabarti, C.N. Rao and D.P. Sen of IIT, Kharagpur, and Dr. D.K. Saha of ISI, Calcutta allowed me to use thin sections and photographs from their unpublished collections. Dr. B. Mishra helped in photomicrography. Drs. Abha Chatterjee and B.P. Sandilya provided translations for the French and German terms respectively. I am grateful to all those mentioned above.

The contents of this book are based on the courses given by me at various levels at IIT, Kharagpur. I recall with appreciation the comments and suggestions received from my students from time to time. Some of my M. Tech students, particularly, Asesh Kumar Maji, assisted in copy editing, indexing and computational work at different stages. My wife Ila (Sujata) took care of my mundane problems thereby leaving me free to work on the manuscript.

A large part of the cost of preparation of the manuscript was defrayed by the Curriculum Development Cell of IIT, Kharagpur under the Quality Improvement Programme (QIP). It is a pleasure to acknowledge the help and co-operation received from Prof. G.L. Datta and the QIP staff at various stages of preparation of the manuscript. Mr. Subrata Giri typed the final version of the manuscript. Mr. Tapan Sarkar drew most of the text figures. Mr. S. Ghosh provided photocopies of the halftone illustrations. But for the persuasion and interest of the staff of M/s Oxford & IBH Publishing Co. Pvt. Ltd., this work would not have been undertaken. I thank all of them.

Department of Geology & Geophysics
IIT, Kharagpur

Supriya Sengupta

CONTENTS

1

INTRODUCTION

SCOPE OF SEDIMENTOLOGY

Sedimentology deals with the processes and products of sedimentation. Sediments are produced either by disintegration and alteration of pre-existing rocks or by precipitation from solution. The particles ejected out of volcanoes and the dust particles of cosmic origin also add to the sediment mass of the earth in a limited way. Sediments are transported by running water, wind or moving ice to various depositional environments. The process of sediment transportation is often accompanied by the production of rhythmic bedforms or more complex structures which, when well preserved, provide clues to the palaeoenvironment and palaeocurrent.

Sediments produced out of mechanical or chemical processes are consolidated into sedimentary rocks by the pressure of overburden, recrystallisation and cementation. Textures of sedimentary rocks bear the imprint of the nature of the changes (diagenesis) undergone during consolidation of sediments into sedimentary rocks. The whole process of rock decay, sediment transportation, deposition, precipitation and diagenesis takes place at or near the surface of the earth at normal pressure-temperature conditions (Fig. 1.1). This distinguishes the sedimentary processes from the igneous and metamorphic processes where a higher order of temperature and pressure is involved. Sedimentary rocks cover wide areas of the earth's surface although they account for only about 5% of the crustal volume. Thus they form only a thin veneer on the outermost part of the earth's surface. The average thickness of the sediment cover on the continental crust is about 2 km but in the ocean basin only about 1 km.

Sediments laid down in layers within depressions (basins) in the earth's crust are preserved to constitute a stratigraphic record. The ultimate aim of a sedimentological study is to unravel the chain of events responsible for production of particular stratigraphic sequences.

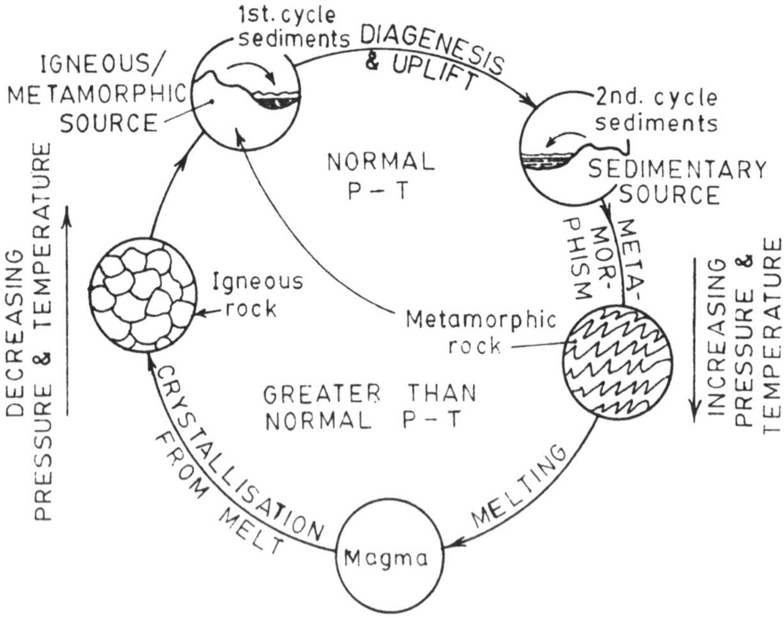

Fig. 1.1: Modern version of the fundamental rock cycle conceived by James Hutton nearly 200 years ago (modified and redrawn after Siever 1983).

HISTORY OF DEVELOPMENT

The Early History

The pioneers responsible for laying the foundation of sedimentology were not geologists by profession. The basic principles of stratigraphy and sedimentation developed out of studies by naturalists belonging to different disciplines. In the mid-17th century, a Danish physician-cum-clergyman, Nicolaus Steno (1638–1687), noticed that layers of sediments are always laid down in water in a sequence in which the oldest one lies at the bottom and the youngest one on top. This observation led to the formulation of a very fundamental rule known as the Law of Superposition. Steno's second law, called the Law of Original Horizontality, states that the primary bedding of the layers of sediments laid down in water always parallels the surface of the earth. His third law, the Law of Original Continuity, states that all water-laid strata, must continue laterally. A corollary to this law, which came to be recognised by the eighteenth century, states that a truncation of an original sedimentary layer implies removal of the original sediments, either by erosion or due to faulting.

These laws, which for centuries provided the basis for geological mapping and stratigraphic correlation, no doubt have their limitations. For example, ripple migration on a sediment bed may cause the original layers to be deposited at an angle to the horizontal plane. Similarly, variations in depositional condition at the time of sedimentation (facies change) can interrupt the lateral continuity of sediment layers.

Surprisingly, Steno's fundamental laws did not find practical application for nearly a century. Johann Gottlob Lehmann is believed to be one of the first to have applied the Law of Superposition to large-scale geological mapping in parts of Germany in the mid-18th century. In 1815, a British civil engineer, William Smith (1769–1839), produced the first stratigraphic section from the records maintained at the construction site. His field studies also led to the production of the first geological map of Britain.

In 1785, James Hutton (1726–1797), a Scottish physician, recognised the cycle of weathering, erosion and transportation by running water (see Fig. 1.1). He also observed that after consolidation the sediments laid down in the sea produce stratified deposits. According to many, Hutton's appreciation of the immensity of geological time and his rejection of supernaturalism in providing explanations for geological phenomena, marked the beginning of rational thinking in geology. Hutton's ideas were publicised during his lifetime by his friend, John Playfair, and nearly half a century later, by Charles Lyell (1797–1875). Lyell, who expanded and modified the Huttonian thesis, was quick to recognise that the changes that have taken place in the geological past can be explained by the processes that are taking place today. In other words, the laws of nature guiding the earth processes have remained invariant, although the earth itself has changed with time. This concept of uniformity in the earth processes came to be known as the Principle of Uniformitarianism.

With time, many corollaries of uniformitarianism like 'the present is a key to the past' (a phrase coined by Geikie in 1882), developed from the original Huttonian-Lyellian percept. Some of these, such as 'the rates of the earth processes have remained constant', or that uniformitarianism should be called 'actualism' because it refers to actual or real events, have been challenged by later workers on the ground that they are fallacious. In spite of these criticisms it must be remembered that the Huttonian-Lyellian concept has been the basis for most of our geological thinking. Many of the so-called fallacious concepts developed from the beliefs of later workers and can hardly be traced to the original writings of James Hutton or Charles Lyell.

James Hutton was dedicated to the 'volcanist' idea that the internal heat of the earth is wholly responsible for the earth processes. This was in marked contrast to the 'Neptunist' concept, patronised by Abraham Gottlob Werner (1749–1817), that all rocks except recent volcanic lavas are produced from chemical or mechanical deposits in a universal ocean. So profound was the influence of the latter school on contemporary geologists that the idea of lateral change in depositional events (facies change), introduced by Gressly in 1833, did not find easy acceptance either in Europe or in North America for a long time. Geologists were still under the spell of the Neptunist belief that all rock layers initially laid down in a universal ocean, must also be continuous, without change.

An important event in mid-19th century was the observation by James Hall in the Appalachian mountains of North America that the thickest sediments accumulate in long, linear belts of subsidence in the earth's crust. This observation eventually led to the formulation of the concept of geosyncline, a model that dominated the sedimentological world for nearly a century.

Developments Since Late 19th Century

The late nineteenth and early twentieth centuries witnessed rapid developments. Studies by pioneers such as Henry Clifton Sorby (1826–1908) in England led to the foundation of sedimentology as we know it today. Sorby was not only interested in the study of sedimentary structures and their implications, but was also one of the first to make extensive use of the petrographic microscope in the study of sedimentary rocks. He produced two classical volumes on sedimentary petrography and also a monograph on chertified limestones. A few years later, another important study on lithogenesis, by Johannes Walther, appeared in Germany. Walther also formulated the Law of Facies Succession which still bears his name (see Chapter 8).

The first half of the twentieth century witnessed the initiation of specialised studies in many different branches of sedimentology. Systematic studies on the processes of sedimentation and their bearing on stratigraphy were initiated in North America by Grabau, Barrell and Schuchert. Studies by W.M. Davis (1850–1934) and G.K. Gilbert (1843–1918) and later by Twenhofel (1875–1957), established a close link between surface processes of the earth, sedimentation and stratigraphy. This marked the beginning of the use of the present as a key to the past. In 1944 sedimentologists' interests in modern sediments were revived with the publication of studies on the Mississippi delta by H.N. Fisk and his associates. Classical studies on modern eolian deposits (Bagnold 1941; McKee 1979), modern fluvial deposits (Hjulström 1935; Sundborg 1956; Leopold, Wolman and Miller 1964), and modern tidal sediments (Reineck since 1952, see Reineck and Singh, 1980; van Straaten 1954, 1954a; Ginsberg 1975) have been emulated by sedimentologists throughout the world for nearly half a century now (see Chapter 6).

Petrographic studies initiated by Sorby in England and Walther in Germany were continued, among others, by Grabau, Krynine and their students in North America (Chapter 4). The basic ideas on textural parameters of sediments were formulated during the first three decades of the present century (Udden 1914; Wentworth 1922; Wadell 1932, 1935; Zingg 1935;

see Chapter 3). This was also the time when many of the procedures for sedimentological analyses were standardised.

The trend of laboratory simulation of sedimentary processes, initiated in France by Daubrée around 1870, was revived by G.K. Gilbert (1914) in North America. Since that time many of the hydraulic laboratories engaged in experimental studies on sedimentary processes have contributed greatly to our understanding of the basic processes involved in sedimentation. That such studies can be of importance to the perception of many fundamental geological problems was demonstrated in the Netherlands by Kuenen and Migliorini (1950). This was followed by many more contributions to experimental sedimentology throughout the world (see Chapter 5).

Vant Hoff's experiments on marine carbonates in the early years of this century and Correns' experimental studies around 1920 in Germany, marked the beginning of sedimentary geochemistry. This field of study is presently actively pursued, particularly in North America (see Krumbein and Garrels 1952; Garrels and MacKenzie 1971).

A major event of the present century was the revival of interest in palaeocurrent research, initiated in the nineteenth century in England by H.C. Sorby. Palaeocurrent studies by Brinkmann and also by Cloos in Germany around 1930 were rejuvenated in North America (see Pettijohn 1962), to be followed by development of many sophisticated techniques in palaeocurrent research (see Chapter 9).

The studies on tectonics and sedimentation, initiated in North America by James Hall around 1850, were pursued by Shvetsov in Russia, Bertrand in France and many others throughout the continent. Oceanographic expeditions, mainly by the Dutch sedimentologists and geophysicists in the East Indies provided valuable input for refinement of the geosynclinal model—ideas which were pursued and elaborated in North America (see Kay 1951). These ideas, together with the classical facies concept, which found acceptance in North America around 1930 and the petrographic studies initiated by Krynine 1948, provided the background for much of what was achieved during the first half of the twentieth century.

Recent Trends in Sedimentology

Developments until about the first half of the present century were mainly the outcome of individual efforts. The end of the Second World War saw a major change in this trend. Large-scale oceanographic expeditions by the major nations in the post-war period, development and deployment of submersibles, deep-toe equipment, side-scan sonar and undersea TV camera in subsea exploration, large-scale drilling operations at sea and refinement of core recovery techniques since 1950, have provided information which was heretofore unobtainable. Much of what is known today about the siliciclastic and carbonate deposits of shallow seas has been revealed by these explorations.

Discovery of giant petroleum reserves in carbonate rocks provided impetus for the study of carbonate petrography and carbonate depositional environments. Intensive research on carbonates, initiated by the major oil companies, led to the discovery of what is known today about carbonate petrography and facies. An equally intensive investigation of the present-day carbonate depositional environments in the Bahamas, Florida and the Persian Gulf also took place.

Exploration for petroleum opened up new dimensions in organic geochemistry also. Techniques were developed for working out the maturation and thermal history of sedimentary basins. Deployment of sophisticated equipment such as X-ray diffraction (XRD), X-ray fluorescence (XRF), scanning electron microscope (SEM), electron microprobe (EM), cathodo-luminescence (CL) made investigation of fine-grained sediments and sedimentary rocks with a high degree of precision possible. Isotope studies (both stable and radio isotopes) provided clues to environmental interpretation. Application of mathematical and statistical techniques led to the investigation of subtle interdependence between lithological units and chemical constituents within stratigraphic sequences.

The ideas developed during the last few decades have been revolutionary in many ways. Information available from hundreds of thousands of boreholes drilled in search of hydrocarbon has diverted the attention of sedimentologists to the subsurface. Techniques have been developed for recognition of sedimentary facies from well logs, cores and well cuttings. Developments in the field of trace fossil study (ichnology) now allow interpretation of depositional environment from information which, until recently, was considered insufficient. The current emphasis is on facies modelling from both surface and subsurface data. Developments in the field of seismic exploration have introduced concepts such as seismic and sequence stratigraphy (see Chapter 8). New concepts such as allostratigraphy and autostratigraphy have also been introduced. Simultaneously, the traditional ideas on basin development are being modified from the point of view of plate tectonics.

Source

Albritton 1963; Dunbar and Rodgers 1957; Gilluly, Waters and Woodford 1975; Hubbert 1967; Mather and Mason 1939; McIntyre 1963, Miall 1978; Middleton 1973, 1978; Nelson 1985; Shea 1982, 1983; Zenger 1986.

LITERATURE ON SEDIMENTOLOGY

In the early part of the present century there were hardly any text and reference books on sedimentology available. Since 1950 the figure has been increasing almost exponentially. Table 1.1 lists the more important text and reference books on sedimentology under three heads: 1) those dealing with general aspects of sedimentary processes and products, 2) those dealing with specialised sedimentological topics and 3) those dealing with methodology for sedimentological research. All the texts mentioned in this table were written in English. Needless to say, many other well-written texts on sedimentology exist, particularly in the various European languages.

The results of current research on sedimentological topics are available in periodicals. Of these, the following three, published in the English language, are devoted wholly to sedimentology: Journal of Sedimentary Petrology (published since 1931 by the Society of Economic Paleontologists and Mineralogists (renamed in 1989 as the SEPM Society of Sedimentary Geology), Tulsa, Oklahoma, USA; Sedimentology, the Journal of the International Association of Sedimentologists (I.A.S.), (published since 1962) and Sedimentary Geology (published since 1967 by Elsevier Publishing Company, Amsterdam). These three journals together publish more than 200 large research papers on sedimentological topics annually.

Sedimentological research papers are also regularly published by many other geological journals. Notable among them, in the English language, are: Bulletin of the American Association of Petroleum Geologists (A.A.P.G.), Bulletin of the Geological Society of America, Marine Geology (Elsevier, Amsterdam), Journal of Geology (Chicago), Geological Journal (Wiley), Lithos (Oslo), Journal of Petroleum Science and Engineering (Elsevier), Palaeogeography, Palaeoclimatology, Palaeoecology (Elsevier), Basin Research (Blackwell), Canadian Journal of Earth Science (Ottawa) and Geology Today (Blackwell). Several memoirs, special publications, short course notes, reprint series of A.A.P.G., S.E.P.M., and I.A.S. and the Developments in Sedimentology Series of Elsevier are devoted wholly to sedimentological topics. The Lecture Notes Series (Springer) are also often of interest to sedimentologists.

Except for the occasional publications of the Indian Association of Sedimentologists there is no Indian journal devoted wholly to sedimentology. All the Indian geological journals, however, frequently publish articles on sedimentology. Notable among them are: Indian Journal of Geology (formerly, Quarterly Journal of the Geological, Mining and Metallurgical Society of India) published from Calcutta for over 60 years; Journal of the Geological Society of India (published from Bangalore since 1960) and Indian Journal of Earth Sciences (published since 1974 from Calcutta).

Table 1.1. Important text and reference works on sedimentology

			1910
		Hatch & Rastall (1913)	
			1920
	Milner (1922)		
		Twenhofel (1928)	
			1930
		Twenhofel (1932)	
	Krumbein & Pettijohn (1938)		1940
		Shrock (1948) Pettijohn, 1st ed. (1949)	
Kuenen (1950)		Twenhofel (1950)	1950
		Krumbein & Sloss, 1st ed. (1951) Dunbar & Rodgers (1957) Pettijohn, 2nd ed. (1957)	
Carozzi (1960)			1960
	Milner, 2 vols. (1962)		
Potter & Pettijohn, 1st ed. (1963)		Krumbein & Sloss, 2nd ed. (1963)	
	Griffiths (1967)		
		Folk (1968)	
	Bouma (1969)		
Kukal (1970)			1970
	Carver (1971)		
Garrels & MacKenzie (1971) Pettijohn, Potter & Siever (1972) Bathurst (1975), Wilson (1975) Potter & Pettijohn, 2nd ed. (1977) Reading, 1st ed. (1978) Walker, 1st ed. (1978)		Blatt, Middleton & Murray, (1972) Reineck & Singh, 1st ed. (1973) Pettijohn, 3rd ed. (1975) Selley (1976) Friedman & Sanders (1978)	
		Reineck & Singh 2nd ed. (1980) Galloway & Hobday (1983)	1980
Miall (1984) Walker, 2nd ed. (1984) Reading, 2nd ed. (1986) Shannon & Naylor (1989) Miall, 2nd ed. (1989)			1990
Syvitski (1991)			
Text on specialised sedimentological topics	Texts on sedimentological techniques	Texts on sedimentary processes and products	

2

PROCESSES OF SEDIMENTATION

SURFACE PROCESSES AND ROCK WEATHERING

Sedimentary processes are largely controlled by just two agencies, water and wind. Thus the hydrological and wind cycles play very important roles in sedimentation. The hydrological cycle is schematically represented in Fig. 2.1.

About two-thirds of the precipitation that takes place on land finds its way back to the atmosphere through evaporation and transpiration. The rest flows back to the sea mostly as surface run-off through streams and partly as ground-water. The actual amount of water involved in the different phases of movement varies from place to place, depending on local geography and climate but by and large a balance in the water budget is maintained.

Water is also responsible to a large extent for weathering and decomposition of pre-existing igneous, metamorphic and sedimentary rocks. The actual amount of sediment yield depends not only on precipitation, but also on surface conditions, types of rocks exposed and density of vegetation cover. Rock weathering can take place by mechanical, chemical and biological actions. Freezing of water percolating into fractures and joints of rock bodies causes expansion and disintegration in cold climate (frost heaving).

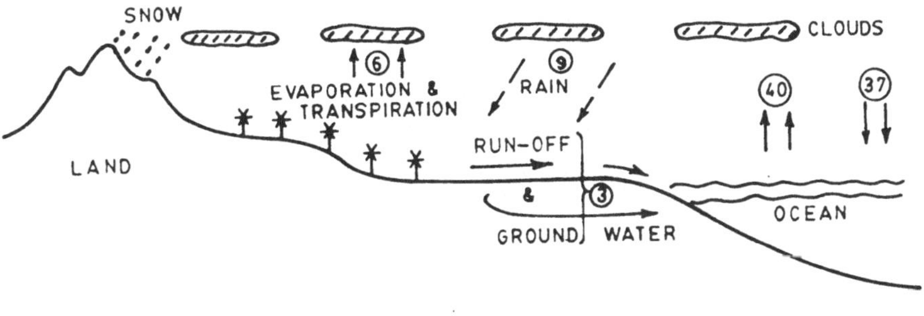

Fig. 2.1: Schematic representation of the hydrological cycle. The balance between evaporation and precipitation in an ideal tropical region is indicated by arrows. The figures within the circles (modified after Siever 1983) give an approximate idea of the proportion of water flow per year.

Earthquakes, the heat of lightning and the impact between large bodies of displaced rock masses are some of the other causes of mechanical disaggregation. In drier climate, crystallisation of soluble salts within cracks and fissures can disrupt rock bodies. Mechanical abrasion and deflation are important erosive processes in deserts.

Much of the actual disintegration of rocks takes place by chemical weathering. The process is particularly effective in tropical, humid climate where rainy seasons follow the summer months. Dissolved carbon dioxide ionises rain-water into carbonic acid. Rain-water, therefore, is generally acidic, with the pH ranging between 4 and 7. Lightning discharges and industrial contamination, particularly sulphurous gases, also produce some acids although in small quantities. The acidic water can take much of the natural carbonate into solution. Humus in the soil also makes natural limestones unstable. Oxidation of organic material in soils also releases CO_2, thereby producing an acidic microenvironment with a pH as low as 2.

Chemical weathering of silicate minerals is a much more complex process. In carbonated water, feldspars break down into clay minerals, silicic acid and carbonates (Holmes and Holmes 1978):

$$\underset{}{6H_2O} + CO_2 + \underset{\text{orthoclase}}{2KAlSi_3O_8} = \underset{\text{kaolinite}}{Al_2Si_2O_5(OH)_4} + \underset{\text{silicic 'acid'}}{4SiO(OH)_2} + K_2CO_3$$

Similarly, albite also produces kaolinite:

$$\underset{\text{albite}}{2NaAlSi_3O_8} + 2CO_2 + 11H_2O = \underset{\text{kaolinite}}{Al_2Si_2O_5(OH)_4} + 2Na + 2HCO_3 + \underset{\text{silicic acid}}{4H_4SiO_4}$$

Kaolinite, in fact, is often the first mineral to form as a product of chemical weathering. Depending on the chemical composition of the source rock being weathered, clay minerals such as smectite, illite, chlorite, and other oxides and hydrates may be produced. Pyrophyllite may be produced by breaking down of plagioclase.

Much of the soda generated by the process of chemical weathering is transported to the sea by running water, but potash remains in the soil to be absorbed later either by plant roots or by clay minerals. The average dissolved silica content of the world's rivers is about 13 ppm. Much of this silica possibly comes from feldspar rather than quartz. Solubility of silica increases with an increase in temperature and alkalinity of the medium. With a decrease in temperature and alkalinity, much of the dissolved silica precipitates in the intergranuiar space of sediments to provide bondage between the grains.

The trend of decomposition of the ferromagnesian minerals can be illustrated by the following example:

$$\underset{\text{diopside}}{H_2O + CO_2 + Ca(Mg{\cdot}Fe)\,(SiO_3)_2} \longrightarrow \underset{\text{silicic 'acid'}}{2SiO(OH)_2} + \text{bicarbonates of Ca, Mg, Fe}$$

Similarly, chemical decomposition of biotite, hornblende and augite leads to the formation of clay and chloritic minerals. Bicarbonate of iron $Fe(HCO_3)_2$ is converted to limonite $Fe_2O_3{\cdot}H_2O$ in the ground-water zone on coming into contact with oxygenated water (see Fig. 2.2A). Ferrous iron released from minerals during weathering is oxidised in soil to a ferric state. Common hydrated iron oxide minerals include limonite ($Fe_2O_3{\cdot}H_2O$), goethite ($FeO{\cdot}OH$), lepidocrocite ($FeO{\cdot}OH$) and anhydrous dimorphs of hematite (n. Fe_2O_3). These minerals, very stable under oxygenated condition, continue to exist in the soil almost indefinitely (Fig. 2.2B). Highly colourful, they impart yellow, orange or brownish red tones to the weathered soil surface. Some of these minerals also provide bondage between loose grains of sediments as ferruginous cement.

Most of the sediments produced on the exposed surface of the earth find their way to the sea through rivers. Chemical analysis of sea-water, however, does not exactly match the composition of average river-water. This is because some of the chemicals arriving in the ocean are precipitated there. The time during which an element stays in the sea-water (called residence time) can be obtained by dividing the amount of an element present in the ocean by the amount of that element (tons/year) contributed by rivers to the sea. Sea-water contains large proportions of sodium chloride because the residence time of Na^+ and Cl^- are of the order of 100 m.y. Since the residence times of Ca^{++} and Mg^{++} are much smaller (1 m.y. and 10 m.y. respectively), these elements are absorbed by marine organisms such as corals to build up their shells and tests. These shells eventually produce extensive coral colonies.

In the deeper parts of the ocean, where the temperature is low, pressure is high and the water is acidic due to dissolved gases, the carbonate is dissolved. This condition is likewise suitable for precipitation of dissolved silica, which has an extremely small residence time. Silica precipitation in deep ocean-waters in the form of chert is aided by microscopic marine organisms such as radiolaria and diatoms. Much of the deep ocean bottom is covered by siliceous ooze and also by extremely fine-grained, muddy sediments derived from land. It must be remembered that chemical precipitation is not confined to oceanic areas only. Natural springs, playas and caverns are some areas on land where chemical precipitation takes place.

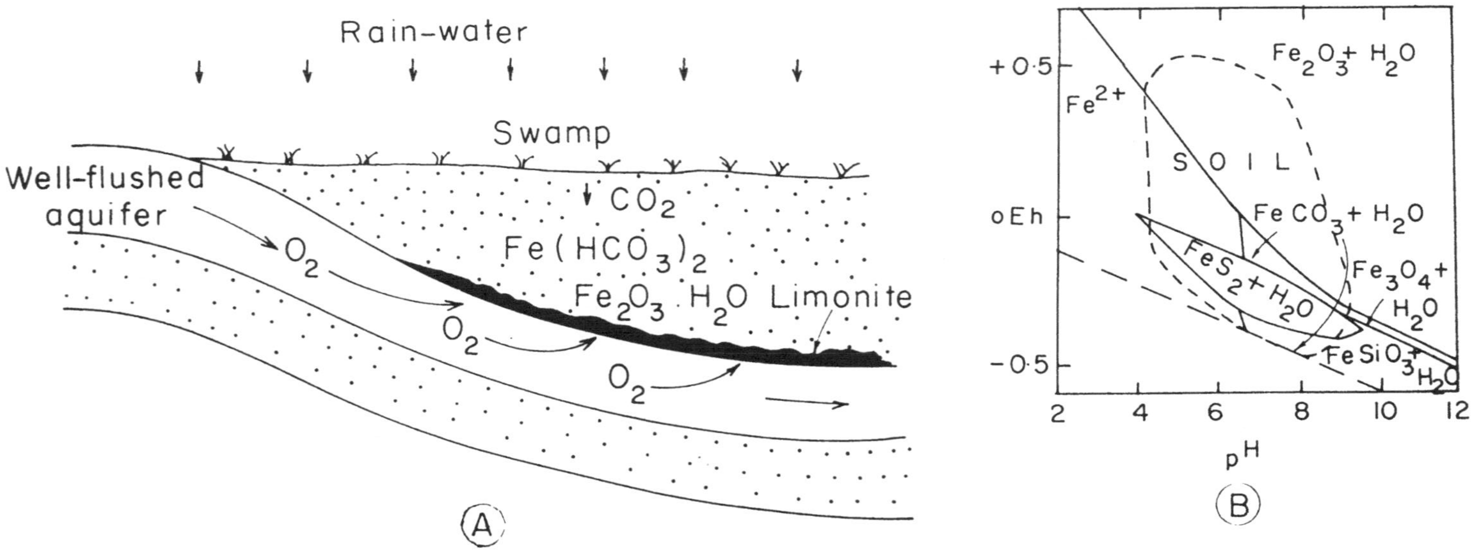

Fig. 2.2: (A) Diagrammatic representation of the process of limonitisation by oxidation of bicarbonate of iron in the ground-water zone (modified after Borchert 1960). (B) Stability fields of iron compounds under surface soil condition (partly redrawn from Garrels and Christ 1965).

Table 2.1. Relative stability of common rock forming silicates (modified after Goldich 1938)

Most stable	Quartz (200)	
	Muscovite (170)	
	Hornblende (160)	
		K-Feldspar
	Biotite (150)	Alkalic plagioclase (150–160)
	Augite (150)	Calcic plagioclase
Least stable	Olivine (133)	

N.B. Figures in parentheses indicate the proportion of anion to a cation content of 100 in each case.

MINERAL STABILITY

The process of weathering effects considerable change in source minerals of sediments. The degree to which a mineral can resist this change determines its stability.

Interestingly, the stability series of rock forming silicates shows a trend almost opposite to that of crystallisation as indicated by Bowen's reaction relation series (Table 2.1). For example, olivine, the earliest to crystallise from a melt, is also the first to be altered during chemical weathering. Quartz, the mineral last to crystallise, can survive chemical decay almost indefinitely (Goldich 1938).

The explanation of this lies in the anion to cation ratios of the minerals. The proportions of negative ions (anion) for the common rock-forming silicates are shown in parentheses in Table 2.1, assuming that the cation content is 100 in each case. The higher the anion proportion of a mineral, the lower its susceptibility to chemical weathering, which essentially is a process of oxidation or hydration (introduction of O^{2-} or OH^-). Thus olivine with a relative anion content of 133 will be the first to be affected by chemical weathering, whereas quartz, with a relative anion content of 200, will not change at all. Under this condition the process of weathering leads to the production of increasingly stable minerals. Two common examples of this process are:

1) olivine (133) → chlorite (180) or serpentine (180) → limonite (200).

2) feldspar (160) → hydromica (200) → clay minerals (225) → bauxite (250–300).

(figures in parentheses indicate proportions of anions relative to a cation content of 100 in each case).

The actual amount of chemical alteration depends not only on the chemical composition and crystal structure of a mineral, but also on local climate, mainly temperature and humidity, as well as the chemistry of the pore fluid.

The problem of mineral stability can also be studied from the point of view of thermodynamics. In a chemical reaction the sum of free energies of formation of all the reaction products minus the free energies of the reactants measures the free energy change of the reaction. The more negative the value of free energy change of a mineral, the more unstable it is. Hence its chances of reaction to produce a more stable product are also higher.

In the stability series obtained by Curtis (1976), according to this principle olivine (most negative) will alter first and muscovite (least negative) will resist alteration for a long time. The common rock-forming silicates, arranged in order of decreasingly negative free energy, are as follows: olivine, pyroxene, amphibole, plagioclase, microcline, muscovite. The trend is the same as that obtained by Goldich (1938).

While free energy indicates the susceptibility of a mineral to chemical reaction, the physical stability of a mineral can be expressed in terms of its resistance to abrasion. The 'durability' of a mineral has been experimentally shown to be related to its hardness. The scale of resistance to abrasion, as worked out by Thiel (1945), for example, is as follows (starting with the mineral having least resistance): barite, siderite, fluorite, goethite, enstatite, kyanite, hematite, augite, apatite, hypersthene, rutile, hornblende, zircon, epidote, garnet, staurolite, microcline, tourmaline, quartz.

Many minerals in this list are 'heavy minerals', that is, they have a specific gravity higher than that of quartz ($\rho = 2.65$) or feldspar ($\rho = 2.54-2.76$). For practical purposes, all minerals having a specific gravity higher than that of bromoform (2.87 at 20°C) are listed as heavy minerals. According to stability the heavy minerals may be grouped as follows (modified by Pettijohn 1975, after Sindowski 1949):
Most stable: zircon, rutile, tourmaline.
Intermediate: staurolite, kyanite, hornblende.
Least stable: (garnet), augite, apatite, olivine.
Because of their stability, zircon, rutile and tourmaline survive prolonged weathering and concentrate in the older soil profiles. Garnet is generally considered to be more stable than that shown here.

SOURCE OF SEDIMENTS

The sediments derived directly from crystalline igneous and metamorphic rocks are the first-cycle sediments. Sediments produced out of lithified first-cycle sediments are termed reworked (polycyclic) sediments (see Fig. 1.1). Reworking can be detected from textural evidence. During the process of cementation of a first-cycle quartz sand precipitation of silica on detrital quartz grains may produce overgrowths. If the edges and corners of the overgrowth on a quartz grain of first-cycle origin are found to be well rounded, we know that the grain of quartz is of second-cycle origin.

The source area and the source rock of a sediment and also, according to some authors, the climate, relief and the environment prevailing at the source, are collectively called its provenance (French: *provenier* meaning 'to originate or to come from'). The mineralogical constituents of terrigenous sediments bear the imprint of the source rock and also, to some extent, its climate and relief. Quartz, feldspar and heavy minerals are particularly suitable for deciphering provenance.

Quartz and Feldspar as Provenance Indicators

Quartz, the most abundant detrital constituent in sedimentary rocks, is often used as an indicator of provenance, but opinion differs about the specific characters of a quartz grain which are indicative of its origin. Undulatory extinction ('strain shadows'), traditionally taken to be indicative of a metamorphic source, has been found in quartz grains of plutonic origin also (Blatt and Christie 1963). Acicular inclusions, long believed to be a characteristic of quartz of granites, have also been detected in metamorphic rocks. No significant difference between shapes of quartz grains of igneous and metamorphic origin have been statistically established, although this was, at one time, believed to be an important discriminating criterion. A group of geologists at Indiana University, after a thorough review of the problem, came to the conclusion that strongly undulose and finely polycrystalline quartz grains are indicative of metamorphic source rocks, but the quartz grains of plutonic and volcanic origin are characteristically strain free (Basu 1985). The Indiana group developed a method of discriminating sedimentary rocks of plutonic, low-, and high-rank metamorphic parentage by plotting the proportions of the following four parameters in a 'diamond diagram': undulatory quartz, non-undulatory quartz, polycrystalline quartz, and the number of crystal units per polycrystalline grains (Fig. 2.3).

As it is very stable, quartz survives prolonged weathering and erosion. The ratio of quartz to feldspar (+ unstable rock fragments) can therefore be used as an index of mineralogical maturity of a sediment of mechanical origin. Comparative studies of sand populations transported under various climatic and environmental regimes have shown that a unique combination of extreme climate, relief, transportation and rate of sedimentation is necessary for elimination of all the unstable fractions leading to a highly matured, first-cycle quartz sand. This imprint of climate, although preserved for the first 75 km of transportation in high-gradient streams, is rapidly destroyed as soon as a high-energy marine environment is reached. For this reason, palaeoclimatic interpretation from mineralogical composition is feasible only for terrigenous, first-cycle sand (Suttner et al. 1981).

Feldspar, which ranks only next to quartz in abundance, may also act as an indicator of provenance. The chemical composition of feldspar may be useful for this purpose. Homogeneous alkali feldspars more sodic than Ab_{50} are invariably of volcanic origin. Alkalic feldspars more potassic than Or_{88} are derived from plutonic and metamorphic sources (Trevena and Nash

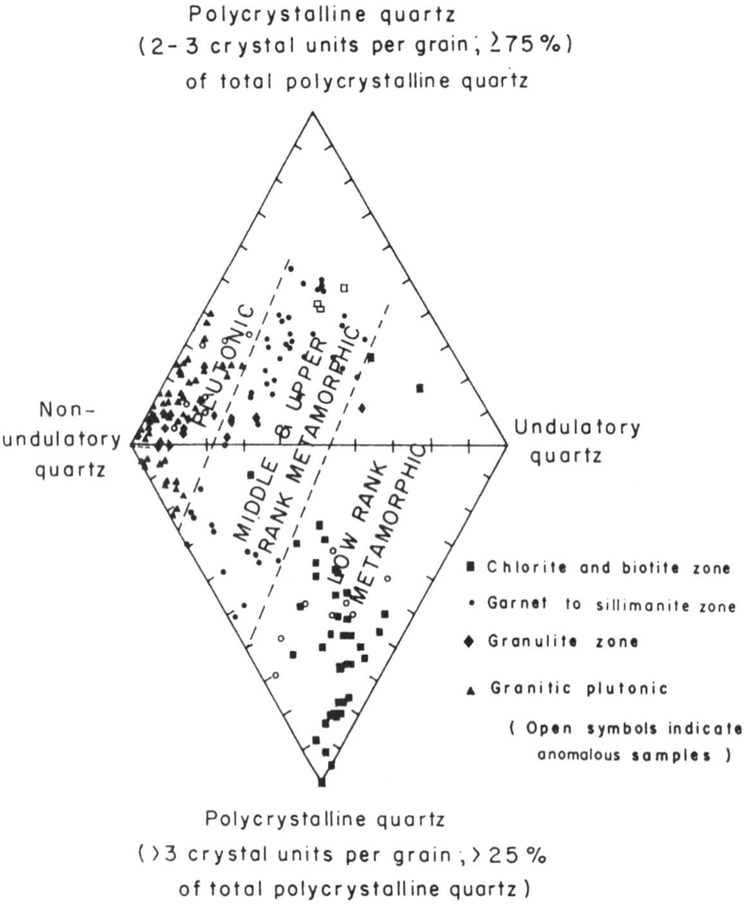

Fig. 2.3: Diamond diagram showing distribution of quartz grains of different varieties with respect to plutonic igneous and metamorphic sources (after Basu et al. 1975; with permission of SEPM, Society for Sedimentary Geology)

1981). Identification of the structural states of detrital alkali feldspars may also lead to recognition of the parent rock. Detrital feldspars from metamorphic sources are mostly microcline, while those from volcanic sources are generally sanidine. Alkali feldspars with relatively higher Al/Si structural states are likely to be associated with rapidly cooling outer shell of a pluton rather than its core. This change in the degree of ordering of feldspar is reflected in the stratigraphic sequence produced out of gradual unroofing of batholiths that have served as a source of sediments (Basu 1977). A quick method for the determination of the structural states of alkali feldspars for this purpose was developed by Basu and Suttner (1975).

In the modern big rivers of the world the proportion of quartz to feldspar increases when the climate is tropical to subtropical, suggesting a lesser chance of survival of feldspar in humid climate (Potter 1978, 1978a). The ratio of quartz to feldspar in the first cycle fluvial sand is > 1 in the humid Appalachians and < 1 in the relatively arid Rocky mountains of the United States (Basu 1976). The foregoing suggests the lesser chance of survival of feldspar in humid climate but it may not be altogether impossible for feldspar to survive prolonged transportation in humid climate although its size decreases by cleaving. The latter conclusion is reached from the finding that irrespective of climate, the proportion of feldspar increases with a decrease in grain-size. A certain amount of climatic rigour, either arid or cold, was long believed essential for the preservation of fresh feldspar grains in sediments, but a later discovery of fresh feldspar under humid, tropical, South American climate led to the belief that short-distance transportation followed by quick burial is the key to the preservation of fresh feldspars in nature (Pettijohn 1975). This implies sudden uplift of the source area and quick subsidence of the basin of accumulation, but the effects of climate and tectonism need not be mutually exclusive in the preservation of fresh feldspars in sediments.

According to current thinking, the proportion of quartz and the relatively unstable elements such as feldspar and rock fragments in a sediment population is dependent not only on the source rock and climate, but also to some extent on the degree of crustal stability (Dickinson and Suczek 1979). These ideas will be discussed later, under tectonics and sedimentation (Chapter 7).

Heavy Minerals as Provenance Indicator

While minerals such as quartz and feldspar can be derived from a variety of rocks, the heavy minerals, or more particularly, assemblages of certain suites of heavy minerals, are restricted to some distinctive source rocks only. Heavy mineral assemblages identified with major source rocks, igneous, sedimentary, and metamorphic, are shown in Table 2.2.

In vertical stratigraphic columns heavy minerals often occur in a reverse order of stability. For example, tourmaline, zircon and rutile, the most stable of the heavy minerals, occur in the older beds, whereas much less stable amphibole, pyroxene and olivine are often found in the younger horizons. As an explanation of this mode of occurrence it has been suggested that the older beds have lost the less stable minerals due to the action of the circulating ground-water or interstratal solutions which have altered or removed these minerals from the rock. If this is actually the case, they can hardly be used as indicators of provenance (Pettijohn 1975, p. 499).

Table 2.2. Heavy mineral assemblages of the major source rocks (condensed from Feo-Codecido 1956)

Heavy mineral assemblage		Source rock
Apatite, biotite, hornblende, monazite, muscovite, rutile, pink tourmaline, zircon	..	Acid igneous rocks
Fluorite, garnet, muscovite, topaz, blue tourmaline, monazite	..	Granite pegmatites
Augite, diopside, hypersthene, magnetite, olivine, chromite	..	Basic igneous rocks
Andalusite, garnet, staurolite, zoisite, epidote kyanite, sillimanite	..	Metamorphic rocks
Barite, iron ores, rutile, tourmaline, zircon (rounded grains common)	..	Reworked sediments

3

SEDIMENTARY TEXTURE

The term texture refers to the pattern formed within a sedimentary rock by the constituting grains. This can best be studied under a microscope. In sedimentary rocks of mechanical origin the framework grains are joined together, either by finer fragmental particles or by chemically precipitated cement. The former is called matrix and the latter, cement. The pattern formed by framework grains, matrix and cement is called clastic texture. Sandstones show this kind of texture.

The texture of a rock of chemical origin is produced wholly by chemically precipitated materials. No intergranular space is to be found in such a rock. Limestones of orthochemical origin show this kind of texture. Allochemical limestones, produced by fragmental pieces of calcareous material joined together either by fine carbonate mud or by calcite spars, show a clastic texture (see Chapter 4 for definition of terms).

Clastic textures are defined mainly by size, shape and roundness of the framework grains. Other features which play important roles in clastic texture are the nature of grain contact (grain-to-grain relationship) and grain packing. Porosity and permeability of a clastic rock depend on these parameters.

GRAIN-SIZE

Concept of Particle Size

The definition of particle size is related, at least in part, to the technique used for measuring it. When a particle is large enough to allow direct measurement of its diameters, say by calipers or by a micrometer scale fitted to the microscope oculars, size can be expressed in terms of the diameters (long, short and intermediate). A problem arises when a grain is too irregular to allow direct measurement of its diameters. In such cases the size is measured indirectly, firstly by measuring its volume, by the process of displacement of water, and then by working out the diameter of the sphere of equivalent volume by the following formula:

$$V = 4/3\pi r^3,$$

where V and r are the volume and radius of the sphere respectively. The diameter thus obtained was named 'true nominal diameter' by Wadell in 1932.

Udden-Wentworth Size Scale

Fragmental particles constituting sedimentary rocks cover a wide range of grain-sizes—from boulder, which by definition is greater than 25 cm, to clay which is less than 0.004 mm in diameter (less than 0.002 mm in Friedman and Sanders (1978) classification). Since handling of such a wide range of grain-size is a difficult problem, a geometric scale is used instead of an arithmetic one. In the so-called grade scale proposed by Udden in 1914, the consecutive figures in the scale increase by multiples of 2 on the higher side and decrease by multiples of 1/2 on the lower side. The various size ranges within the grade scale were named by Wentworth (1922) as follows:

Boulder	|	Cobble	|	Pebble	|	Granule	|	Sand	|	Silt	|	Clay
	256		64		4		2		1/16		1/256 (mm)	

The Udden-Wentworth system not only provided an appropriate scale for representation of grain-size but also standardised the sedimentological nomenclature. Adoption of this system of nomenclature by a committee of geologists and hydrologists dispelled the confusion created by arbitrary assignment of names to different parts of the grade scale by various authors (Lane et al. 1947). The Udden-Wentworth grade scale and nomenclature system are universally used today.

Phi Scale

In the early days of sedimentology, when mechanical or electronic computing facilities were not readily available, handling of fractional values of the Udden grade scale involved tedious computations. As a solution to this problem, Krumbein (1934) introduced a method of logarithmic transformation as follows:

$$\phi = -\log_2 d,$$

where d = diameter of the grain in millimetres.

The use of 2 as the base of the logarithm allows direct conversion of Udden's grade scale to Krumbein's phi scale. Moreover, inversion of sign converts the fractions to positive, round integers, thereby making easier computations of the finer grains (sand, silt and clay), which constitute the bulk of sediments. These advantages popularised the phi scale throughout the world. The combined Udden-Wentworth-Krumbein grain-size scale is shown in Table 3.1.

Table 3.1. Grain-size classification and nomenclature

Wentworth nomenclature	Boulder	Cobble	Pebble	Granule	Sand	Silt	Clay
Udden's grade scale	256 (2^8)	64 (2^6)	4 (2^2)	2 (2^1)	1/16 (2^{-4})	1/256 mm (2^{-8}) mm	
Krumbein's phi-scale	−8	−6	−2	−1	+4	+8 ϕ	

Sand					
	V. Coarse	Coarse	Medium	Fine	V. fine
−1	0.0	+1.0	+2.0	+3.0	+4 ϕ

GRAIN-SIZE MEASUREMENT

The techniques commonly used for grain-size analysis are: (i) direct measurement (in macro- or microscale), (ii) sieving and (iii) settling. Grains larger than sand can be measured directly by means of calipers, while an optical device, varying from a hand lens to a microscope, is needed for measuring finer particles. Sieving is generally used for particles ranging from granule to fine sand. Settling techniques are used for sizes finer than sand (silt and clay). More sophisticated methods such as image analysis of SEM pictures (without intermediary photographs), electrical resistive pulse and light-blockage counters, disc centrifuge and laser scattering devices are available now (see Tables 3.2 and 3.3). The principles behind the techniques traditionally used for grain-size measurement are briefly discussed below.

Grain-size measurements conducted under a microscope are influenced by the shape as well as the orientation of the grains. When the grains are oriented at random, the 'maximum horizontal intercepts' through the grains are measured. The sizes thus obtained are obviously smaller than the actual sizes because seldom does the line of traverse pass through the grain centre. Theoretical studies as well as experimental observations have shown that the data obtained by this method are some 24% smaller than the actual in most cases. Thus the average radius of a sphere, computed on a random section by the techniques of maximum intercept measurement, is 0.736 of the actual radius of the spheres (Krumbein and Pettijohn 1938, p. 130).

To take care of the shape variation of irregular grains, Wadell (1935) recommended computation of the area of the grain by a planimeter on an enlarged picture. The nominal sectional diameter, which is the diameter of a circle of the same area, can be obtained from this measurement. For elongate grains the three diameters – a, b, and c – must be measured separately

Table 3.2. Techniques commonly used for grain-size analysis (modified after McCave and Syvitski 1991)

Altered Atterberg	Modified Udden-Wentworth	Classical method of analysis	Modern method of analysis
Gravel	Gravel	Count	Field sieve
______ 2 mm			
Sand (3 grades)	Sand (5 grades)	Sieve	Settle Count Scatter light
______ 63 μm			
Silt (3 grades)	Silt (5 grades)	Settle	Count Settle Scatter light
______ 2 μm*			
Clay	Clay	Settle	Settle Centrifuge
______ molecular			

*Modified silt-clay boundary after Friedman and Sanders 1978.

Table 3.3. Techniques of grain-size analysis applicable to different size ranges (modified after McCave and Syvitski 1991)

Range	Principle					
	Count	Sieve	Settle from: Top	Settle from: Suspension	Laser scatter	Photon correlation spectroscopy
Gravel	/a	/	X	X	X	X
Sand	/b	/	/f	X	/i	X
Silt	/c	/e	/g	/h	/i	X
Clay	X[d]	X	/g	/h	X[i]	/j

Symbols: / = technique applicable; X = technique of limited applicability or impossible.

a. Measurement on outcrops and photographs.
b. Measurement on thin section or grain mount either directly or by image analyser.
c. Image analysis of SEM photographs or by pulse and light blockage counters.
d. Electrical resistive pulse method measures down to ~ 0.5 μm.
e. Commonly measures down to 44 μm, in special cases to 20–25 μm or even down to 5 μm by special Fritsch sieve shaker.
f. Settling tubes measure weight, differential pressure on pan, visual accumulation or light attenuation method of sensing.
g. Disc centrifuge method. Sediment is introduced via a density gradient and sensed by light attenuation. Discrimination down to 0.01 μm is claimed.
h. Homogeneous settling suspension. Measurement by simple pipette method, sedimentation balance, withdrawal from bottom, light or X-ray attenuation techniques.
i. Laser forward scattering devices.
j. Coulter and many other similar counters.

and the mean diameter then computed by one of the following formulae:

arithmetic mean, $d_a = (a + b + c)/3$
harmonic mean, $d_h = (3\,abc)/(ab + bc + ca)$
geometric mean, $d_g = \sqrt[3]{abc}$

Computation of the arithmetic mean suffices in most cases. On average, the harmonic mean lies within about 6% of the arithmetic mean.

Sieving Technique

Sieves are screens of standard size. In the commonly used sieves specified by the American Society for Testing Materials (A.S.T.M.), the openings increase uniformly in multiples of $\sqrt[4]{2}$, starting from an opening of 0.0029 inch (200 mesh). The millimetre equivalents of these openings lie quite close to the Wentworth grade limits and can be used for sedimentological purposes.

In the sieving technique a known weight of sediment sample consisting of loose grains is placed on top of a stack of screens arranged from bottom to top in order of increasing sieve diameter. When the stack is shaken, the particles divide into several fractions, depending on the number of sieves in the stack. The fraction lying on each sieve is collected separately and weighed.

The theory of sieving assumes that the weight of material lying on a particular screen is proportional to the weight of the original material of that particular size in the sample. Moreover, the sieve diameter of a grain is assumed to be the same as its nominal diameter. The theory of sieving, based on these assumptions, has many drawbacks. An elongate grain can pass through a square or round hole screen vertically. The smallest diameter of the grain involved in this case, may not even be remotely related to the nominal diameter. Moreover, a grain having a long diameter 1.4 times the sieve opening can pass through a square mesh sieve diagonally, the diagonal of a square of unit length being $\sqrt{2} = 1.41$. Common manufacturing defects of sieves also add to the problem. Square mesh sieves often stretch into rhombic shapes after prolonged usage, thereby increasing the effective diameter. Moreover, in this method, the grains are sorted according to their shapes as well as sizes. Flakes of mica, for example, lie flat on screens far bigger than their nominal diameter, while elongate grains, passing vertically, accumulate on the finer sieves.

Interpretation of the results of sieving is also not straight forward. The conventional techniques of statistical analysis demand measurement of frequency (number counts), whereas in sieving the total weights of grains lying on each sieve are measured instead of numbers. This weight is represented by a complex parameter, frequency (n) multiplied by the mass of each grain on the sieve (M), or

$$W = n \cdot M,$$

where $M = (4/3\pi\ r^3)\ \rho$, $4/3\pi\ r^3$ being the volume of a spherical grain of radius r and density ρ. This leads to additional complications because density of the grains influences size measurements. Moreover, r, the sieve opening, is *assumed* to be the same as the nominal grain diameter.

In view of these complications, parameter W, the total weight of grains on each sieve, is designated as 'weight frequency' to distinguish it from true frequency (n). Appropriate techniques for analysis of weight frequency data not being available, the conventional techniques for analysis of frequency data are used for computing statistical parameters of grain-sizes from sieve data. Some attempts have been made to formulate new techniques of computation of mean and variance of the weight frequency data (Adhikari, Roy and Sengupta 1981), but the methods developed are too sophisticated to be discussed here.

Settling Techniques

In this method the particle concerned is allowed to fall through a column of water and the size of the grain is worked out from its fall velocity (v), using Stokes' equation:

$$v = \frac{2\,(\rho_1 - \rho_2)\, gr^2}{9\eta},$$

where ρ_1 and ρ_2 are the densities of the grain and the fluid respectively, η = coefficient of viscosity, g = acceleration due to gravity and r = radius of the grain concerned. The Stokes' equation can therefore be simplified as follows:

$$v = C \cdot r^2,$$

all parameters, except the grain diameter (r) and fall velocity (v), being constant for a particular experiment. Hence the radius of the particle can be worked out from its settling velocity.

The type of equipment commonly used for measuring grain settling velocity is illustrated in Fig. 3.1. The particles concerned are allowed to fall through a tall (about 2 m) column of water and the weight of the grains accumulating on the balance pan immersed in water is recorded automatically against time. The advantage of this system is that each grain of the sample is allowed to fall through the whole column of water and hence the curve obtained is a direct record of the cumulative weight percentage of each size fraction. Various other models of settling tubes, measuring fall velocities of fine particles from suspension densities, are available today. Modern versions of settling tubes use pressure-sensitive devices or light attenuation techniques for measurement. Computerised systems in modern settling tubes take care of such variables as water temperature, grain shape, grain density, providing grain-size distribution patterns of the samples directly.

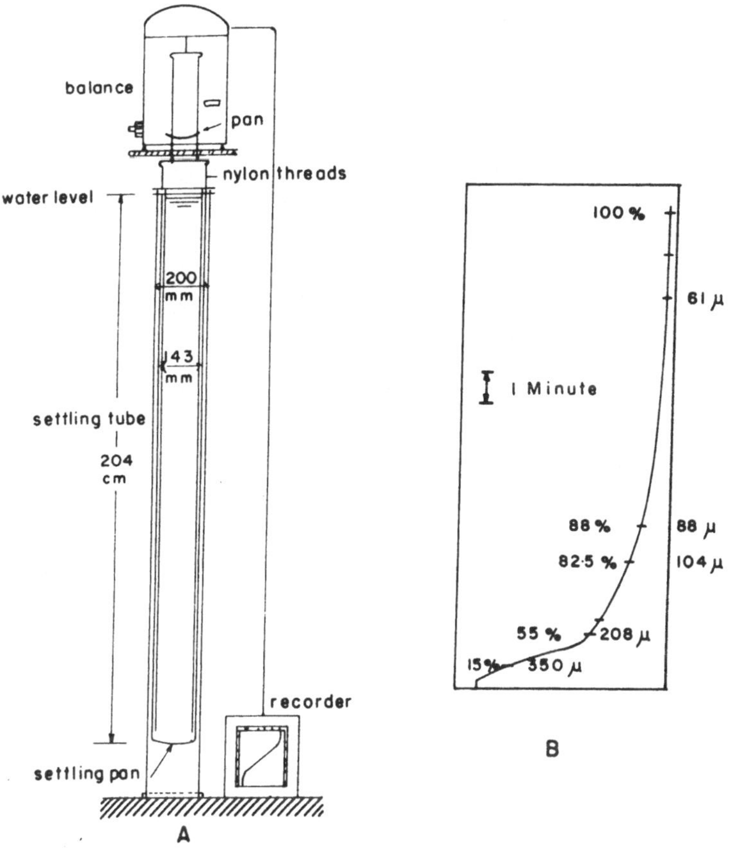

Fig. 3.1: (A) Schematic representation of the Groningen sedimentation balance for grain-size measurement by settling technique. (B) Sketch of a grain-size distribution chart obtained from the Groningen balance (modified after Sengupta and Veenstra 1968 with permission of Elsevier Science Publishers B.V.).

In spite of these developments the settling method cannot be used for measuring grains of all sizes because of the preconditions necessary for Stokes' law to be valid. The more important of these are that: 1) the particle concerned is spherical, 2) the fluid medium is homogeneous, 3) the medium is of unlimited extent, 4) a constant settling velocity is attained during fall of the particle through the fluid column and 5) the settling velocity of the grain is not too high.

The lower size limit for which the above-mentioned conditions are valid lies between 0.15 and 0.5 µ. On the coarser side, 60 µ is believed to be the limit (some would extend it to 80 µ). Free fall of grains larger than this size is hindered by generation of wakes (low pressure zones), thereby invalidating Stokes' law. These limitations confine the applicability of Stokes' law to the silt and clay size grains only. Equations were developed by Oseen in 1910 and Rubey in 1933, to extend the limit of settling analysis to coarse silt. The rates of settling of grains coarser than this size vary as the square root of grain diameter (see Krumbein and Pettijohn 1938).

In view of the difficulties involved in converting settling time to grain diameter, theoretical or empirical curves relating settling velocity to grain diameter are often used for conversion (see, for example, Fig. 3.2).

A comparative study of the different techniques of grain-size measurement is presented in Table 3.4. In spite of its drawbacks, the sieving method is still very popular for grain-size measurement because the equipment is relatively inexpensive and the results are replicable, that is, overall reproducibility is good.

Some techniques for conversion of sieve ('weight frequency') data to thin section (number frequency) data are available (Friedman 1958, 1962; Sahu 1964). Friedman (1962) empirically established a linear correlation between plots of thin-section quartile parameters and sieving quartile parameters (Fig. 3.3). This provides an easy conversion of one to the other. Mathematical models for comparison of sieve and thin-section data are also available (Burger and Skala 1976).

ANALYSIS AND INTERPRETATION OF GRAIN-SIZE DATA

Grouped data of grain-sizes, obtained by any of the methods described above, are represented in tabular form. The attributes of grain-size frequency distribution—central tendency, dispersion, symmetry and peakedness—are then computed, either graphically or by using the technique of moment measures. Moments are preferred to graphic plots because they take care of every part of the frequency distribution. Following the procedure given in the Appendix, the first four moments can be readily calculated using a desk calculator (see Tables 3.5–3.8). Using a suitable programme, the results can be obtained in seconds on an electronic computer. It is a good practice to plot the results also as histogram, frequency polygon and cumulative frequency curve (Fig. 3.4). The visual impression created by these plots often helps in arriving at geologically meaningful conclusions. Before the modern electronic computers came into use it was a common practice to compute the statistical parameters in terms of phi-units with the help of the percentile values read off the cumulative frequency curve. The 'inclusive graphic measures' proposed by Folk and Ward (1957) were very popular for computational purposes for many years (Table 3.9).

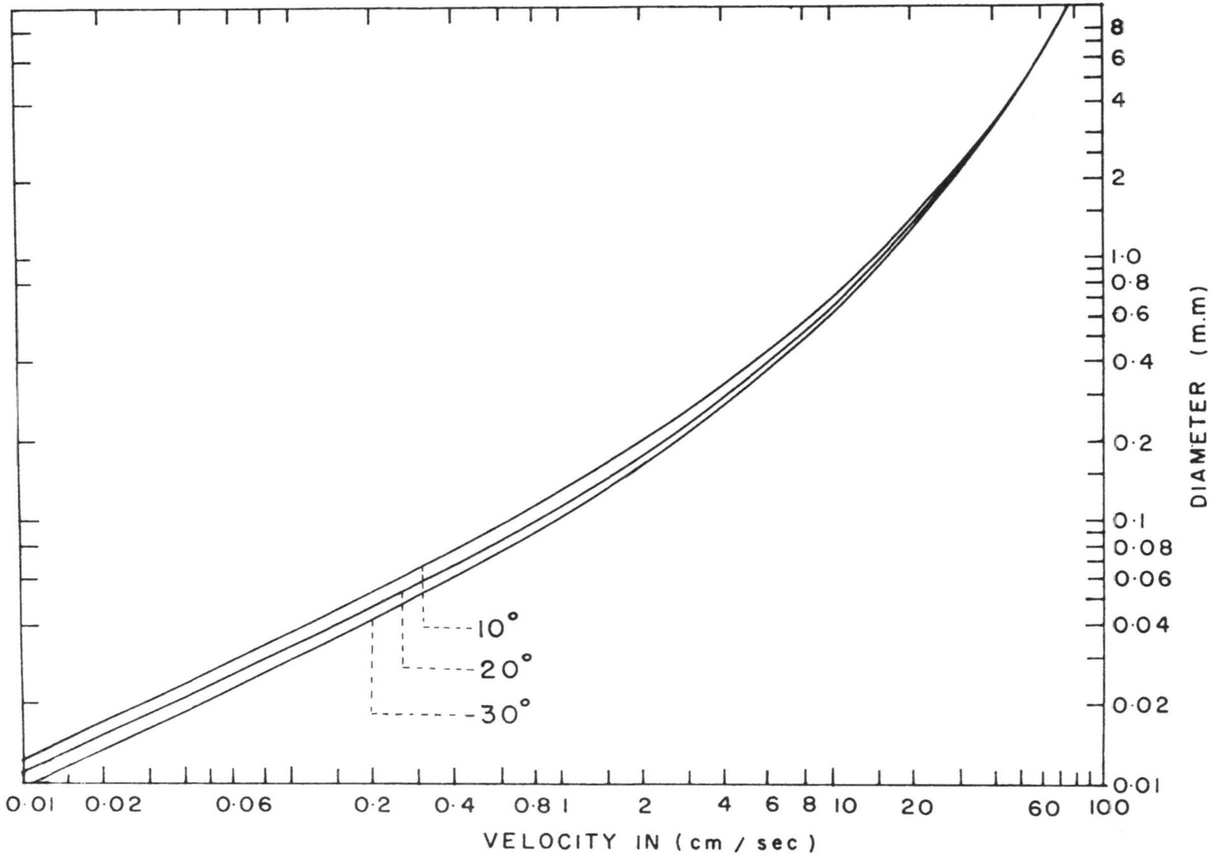

Fig. 3.2: Terminal fall velocities of quartz spheres of different diameters in water at different temperatures, as determined experimentally by the U.S. Department of Agriculture, Soil Conservation Service (modified and redrawn after Vanoni, Brooks and Kennedy 1961).

Table 3.4. Advantages and disadvantages of various methods used for grain-size measurement

Technique of measurement	Advantages	Disadvantages
Direct method:		
calipers or microscope (ocular micrometer)	Observations recorded as frequency (number counts), hence statistical computations are easy. Shape and roundness can also be measured together with grain-size.	Tedious and time-consuming. Measurements obtained are less (by about 24%) than actual value. Conversion to 'weight frequency' is required for comparison with sieve data.
Sieving	Easy to use. Equipment inexpensive. Overall reproducibility of results is good. This technique is widely used. Hence comparison with results obtained elsewhere is possible.	Particle shape and density influence measurements. Square hole sieves create special problems. Manufacturing defects are common in sieves. Procedure, though not as tedious as microscope measurements, is time-consuming. Measures 'weight frequency', a parameter for which the statistical method of analysis is not yet fully developed.
Settling method	Rapid. Easy to use. Continuous record of settling time (velocity) available automatically by modern equipment. Shape, volume, specific gravity of grains are taken into account in settling velocity. Transportation and deposition in nature is controlled by grain settling velocity; hence this is the most dependable parameter for hydrodynamic interpretation.	Shape and density of grains influence settling velocity but not exactly in the same way as in sieving. Hence comparison with results of sieving is not easy. Conversion of settling velocity to grain-size has to be done by formulae which have limited applicability, or with the help of empirically derived curves based on observations on quartz spheres. Hence interpretations are of limited value. Settling tubes of extraordinary length are required for handling the whole range of sediment size. Natural samples tend to fall as one unit, thereby artificially increasing settling velocity.

The particles produced out of the natural processes of weathering and erosion of a source rock are further sorted during transportation and deposition. The mechanism of grain sorting, which involves selective elimination of certain grain-sizes, varies from environment to environment. Grain-size parameters of sediment populations from various modern environments have been studied intensively for more than half a century with the hope of establishing a direct correlation between the two. The techniques used in most cases involved plotting of one of the parameters of grain-size dis-

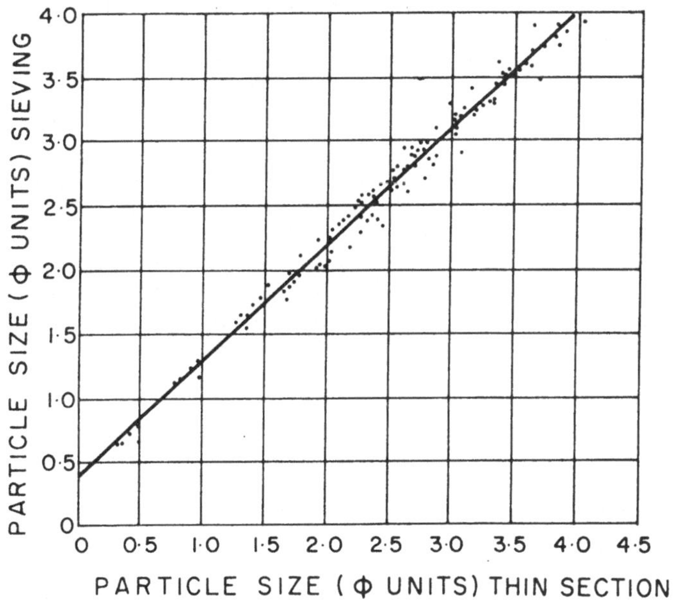

Fig. 3.3: Single overall regression line for determining sieve-size distribution from thin section analysis (after Friedman 1958, fig. 4, with permission of the author and University of Chicago Press).

tribution against another with the intention of discriminating between different environments (Friedman 1961, 1967; Passega 1957, 1964; Sahu 1964a; Moiola and Weiser 1968). Friedman tried combinations of various statistical parameters and found plots of third moment ('mean cubed deviation') against standard deviation (square root of second moment) to be one of the most efficient methods for environmental discrimination (Friedman 1979).

The principle involved in this discrimination can be demonstrated by referring to Friedman's earlier publications (Friedman 1961; also see Figs. 3.5 and 3.6 of this text). River sand shows poorer sorting than beach sand because of the lower energy conditions prevailing in rivers. River sands are also distinguished by positive skewness because of the presence of fines. The finer particles are washed away in the beach, leading to negative skewness. A perfect discrimination between beach and river sands can thus be obtained by plotting standard deviation (sorting) against skewness (Fig. 3.5). In Fig. 3.6 the (eolian) dune sand is finer than the beach sand, and is also positively skewed because the fines are retained in it. Beach sand is coarser because the competence of the ocean wave is greater than that of the wind. Plotting of mean size against skewness therefore discriminates beach from eolian dune sand.

Table 3.5. Results of sieving analysis. Sample No. USR 23-A. Initial weight of sample = 50.292 gm

ASTM No.	Sieve (mm)		Opening (ϕ)		Midpoint (ϕ)	Gross wt. (gm)	Empty watch glass wt. (gm)	Weight f (gm)	Weight (%)	Cumulative weight frequency
10		>2.00	< −1.00		—	30.950	29.526	1.424	2.823	2.823
14	2.00	1.41	−1.00	−0.50	−0.75	18.721	14.763	3.958	7.847	10.670
18	1.41	1.00	−0.50	0.00	−0.25	37.015	29.526	7.489	14.847	25.517
25	1.00	0.71	0.00	0.50	0.25	36.013	29.526	6.487	12.861	38.378
35	0.71	0.50	0.50	1.00	0.75	39.562	29.526	10.036	19.897	58.275
45	0.50	0.35	1.00	1.50	1.25	38.622	29.526	9.096	18.033	76.308
60	0.35	0.25	1.50	2.00	1.75	38.959	29.526	9.433	18.701	95.009
80	0.25	0.177	2.00	2.50	2.25	31.563	29.526	2.037	4.039	99.048
120	0.177	0.125	2.50	3.00	2.75	15.201	14.763	0.438	0.868	99.916
170	0.125	0.088	3.00	3.50	3.25	14.794	14.763	0.031	0.061	99.977
230	0.088	0.062	3.50	4.00	3.75	14.770	14.763	0.007	0.014	99.991
PAN	<0.062		>4.00		—	14.767	14.763	0.004	0.008	99.999
							Total weight	50.440		
							Initial weight of sample	50.292		
							Gain in weight	0.148		
							Error = +0.294%			

Table 3.6. Computation of first and second moments. Sample No. USR 23-A

Size class (ϕ)		Midpoint m_i(ϕ)	Weight frequency f_i	$f_i m_i$	$f_i m_i^2$
< −1.00		—	(1.424)	—	—
−1.00	−0.50	−0.75	3.958	−2.968	2.226
−0.50	0.00	−0.25	7.489	−1.872	0.468
0.00	0.50	0.25	6.487	1.622	0.406
0.50	1.00	0.75	10.036	7.527	5.645
1.00	1.50	1.25	9.096	11.370	14.212
1.50	2.00	1.75	9.433	16.508	28.889
2.00	2.50	2.25	2.037	4.583	10.312
2.50	3.00	2.75	0.438	1.204	3.311
3.00	3.50	3.25	0.031	0.101	0.328
3.50	4.00	3.75	0.007	0.026	0.098
>4.00		—	(0.004)	—	—
			$N = \Sigma f$ = 49.012	Σfm = 38.101	Σfm^2 = 65.895

Note: Figures in parentheses are omitted in computation.

First moment (arithmetic mean): $\mu_1 = \Sigma(fm)/\Sigma f = 38.101/49.012 = 0.777\phi$

Second moment (variance): $\mu_2 = \sigma^2 = \{\Sigma(fm^2) - (\Sigma fm)\mu_1\}/\Sigma f = 0.740\ \phi$

Standard deviation: $\sigma = 0.860\ \phi$

Table 3.7. Computation of α_3 and α_4. Sample No. USR 23-A

Size class (ϕ)		Midpoint m_i (ϕ)	'Weight frequency' f_i	$(m_i - \mu_1)$	$(m_i - \mu_1)^3$	$(m_i - \mu_1)^3 \cdot f_i$	$(m_i - \mu_1)^4$	$(m_i - \mu_1)^4 \cdot f_i$
< −1.00			(1.424)	—	—	—	—	—
−1.00	−0.50	−0.75	3.958	−1.527	− 3.560	−14.093	5.437	21.519
−0.50	0.00	−0.25	7.489	−1.027	− 1.083	− 8.112	1.112	8.331
0.00	0.50	0.25	6.487	−0.527	− 0.146	− 0.949	0.077	0.500
0.50	1.00	0.75	10.036	−0.027	0.000	0.000	0.000	0.000
1.00	1.50	1.25	9.096	0.437	0.106	0.963	0.050	0.455
1.50	2.00	1.75	9.433	0.973	0.921	8.689	0.896	8.455
2.00	2.50	2.25	2.037	1.473	3.196	6.510	4.708	9.589
2.50	3.00	2.75	0.438	1.973	7.680	3.364	15.153	6.637
3.00	3.50	3.25	0.031	2.473	15.124	0.469	37.402	1.159
3.50	4.00	3.75	0.007	2.973	26.277	0.184	78.123	0.547
>4.00			(0.004)	—	—	—	—	—
			$N = \Sigma f$ = 49.012			$\Sigma(m - \mu_1)^3 \cdot f$ = − 2.970		$\Sigma(m - \mu_1)^4 \cdot f$ = 57.192

Note: Figures in parentheses are omitted in computation.

Computation of α_3 and α_4:

$$\alpha_3 = \{\Sigma(m_i - \mu_1)^3 \cdot f_i\}/N \cdot \sigma^3 = -0.095$$

$$\alpha_4 = \{\Sigma(m_i - \mu_1)^4 \cdot f_i\}/N \cdot \sigma^4 = 2.133$$

Table 3.8. Statistical parameters of grain-size distribution. Sample No. USR 23-A

	Mean (ϕ)	Standard deviation (ϕ)	Skewness	Kurtosis
Graphic measures (Folk and Ward 1957):				
M_z	0.704			
σ_1		0.865		
SK_1			−0.172	
K_G				0.965
Moment measures:				
μ_1	0.777			
σ		0.860		
α_3			−0.095	
α_4				2.133

Table 3.9. Folk and Ward (1957) grain-size measures

Parameter	Formula
Graphic mean	$(M_z) = \frac{\phi_{16}+\phi_{50}+\phi_{84}}{3}$
Inclusive graphic standard deviation	$(\sigma_1) = \frac{\phi_{84}-\phi_{16}}{4} + \frac{\phi_{95}-\phi_5}{6.6}$
Inclusive graphic skewness	$(SK_1) = \frac{(\phi_{84}+\phi_{16}-2\phi_{50})}{2(\phi_{84}-\phi_{16})} + \frac{(\phi_{95}+\phi_5-2\phi_{50})}{2(\phi_{95}-\phi_5)}$
Graphic kurtosis	$(K_G) = \frac{(\phi_{95}-\phi_5)}{2.44(\phi_{75}-\phi_{25})}$

In another line of approach popular among sedimentologists, the different modes of sediment transportation are identified by plotting the cumulative frequency distributions of grain-sizes on probability paper (Fig. 3.7). A perfectly symmetrical ogive, which is the cumulative plot for a normal distribution, plots as a straight line on log-probability paper. When several such straight lines show up on the log-probability plot, as in Fig. 3.7, it is assumed that the sediment population has been transported by different mechanisms, each of which has led to a normal distribution. The net distribution is then divided into traction load (coarsest fraction), suspension load (finest fraction) and saltation load (medium fraction). Using this technique, Visher (1969) identified a variety of transportational-depositional conditions on the assumption that specific combinations of traction, suspension and saltation load characterise particular environments.

The one-to-one correlation between segments of log-probability plots and specific modes of sediment transportation assumed in this type of interpretation, has raised doubts in some minds (Komar 1986). Sediment populations transported either wholly as bed load or wholly as suspended load in experimental channels have been found to be segmented in log-probability plots (Sengupta 1975, 1979), suggesting that these are not necessarily the

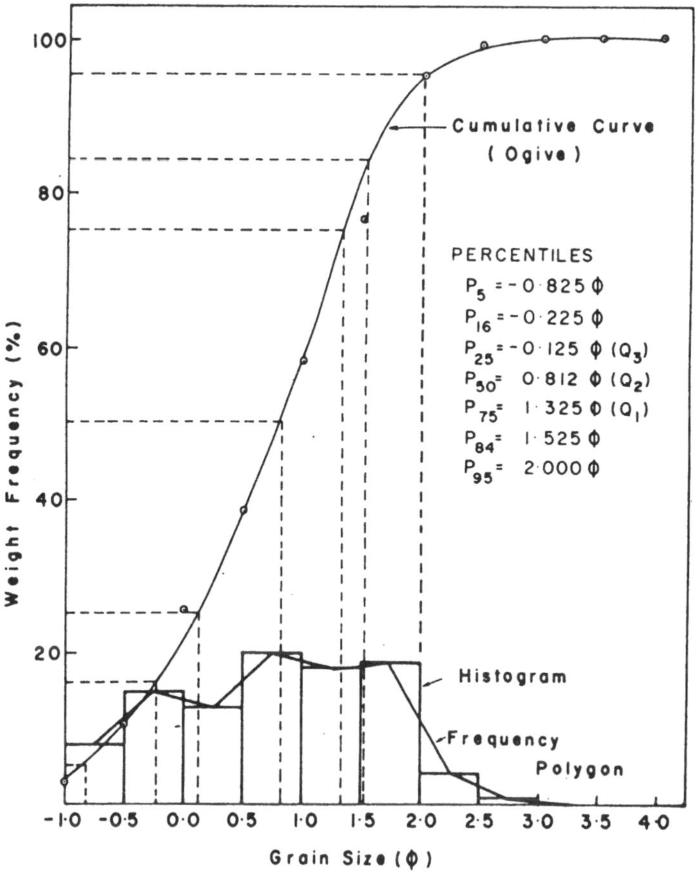

Fig. 3.4: Graphic representation of grain-size frequency data. Histogram, frequency polygon and cumulative curve for the data shown in Table 3.5.

result of transportation by more than one mechanism. The degree of log normality attained by a sediment population is perhaps a measure of the flow parameters rather than depositional environment. In another study segmented log-probability plots, commonly interpreted as mixtures of several log-normal populations, have been found to fit log-hyperbolic better than log-normal distributions (Christiansen, Blæsild and Dalsgaard 1984). The recent finding that log normality may develop out of a log-hyperbolic distribution when the sorting process is nearly completed by a high velocity current, has provided new clues to the interpretation of grain-size data (Sengupta, Ghosh and Mazumder 1991).

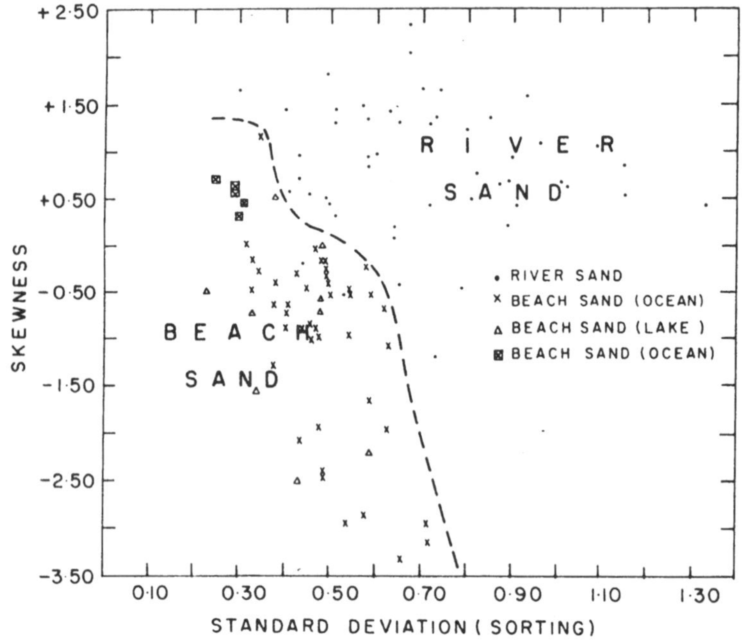

Fig. 3.5: Plots of sorting vs. skewness of grain-size distribution for discrimination of river and beach sands (after Friedman 1961, fig. 4, with permission of the author and SEPM, Society for Sedimentary Geology).

ROUNDNESS AND SHAPE

Roundness and shape of clastic particles are two separate parameters, independent of each other. Roundness implies rounding of the edges and corners of a clastic grain. Shape is expressed as the ratio between the three measurable diameters of an irregular grain: length, breadth and width. Other measures of shape, including sphericity and flatness, are also used.

Roundness

Following Wadell, roundness is defined as the ratio of the average radius of curvature of the edges and corners of a grain to the curvature of the maximum inscribed sphere. Measurements on spheres being difficult, two-dimensional projections of grains are generally used. Roundness in such a case is expressed as:

$$\text{Roundness} = (\sum_{1}^{N} r_i/N)/R,$$

where $r_i s$ are the radii of the corners of the particles, N = the total number

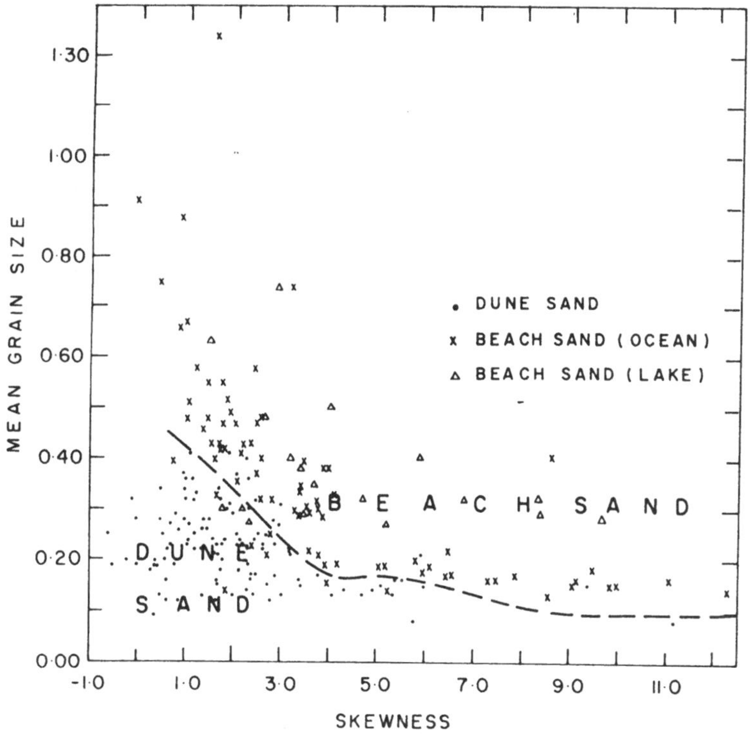

Fig. 3.6: Plots of skewness vs. mean grain-size for discrimination of beach and (eolian) dune sand (after Friedman 1961, fig. 2, with permission of the author and SEPM, Society for Sedimentary Geology).

of corners measured, and R is the radius of the largest possible inscribed circle (Fig. 3.8).

According to Wadell's definition, the roundness of a sphere is 1.0. The nearer a natural particle is to this value, the more rounded it is. A chart giving photographs of pebbles of different roundness values was published by Powers (1953) to provide a quick aid to roundness determination by visual comparison (Fig. 3.9). Verbal expressions of roundness as given by various authors are shown in Table 3.10.

Rounding of sand-size particles of quartz and feldspar is negligible during fluvial transport. Whatever rounding is attained during transportation is confined only to the first few kilometres and that, too, observable only in the pebble-size grains (Fig. 3.10). Beach action is an effective agent for rounding of pebbles but not for sand. Rounding of a sand grain is believed, on the basis of experimental studies, to be essentially due to eolian action. The

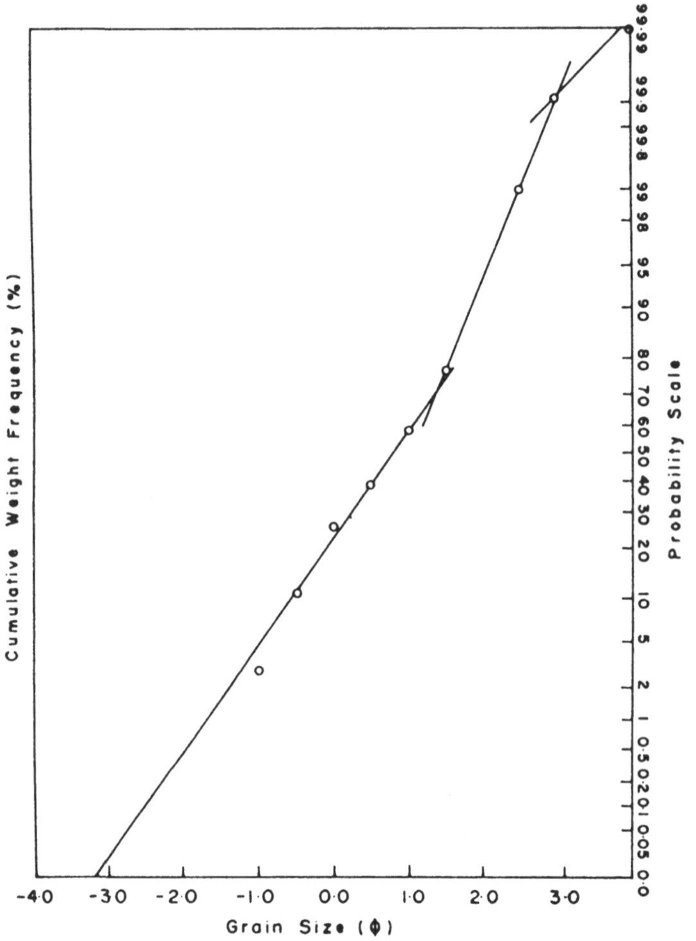

Fig. 3.7: Probability plot of grain-size distribution of sand Sample No. USR 23-A from the Usri River, Bihar, India (data shown in Table 3.5).

effect of solution in the rounding of quartz grains has been shown to be negligible (Kuenen 1960a and b, 1964).

For grains larger than sand size and roundness generally show a positive correlation. This is so because the smaller grains, which are produced by chipping of the corners of a large pebble, are more angular than the remaining central part. The fact that pebble roundness is controlled by grain-size rather than distance of transportation puts a serious limitation to the use of pebble roundness as a palaeocurrent indicator. Size roundness relation in sand sizes is more complicated. For example, Banerjee (1964) showed a

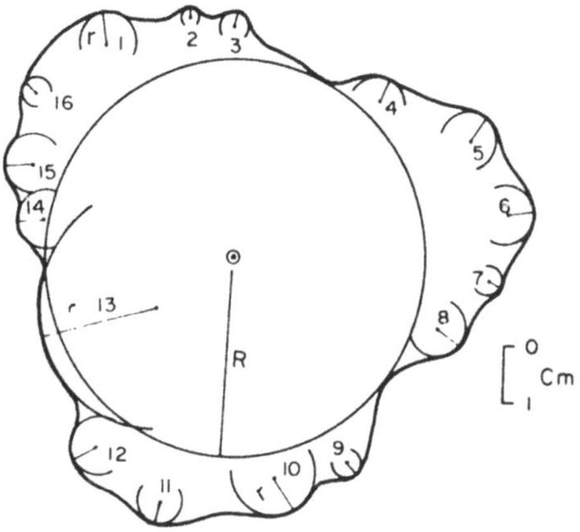

Fig. 3.8: Diagram illustrating technique of measurement of grain roundness on an enlarged projection. Roundness = $(\sum_{1}^{N} r_i/N)/R = (81.1/16)\text{mm}/34.5\text{ mm} = 0.15$.

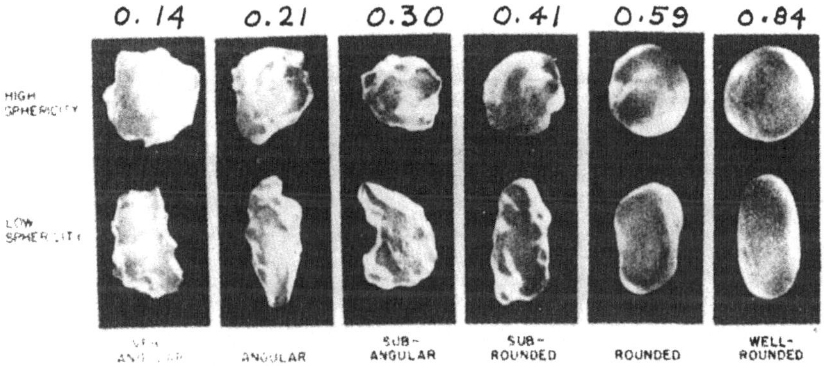

Fig. 3.9: Chart for estimation of roundness of sand grains of high and low sphericity by visual comparison (after Powers 1953, fig. 1, with permission of SEPM, Society for Sedimentary Geology).

negative correlation between size and roundness of sand and explained this as being the result of mixing of first-cycle and second-cycle sand grains.

Shape

Shapes of sedimentary particles can be defined in many ways. A simple shape classification, introduced more than half a century ago (Zingg 1935), used the ratio of breadth to length (b/a) and thickness to breadth (c/b) of the particle concerned. Four different shape nomenclatures—tabular, equant,

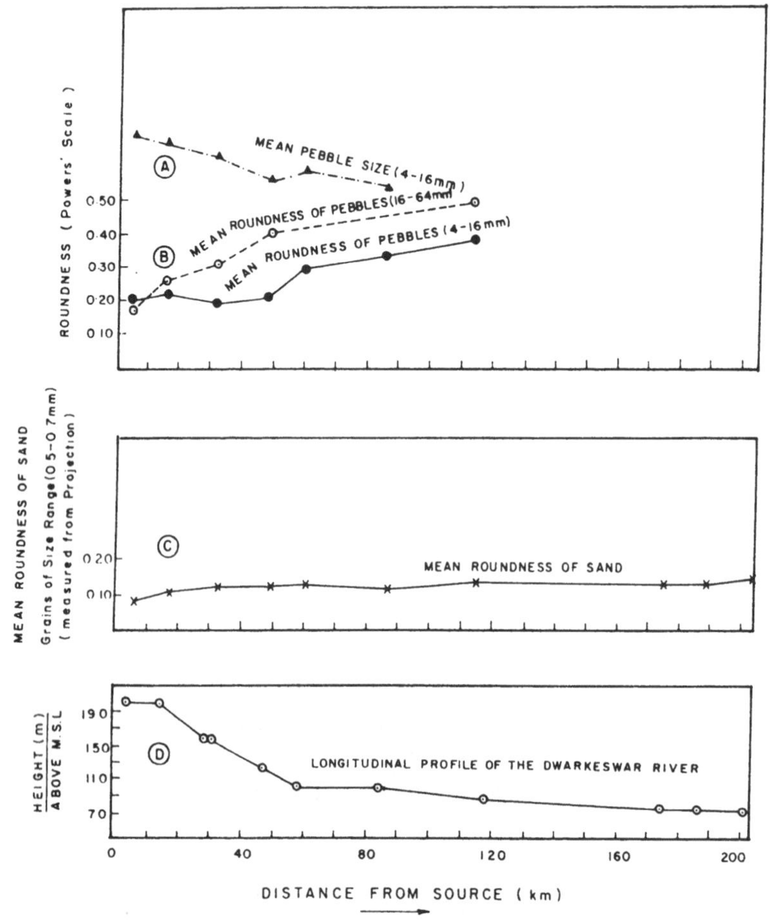

Fig. 3.10: Downstream change in roundness of pebbles and sand from the Dwarkeswar River, W. Bengal, India (after A.K. Mukhopadhyay 1988, unpublished M.Sc. thesis, IIT, Kharagpur).

bladed and prolate—were proposed on the basis of these ratios (Fig. 3.11).

In another approach to shape measurement the closeness of a grain to the shape of a sphere ('sphericity') is determined by the ratio s/S, where 's' is the surface area of a sphere of the same volume as the particle having a surface area of 'S'. For an ideal sphere this ratio is 1. For practical purposes, following Wadell, the diameter of a sphere (d_n) having the same volume as the particle in question is compared to the diameter (D_s) of the smallest sphere circumscribing the particle of irregular shape. The Wadell sphericity index in that case is represented as d_n/D_s. Since measurements of spheres are difficult, the diameter of a particle of irregular shape and that

Table 3.10. Comparison of roundness values of Pettijohn and Powers

Roundness terms	Pettijohn (1975)		Powers (1953)
	class limits	midpoints	midpoint
Very angular	-	-	0.14
Angular	0.00–0.15	0.125	0.21
Subangular	0.15–0.25	0.200	0.30
Subrounded	0.25–0.40	0.315	0.41
Rounded	0.40–0.60	0.500	0.59
Well rounded	0.60–1.00	0.800	0.84

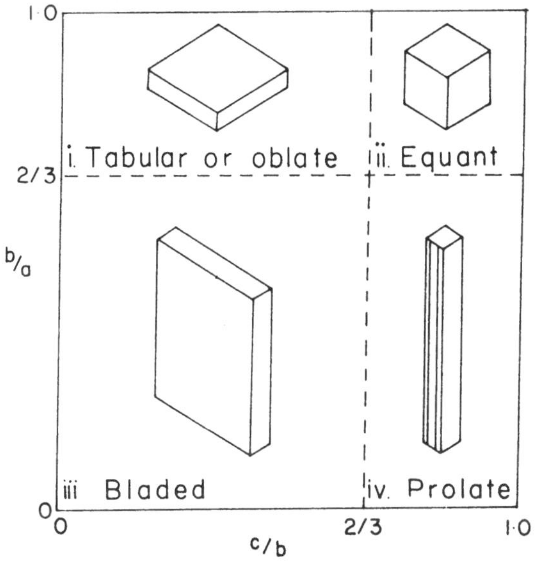

Fig. 3.11: Zingg's (1935) classification of pebble shapes (redrawn from Pettijohn 1975, fig. 3-18).

of the circumscribing sphere are compared with the help of two-dimensional projections. It is difficult to establish a direct relationship between Wadell's sphericity index and Zingg's shape for a sedimentary particle. Pebbles of entirely different Zingg shapes (oblate and prolate) may have the same or similar sphericity.

A more modern measure of equidimensionality of grains was introduced by Sneed and Folk (1958) for analysis of forms of particles. The triangular diagram (Fig. 3.12) used for this purpose has three end members: compact or spherical (all axes equal), platy or oblate spheroid (two long and one short axes) and elongate or prolate spheroid (one long and two short axes).

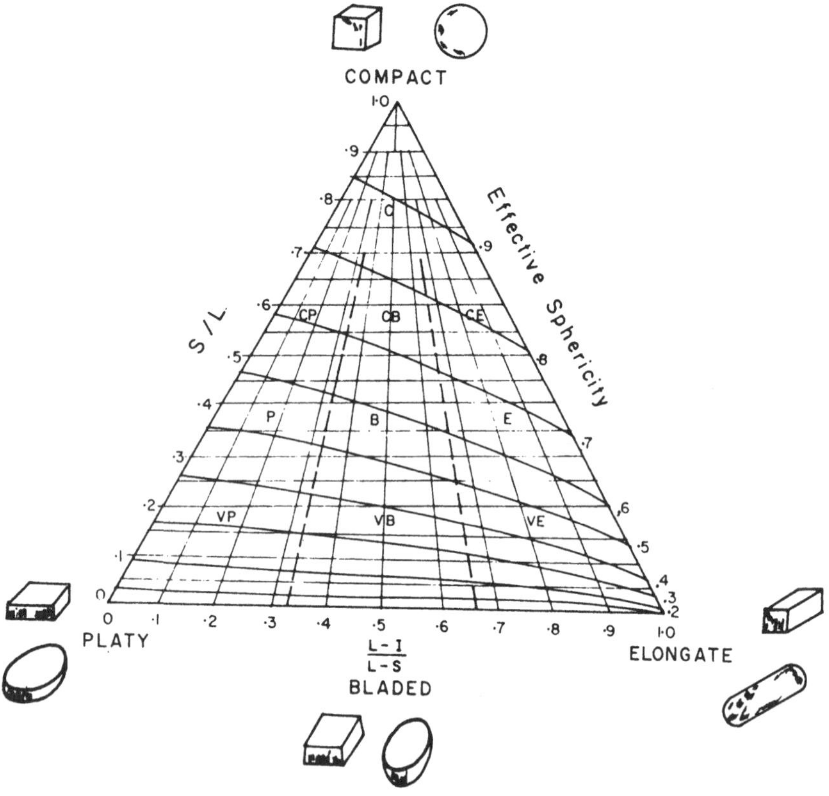

Fig. 3.12: Graphic method of determination of sphericity after Sneed and Folk (1958, fig. 2). Procedure: compute S/L and $(L - I)/(L - S)$. The sphericity of the point at the intersection of these two parameters is obtained by interpolating between the curved lines. L, I, and S represent the long, intermediate and short diameters of the grain respectively (with permission of the University of Chicago Press). Shapes of the grains of the end members are represented after Fritz and Moore 1988.

A given particle, with all three axes unequal, can occupy any field within this triangle, depending on $(L - I)/(L - S)$ ratio, where L, I and S represent the long, intermediate and short diameters respectively. The form classes obtained by this method give a better quantitative estimate of the pebble shape than the Zingg method.

A new shape property, 'rollability', was proposed by Winkelmolen (1971). Rollability is measured by the time taken by a grain, put into a slightly inclined, revolving drum, to roll down to the bottom of the drum. The average rollability value of a given sample is calculated per sieve fraction. The shape distribution character (SDC) obtained by plotting the relative rollability values against the size fractions, is claimed by Winkelmolen (1972) to have genetic significance. Shelf, shore and alluvial channel deposits of the Lower Oligocene of Belgium were shown to have distinctly different SDC characteristics.

The shape of a mineral or rock fragment, often controlled by its cleavage or fracture, is rarely modified by prolonged transportation. Analysis of particle form led Sneed and Folk (1958) to conclude that sphericity depends mainly on the inherent abrasional properties of different rock types and is a function of size as well as distance of transportation. It is little affected by selective sorting. In the course of another study, Dobkins and Folk (1970) concluded that roundness of isotropically wearing basalt pebbles increases from rivers to beaches, while sphericity decreases and the oblate-prolate index becomes progressively negative, i.e., disc-like. The production of discs, which abound on beaches, is the result of abrasion caused by pebbles sliding back and forth over sand or smaller pebbles in the surf zone. Selective shape sorting is also important, at least in some cases, as sandy beaches are capable of trapping the discs.

GRAIN-TO-GRAIN RELATIONSHIP

Packing

Arrangement of the framework grains within a clastic sedimentary rock and also the nature of contact between the adjacent grains, control what is commonly termed 'packing'. Packing of the grains in turn controls porosity and also, indirectly, permeability. Point contact, long contact, concavo-convex contact and sutured contact are some of the terms commonly used in describing the nature of grain contacts (Fig. 3.13). When the matrix proportion is very high, grains simply 'float' in the matrix. Convcavo-convex contact generally implies solution effect, whereas solution, together with pressure, produces sutured grain contacts. In most other cases, when the rock is unaffected by solution or pressure, point, tangential or long contacts are found.

The index of packing of the framework grains may be determined by counting the number of grain-to-grain contacts in a randomly cut section of the rock. The result may be expressed either as 'packing proximity' or as 'packing density'. The former is defined as the ratio of number of grain-to-grain contacts to the total number of grains counted in the traverse. The latter is the ratio of the sum of all grain intercepts to the total length of the traverse. This, as Pettijohn (1975, p. 75) has pointed out, will have an inverse relationship to porosity. In this method of computation it is assumed that there is no matrix or cement present in the rock. A comparison by Pettijohn shows that packing (number of grain contacts) is higher in orthoquartzites than subgreywackes. Loose sand and calcareous oolites have

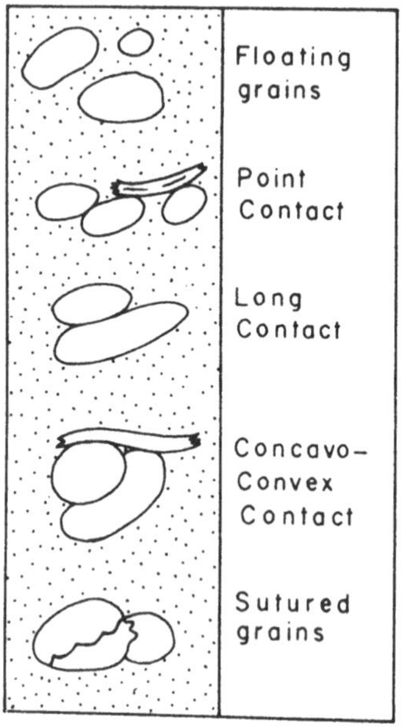

Fig. 3.13: Common types of grain contacts found in clastic sedimentary rocks (based on Pettijohn, Potter, and Siever 1987, fig. 3-11).

a much smaller packing index than most sandstones. (See Chapter 4 for explanation of terms.)

Porosity

Porosity or the porportion of voids within a rock is a function of all the textural parameters—grain-size, shape, sorting, packing, as well as the amount of void-filling materials, matrix and cement. The absolute porosity of a rock is given by the ratio—total pore volume/bulk volume of the rock. For practical purposes, for example to assess the usefulness of a rock as a reservoir for petroleum, the effective porosity is more important than the absolute porosity. This is given by the ratio of the interconnected pores to the total bulk volume of the rock. The effective porosity is generally 5 to 10% less than the absolute porosity. It is the effective porosity that determines the space occupied by extractable fluids. However, not all fluids occupying pores can be extracted because in very fine pores the liquid adheres to the pores due to surface tension.

Porosity is determined to some extent by the manner of packing of the grains within a rock. An arrangement of packing which increases stability of the grains invariably decreases porosity. For this reason a stable rhombohedral arrangement of the grains within a rock produces a porosity of only

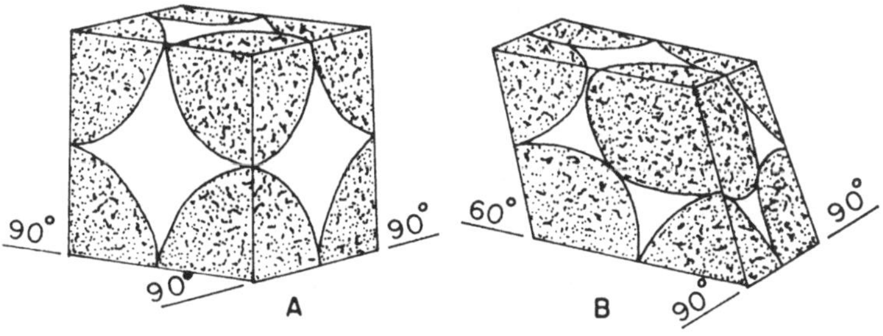

Fig. 3.14: Influence of grain packing on porosity: (A) cubic packing (porosity = 48%), (B) rhombohedral packing (porosity = 26%) (after Graton and Fraser, 1935 with permission of the University of Chicago Press).

26%, whereas an unstable cubic arrangement produces a porosity of 47.6% (Fig. 3.14). Porosities of reservoir rocks for petroleum are graded as follows: 5–10% (poor), 10–15% (fair), 15–20% (good), 20–25% (very good) (Levorsen 1958, p. 98).

Porosity can be determined under the microscope in thin sections of rock specimens. Cores or well cuttings collected from boreholes are invariably studied under a binocular microscope for estimation of porosity. The concerned rock is then qualitatively classified as tight, dense, porous or cavernous. Porosity may also be estimated from such parameters as speed of drilling. A sudden decrease in drilling speed may indicate an increase in porosity of the rock formation because loosely packed, porous materials can only be drilled slowly. *In situ* determination of porosity can be done in the subsurface using various geophysical logs. These methods have been discussed in the section on wireline logs in Chapters 8 & 9.

Permeability

Permeability is the property allowing passage of fluid through a rock. The unit of permeability is a darcy. According to the American Petroleum Institute, a porous medium is said to have a permeability of one darcy 'when a single phase fluid of one centipoise viscosity that completely fills the voids of the medium will flow through it under conditions of viscous flow at a rate of one centimeter per second per square centimeter of cross-sectional area under a pressure of equivalent hydraulic gradient of one atmosphere (76.0 cm of Hg) per centimeter'.

The rate of flow of a fluid through a porous medium is dependent on hydraulic gradient. All rocks become permeable to some extent when the pressure difference is sufficiently high or the fluid viscosity is sufficiently low. Temperature therefore exerts an influence on permeability. Higher temperature lowers fluid viscosity thereby making an otherwise non-permeable rock permeable. Grain shape and packing are the two other important factors controlling permeaility of a formation. A decrease in grain-size lowers permeability by increasing capillary pressure. Moreover, permeability in many cases is an anisotropic property controlled by features such as bedding planes, grain orientation and fractures.

In the laboratory permeability is determined by forcing a liquid through a core sample of the rock concerned by means of a pump and measuring the pressure drop across the sample on a manometer. The equipment used is called a permeameter. *In situ* measures of permeability can be obtained in boreholes by means of logging techniques.

Uncemented gravel and sand with a permeability of $10–10^5$ darcy are good aquifers. Those with lower darcy values are poor aquifers. Clays with darcy values of less than 10^{-5} are impervious. In the petroleum industry the unit of permeability used is a millidarcy (md = 0.001 darcy). Rocks are classified as follows according to their permeability: fair (1.0–10 md), good (10–100 md), very good (100–1000 md) (Levorsen 1958).

APPENDIX

Technique for Granulometric Analysis

Determination of Grain-Size Frequency ('Weight Frequency') Distribution by Sieving

1. Grain-size frequency ('weight frequency') distributions of loose sand-size grains can be determined by the following method. Well-cemented sandstones must be disaggregated (without breaking the framework grains) before sieving. Sandstones with carbonate cement can be easily disaggregated by boiling chips of the rock in dilute HCl. Ferruginous cement in sandstones can be dissolved by boiling chips of the rock in dilute HCl together with granules of tin (Sn). Nascent Cl^- liberated by the process converts oxides of iron into soluble ferric chloride. Sandstones with siliceous cement cannot be disaggregated without damaging the framework grains. Grain-size distributions of such rocks must be determined under the microscope using a calibrated eyepiece micrometer.

2. Sand samples must be thoroughly dried in hot air oven before sieving. To eliminate organic materials treat the sample with H_2O_2.

3. When the original sample size is too large for sieving a subsample of desired weight must be obtained by splitting. The cone-and-quartering method of sample splitting usually works well but a mechanical sample splitter should preferably be used. Too large a sample clogs the sieves; too small a sample causes delays because of the extreme care needed for collection and weighing.

Recommended initial weights of samples to be placed on the topmost sieve (after Vanoni et al. 1961):

Sand size	Frame diameter of sieves	
	8-inch (20.3 cm)	3-inch (7.6 cm)
Medium sand size	25–50 gm	3–7 gm
Fine sand size	10–20 gm	1.5–3 gm

4. The sieves should be thoroughly cleaned and arranged in a stack from bottom to top in order of increasing sieve diameter in a mechanical ('Ro-Tap') sieve shaker. Ten minutes of shaking is recommended normally but the time may be increased when the sample size is large.
5. All the materials accumulated on each sieve are to be collected and weighed. Gentle tapping and sweeping by a camel hair brush may be necessary for collection of the grains.
6. The results of sieving analysis are to be tabulated in the manner shown in Table 3.5. Net weight of material corresponding to each size class, weight % and cumulative weight % are to be calculated.
7. The results of sieving analysis are to be presented graphically and statistical parameters of grain-size distributions are to be worked out as follows.

A. Graphic Presentation

This involves presentation of the results of sieving in the form of a histogram, frequency polygon and cumulative curve (see Fig. 3.4). The cumulative curve is the best-fitting smooth curve through the cumulative weight frequency plots. Successive points on the graph must *not* be joined by straight lines.

B. Computation of Statistical Parameters

This is done either (i) graphically or (ii) analytically.

i) In *the graphic method* percentiles are computed from the cumulative curve of weight frequency and the values obtained are substituted in appropriate formulae, e.g., the Folk and Ward formulae (Table 3.9) to obtain the graphic measures for mean (M_z), standard deviation (σ_1), skewness (S_{K_1}) and kurtosis (K_G). The results corresponding to Folk and Ward measures for the sample No. USR 23-A are shown in Table 3.8.

ii) The analytical method : Computation of moment measures

The procedure for computation of moments are given in Tables 3.6 and 3.7. The statistical parameters of grain-size distribution are obtained as follows:

The first moment (arithmetic mean), $\bar{X} = \mu_1 = \Sigma(fm)/\Sigma f$, where m and f denote midpoint of the class and weight frequency of the class respectively.

The second moment, $\mu_2 = \sigma^2 = \{\Sigma(fm^2) - (\Sigma fm)\ \mu_1\}/\Sigma f$ measures variance (see Table 3.6). Standard deviation (σ) is the square root of variance. The third (μ_3) and fourth (μ_4) central moments, are functions of skewness and kurtosis of the distribution respectively. Simpler approximations of skewness and kurtosis are given by α_3 and α_4 (see Table 3.7):

$$\alpha_3 = \{\Sigma(m_i - \mu_1)^3 \cdot f_i\}/N\sigma^3;$$
$$\alpha_4 = \{\Sigma(m_i - \mu_1)^4 \cdot f_i\}/N\sigma^4.$$

Values of μ_1, σ, α_3 and α_4 for the sample No. USR 23-A are shown in Table 3.8.

C. Tests for Normality of Distribution

i) Analytical procedure

Whether or not the frequency is 'normally' distributed can be tested either analytically or graphically. In normal distribution β_2 attains the value of 3 and γ_2 is zero, where β_2 and γ_2 are as follows:

$$\beta_2 = \mu_4/\mu_2^2;$$
$$\gamma_2 = \beta_2 - 3,$$

where μ_4 = 4th central moment; μ_2 = 2nd moment = variance. Elaborate computations are necessary for obtaining β_2 and γ_2. A simpler method is to ascertain whether α_3 and α_4 correspond to zero and three respectively.

ii) Graphic procedure

Sedimentologists are accustomed to testing the normality of distribution by plotting the cumulative grain-size frequency on a 'probability paper' (see Fig. 3.7). A symmetrical ogive, which is the cumulative curve corresponding to a normal frequency distribution, has a perfectly linear appearance on a probability plot. The better the fit of a straight line through the plots, the closer the match to a perfectly normal distribution. When the grain-sizes are plotted on a logarithmic (phi) scale, a straight line fit indicates that the distribution is log-normal.

4

SEDIMENTARY PETROLOGY

After transportation and deposition loose sediments are consolidated into sedimentary rocks by a process called diagenesis. The sedimentary rocks thus formed may be broadly divided into two groups: exogenetic and endogenetic. Exogenetic rocks are the products of fragmentation of the source rock outside the basin of sedimentation. Endogenetic rocks are produced out of precipitation from solution within the basin. The exogenetic and endogenetic rocks can be distinguished by their textures. While the endogenetic rocks show an interlocking or crystalline fabric ('non-clastic' texture), the exogenetic rocks show discrete framework grains separated by voids. The latter texture is called 'clastic' from the Greek *clastos*, meaning broken.

Classification of sedimentary rocks into clastic and non-clastic, though popular among sedimentologists, may at times lead to ambiguity. A chemically precipitated sediment for example, may also be transported from its place of origin after fragmentation and deposited elsewhere to form a sedimentary rock. Such rocks, although essentially endogenetic, may show a clastic texture. To avoid confusion in such cases, Folk (1968) introduced the following terms: terrigenous, orthochemical and allochemical.

Terrigenous rocks are produced from sediments derived from land areas located outside the basin. The process of derivation is purely mechanical but not all terrigenous sediments are the products of weathering and erosion. Fragments of volcanic origin and the products of crushing by glaciers also add to the terrigenous sediments. The orthochemical or truly chemical rocks are produced by chemical or biochemical precipitation within the basin. Evaporites such as halite, anhydrite and gypsum and non-evaporites such as limestone, dolomite and ironstone are examples of such precipitates. Intermediate between the terrigenous and orthochemical rocks are the allochemical rocks (*allo* means false). Produced initially within a basin by chemical precipitation, these rocks are fragmented, transported and redeposited elsewhere. Both terrigenous and allochemical rocks show a 'clastic' texture. Only the orthochemical rocks show a non-clastic texture. Sediments of chemical and terrigenous origin may combine to form rocks of a hybrid nature. Calcareous and carbonaceous shales are examples of such hybrid rocks. While accepting the rationality of Folk's scheme of classification, it must be appreciated that the terms clastic and non-clastic have thoroughly pervaded sedimentological literature and cannot be totally ignored at present. These terms will be used in the present text whenever indispensable.

Of all the major groups of sedimentary rocks found on the earth, only three—sandstone, shale and limestone—together account for about 95% of the total. Theoretical calculations, based on the idea that the average chemical composition of the sedimentary rocks must match the average composition of the igneous rocks from which they are derived, lead to the following figures of abundance of the major rock types: shale (81%), sandstone (11%) and limestone (8%) (Garrels and Mackenzie 1971). Actual measurements of the stratigraphic record, however, show a smaller proportion of shale and a higher proportion of sandstone and limestone (Pettijohn 1975).

ROCKS OF MECHANICAL ORIGIN

Sedimentary particles of mechanical (terrigenous) origin are classified according to their size. Particles larger than 2 mm constitute gravel (Latin—*rudite*), those between 2 and 1/16 mm constitute sand (Latin—*arenite*), while those finer than 1/16 mm constitute silt and clay (Latin—*lutite*). Accordingly, sedimentary rocks of mechanical origin are classified broadly into three groups—rudaceous, arenaceous and lutaceous (also called argillaceous). This scheme of classification is followed here for describing rocks of mechanical origin.

Rudaceous Sedimentary Rocks (Conglomerates and Breccias)

Unconsolidated sediments coarser than 2 mm are grouped under the term gravel. Indurated gravels are called conglomerates. Since gravel is classified as granule, pebble, cobble and boulder (see Table 3.1), these terms are often used as prefixes to indicate the size of the dominant fragments in a conglomerate (e.g., pebble conglomerate, boulder conglomerate). While the particles constituting a conglomerate are essentially rounded, a breccia consists of gravel-size clasts which are angular. The term rubble is applied to an assemblage of loose angular fragments of gravel size. Breccias need not be of sedimentary origin only. For example, fault breccias are produced by tectonic activity. Large, well-rounded particles of volcanic origin (volcanic bombs) constitute a rock called an agglomerate.

Like other sedimentary rocks of mechanical origin the space between the framework grains of a conglomerate is filled by finer particles. These fine particles constitute the matrix of the rock. Gravel and sand fragments can combine in various proportions to produce a conglomerate. The resultant rock is named accordingly.

Conglomerates may be classified either in a purely descriptive way (example: chert conglomerate, limestone conglomerate) or according to the mode of their origin (fluvial conglomerate, glacial conglomerate).

In the comprehensive genetic classification proposed by Pettijohn (1975), conglomerates are grouped into four major categories which are geologically meaningful: epiclastic, pyroclastic, cataclastic and meteoric (see Table 4.1).

Epiclastic conglomerates are the commonest of all conglomerates. These consist of gravels of terrigenous origin. The conglomerates of fluvial origin as also those generated on beaches by wave action are of this type. The intraformational conglomerates, which are also epiclastic, are composed of clasts derived from within the basin. Shale-pebbles, produced out of mud cracks, are a good example of intraformational conglomerates, but the process may be more complicated than that.

Extraformational epiclastic conglomerates include oligomicts, petromicts (polymicts), and diamicts. Oligomictic conglomerates consist of pebbles of homogeneous composition. Petromictic (polymictic) conglomerates consist of pebbles of varied rock types. Diamictic conglomerates are characterised by a high proportion of matrix. These may be of glacial (tills or tillites) or non-glacial (tilloid) origin. Pebbly mudstones (argillites) are also included in this group. Conglomerates found in alluvial fan deposits are good examples of tilloids.

The clast-supported conglomerates belonging to the basal Bababudhan Group of South India may be cited as an Indian example of oligomictic conglomerate of Archaean age. Vein quartz pebbles predominate within these conglomerates but some quartzite pebbles (metacherts?) are also present. These conglomerates are believed to have been derived from a granitic or gneissic source and deposited on a braided stream plain (Srinivasan and Ojakangas, 1986). The Talchir (Lower Permian) diamictites of India are some of the well-known Indian glacial conglomerates (Fig. 4.1). The pebbles within these conglomerates show striations and grooves on the surface and flattening of faces parallel to the longest axis (Fig. 4.2). Talchir diamictites of similar character have also been reported, among others, by Casshyap and Tewari (1982), Sen (1991).

Other important types of conglomerates include pyroclastic conglomerates of volcanic origin, cataclastic conglomerates produced by earth movements and meteoric conglomerates produced by meteoric impact. Pyroclastic conglomerates, produced by agglomeration of pieces of lava which solidified in flight, having been hurled into the air during volcanic eruption, are termed agglomerates.

Table 4.1. Classification of conglomerates and breccias (modified after Pettijohn 1975)

Type of conglomerate	Name of conglomerate	Description	Origin and examples
	Oligomictic conglomerate	Consists of pebbles of same composition. Matrix proportion low.	Derived from vein quartz or a monomineralic rock (most fluvial, wave or beach conglo.).
Epiclastic	Polymictic (Petromictic of Pettijohn) Conglomerate	Consists of pebbles of varied types. Matrix proportion low.	Derived from rocks of varying composition, e.g. granite (fluvial, wave or beach conglomerate).
	Diamictite: Conglomeratic argillite	Laminated argillites with ice rafted pebbles.	e.g. varves.
	Diamictite: Tillites	High matrix conglomerates with striated, faceted, pentagonal boulders.	Glacial.
	Diamictite: Tilloid	Chaotic assemblage of non-glacial boulders in muddy matrix.	Non-glacial (example: conglomerates originating from subaqueous gravity flow as in alluvial fans.
	Intraformational conglomerate	Mud, or shale-pebble intraclasts.	Temporary phases of subaerial erosion.
Pyroclastic	Volcanic Breccia and agglomerate	Blocks of previously deposited volcanic material. Volcanic bombs solidifed in flight.	Volcanic.
Cataclastic	Breccia	Angular fragments, crushed pieces, fault gouge, slickenside blocks.	Slumping, folding, faulting etc.
Meteoric	Impact Breccia	Effects of shock recognisable.	Meteoric impact.

Fig. 4.1: A pentagonal granite gneiss boulder within Talchir diamictite, Saharjuri coalfield, Bihar, India.

Arenaceous Sedimentary Rocks (Sandstones)

Terrigenous sedimentary rocks composed mainly (~70%) of sand-size particles are called sandstones. Sand may be deposited in a variety of environments ranging from marine to non-marine, through mixed (see Chapter 6). Sand-size fragments of quartz, feldspars, micas, heavy minerals and rock fragments constitute the bulk of framework grains of sandstone. Rocks composed of carbonate or pyroclastic grains are not treated as sandstones. The voids between the framework grains of sandstones are filled either by finer grains of terrigenous origin or by chemically precipitated cement. Under a microscope the cement shows a crystalline texture but the matrix is composed of discrete, detrital particles. Matrix is a relative term. Particulate materials (irrespective of their actual sizes) distinctly smaller than the framework grains constitute the matrix of a terrigenous rock. In the case of a conglomerate, sand-size grains constitute the matrix, whereas in sandstones the matrix is composed of fine, silt-size grains only.

Classification of Sandstones

Sandstones are classified by observable petrographic criteria. Proportions of quartz, feldspar, rock fragments and mica are commonly used for this purpose. Being the most stable of all detrital grains, quartz survives

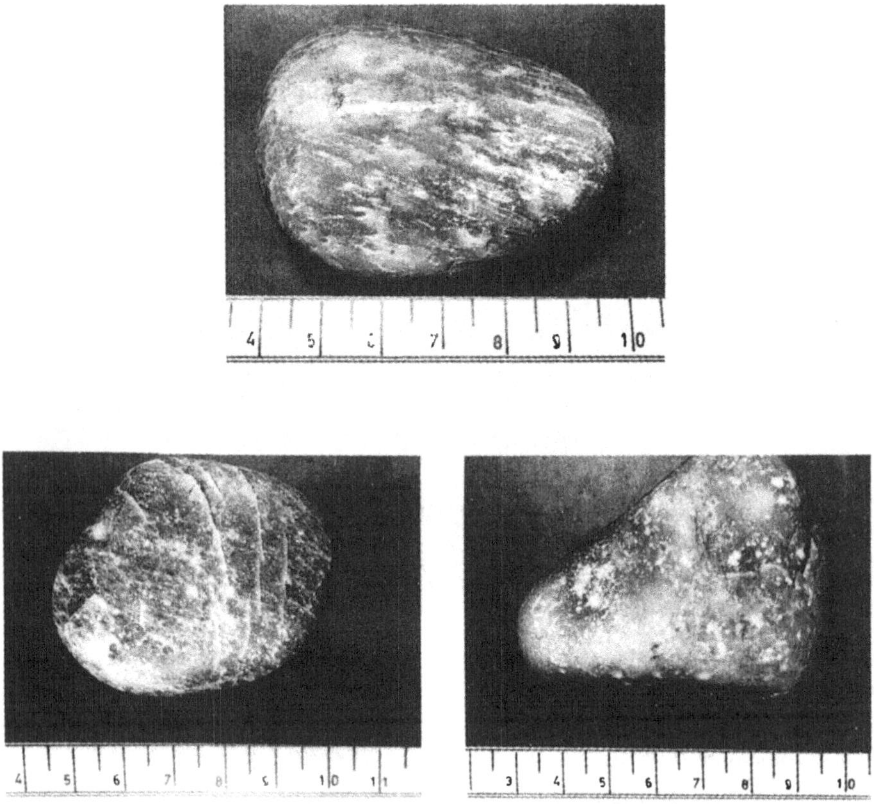

Fig. 4.2: Typical glaciated pebbles from the Talchir Boulder Bed, Giridih, Bihar, India. Note pentagonal shapes and striated, pitted surfaces. Scale in cm.

prolonged weathering and abrasion at the source and also during transportation. A sandstone composed wholly of quartz grains is considered the most matured. For this reason the ratio of quartz to the relatively unstable components, feldspar and rock fragments, is used as an indicator of the mineralogical maturity of a sandstone. The proportion of feldspars and rock fragments is indicative of the nature of the source rock.

When sand is transported in nature by running water, the finer fractions are carried in suspension while the coarser ones are moved along the bottom. The finer fractions eventually contribute to the matrix of the sandstone. The fluidity of a running body of water being inversely proportional to its suspended load, it is argued that a large amount of matrix in a sandstone is indicative of low fluidity of the transporting agency. The matrix content of a sandstone is thus important for its interpretative value (Pettijohn 1975). This idea, however, has been challenged by some authors on the ground that a part or whole of the matrix may be of secondary origin. Increase of matrix content of many sandstones with increase of age or with decrease of framework grain-size led to such a conclusion (Dickinson 1970; Okada 1971).

Many different schemes of classification of sandstones have been proposed from time to time by various authors using the four criteria—quartz, feldspar, rock fragments and matrix. The difficulty involved in plotting four variables on two-dimensional diagrams prompted some authors to use two separate triangles—one devoted to high-matrix sandstones (called wackes) and the other to low-matrix sandstones (called arenites). In each case the end members of the triangles are quartz, feldspar and rock fragments (Gilbert 1954). Arkosic and lithic are the adjectives used for sandstones having high (>25%) feldspathic and lithic contents respectively.

While the schemes using two separate triangles for wackes and arenites have practical advantages, they divide sandstones into two rigid compartments. In reality, however, a sandstone may have any amount of matrix ranging from zero to about 30%. The idea of a continuous change in the matrix content of sandstones was introduced by Dott (1964) who used a three-dimensional prism instead of two fixed triangles. Sandstones of various matrix content were represented by triangular sections of this prism. The boundary between arenites and wackes was arbitrarily placed by Dott at a matrix content of 15%. The end members of each triangle were quartz, feldspar and rock fragments. This technique of representation was a closer simulation of the real situation than that proposed by other authors.

Figure 4.3 gives a modified version of Dott's scheme of classification. The demarcation between quartz arenite and quartz wacke here has been put at $12\frac{1}{2}$% (or 1/8th of the field of view of the microscope, which is easier to estimate than 15%). The boundary between sandstone and mudstone has been put at 30% because a sandstone, by definition, must have about 70% sand-size grains. Quartz and metaquartzite fragments have been grouped together because they behave similarly from the point of view of mineralogical maturity. Chert grains have been included in rock fragments because chert is a distinct indicator of provenance (Okada 1971). For reasons to be discussed later, graywackes have been plotted in a separate triangle at the high matrix end of the prism. The daughter triangle of Fig. 4.3 gives a detailed classification of lithic arenites as proposed by Folk (1968). Sandstones with appreciable quantities of slate, phyllite, mica schist and other phylloid fragments of low-grade metamorphism are termed phyllarenites. Those containing large proportions of detrital limestone fragments are called calclithites. Sandstones with chert or volcanic fragments are named chert and volcanic arenites respectively.

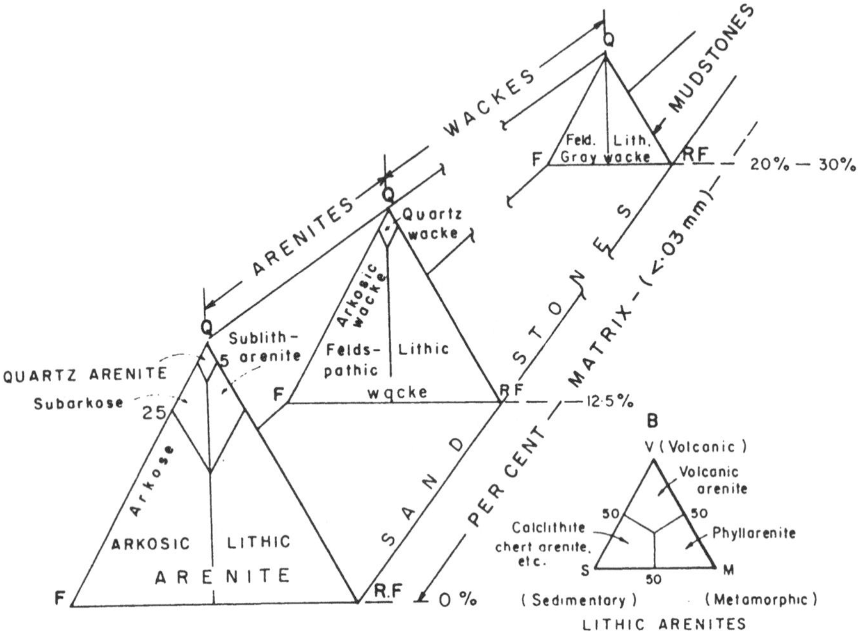

Fig. 4.3: A revised version of Dott's (1964) scheme of classification of sandstones. Q = quartz + metaquartzite, F = feldspars, R.F = rock fragments. The daughter triangle shows subdivision of lithic arenites after Folk (1968).

Description of Sandstones

Quartz Arenite (also called Orthoquartzite)

Quartz arenites consist almost entirely of quartz grains cemented together by silica or welded together by predissolved quartz (see diagenesis). Chemical precipitates other than silica may also serve as cement. The quartz grains constituting the framework are often well rounded and well sorted (Fig. 4.4). A small amount of heavy minerals of stable variety (tourmaline, zircon, rutile, ilmenite) may be present in quartz arenites. The Rewa sandstones of the Vindhyan Supergroup may be cited as an Indian example of quartz arenite. Modal analyses of representative samples of Rewa sandstones are presented in Table 4.2.

The detrital grains of quartz in arenites may be produced by erosion of vein quartz. They may also be derived from one of the following source rocks through prolonged weathering, erosion and transportation under high energy conditions (Folk 1968):

1. Plutonic (granitic) source → arkose → subarkose → quartz arenite
2. Metamorphic source → graywacke → subgraywacke → quartz arenite

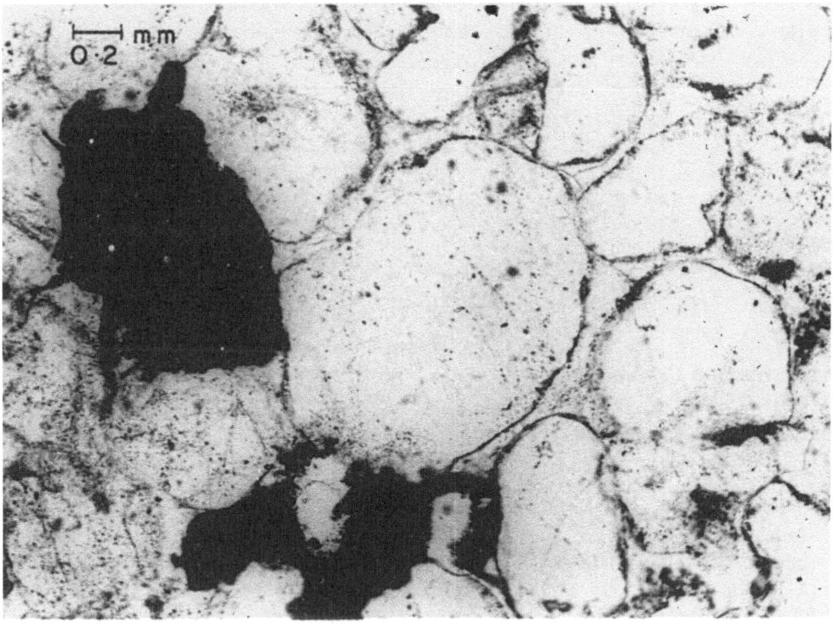

Fig. 4.4: Photomicrograph of a quartz arenite under plane polarised light. Upper Proterozoic Chandarpur Formation near Mahasamund, Raipur, M.P., India (thin section No. SCH-3, courtesy of S.S. Das and C.N. Rao).

3. Sedimentary source → submature orthoquartzite → mature orthoquartzite (quartz arenite).

Chemically, quartz arenites consist of a very large proportion (>95%) of SiO_2 and small amounts of CaO, CO_2, Al_2O_3, FeO + Fe_2O_3. The chemical composition is controlled mainly by the type and proportion of heavy minerals and cementing materials in the rock (see Table 4.3).

QUARTZ WACKE

Sandstones with a fairly large proportion (say > 12.5%) of fine-grained matrix are termed quartz wackes. The presence of a large proportion of matrix, a wide range of grain-size variation, lack of rounding and sorting of the framework grains are indicative of textural immaturity. The presence of feldspars and rock fragments of various types indicates mineralogical immaturity. A fine-grained matrix is the dominant binding material in these rocks. A large amount of matrix in wackes leaves hardly any space for cement (Fig. 4.5). In some quartz wackes replacement of the matrix by chemically precipitated cement may create the illusion of an arenite, but the textural immaturity of these sandstones is clearly indicated by lack of rounding

Table 4.2. Modal analyses of some Indian sandstones

Sample No.	Quartz & Quartzite	Feldspars	Micas	Rock fragments	Cement (calcareous)	Cement (siliceous)	Cement (ferruginous)	Matrix	Total
Quartz Arenites									
20.4*	93.90	—	—	0.80	—	3.45	1.72	—	99.87
20.1*	88.53	—	—	0.18	—	5.83	5.46	—	100.00
20.10*	86.16	—	—	0.21	—	3.04	10.38	—	99.79
Quartz Wackes									
AM 18W+	66.00	3.00	8.00	—	2.00	—	3.00	18.00	100.00
AM 13+	72.00	2.00	3.00	—	—	—	7.00	16.00	100.00
AM 8+	75.00	2.00	2.00	—	—	—	5.00	16.00	100.00
Arkosic Wacke & Calcareous Arkose									
M 30**	52.57	25.11	3.14	—	—	—	4.18	14.99	99.99
M 46**	45.57	20.05	—	—	30.86	—	—	3.52	100.00
M 5**	46.54	20.66	13.90	—	—	—	0.21	18.69	100.00

* Rewa sandstones, Vindhyan Supergroup (Precambrian), Mihar, M.P., India (Basumallick 1962).

+ Sandstones from Ironstone Shale Formation (Permian), Raniganj coalfield, W.B., India (A. Mandal 1985, Unpub. M.Sc. thesis, IIT, Kharagpur).

** Sandstones from Raniganj Formation (Permian), Jharia coalfield, Bihar, India (S. Nayak 1985 Unpub. M.Sc. thesis, IIT, Kharagpur).

and sorting of the framework grains. A high proportion of trapped detrital clay in the interstices of framework grains (protomatrix) reflects an abundance of suspension load in the transporting medium, as in a turbid river during flood, but a certain amount of matrix may also be produced at a later stage by diagenetic alteration of sand-size grains (epimatrix), or deformation and squashing of pelitic fragments (pseudomatrix). Recrystallised material occurring between framework grains constitutes orthomatrix. The sandstones belonging to the Barakar, Ironstone shale and Raniganj Formations of the Indian Gondwana are good examples of quartz wackes. Mineralogical compositions of some of the typical quartz wackes are given in Table 4.2.

Feldspathic Sandstone and Arkose

A sandstone with a high (>10%) proportion of feldspar is designated as a feldspathic sandstone. When the matrix content of such a rock is also high, the rock is called a feldspathic wacke. Low matrix feldspathic sandstones are classified as arkose or subarkose depending on whether the feldspar content is greater or less than 25%. Arkose therefore is the name restricted to arkosic arenites with a high (>25%) proportion of feldspar and a very low amount (<10%) of matrix (Fig. 4.6). The term arkose should

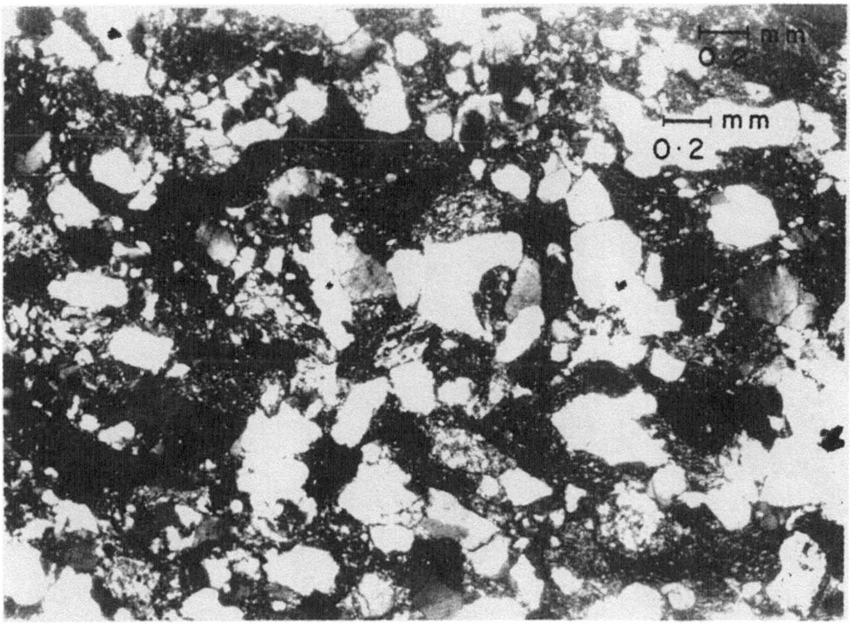

Fig. 4.5: Photomicrograph of a quartz wacke under cross-polarisers. Nahan Group, Koshailia Nala, Kalka, India (thin section No. DP 25/68, courtesy of D.P. Sen).

be used with caution. Many feldspar-bearing sandstones which seem to be arkose at first glance, turn out on detailed scrutiny to be subarkose or arkosic wacke.

Next to quartz and feldspar, the mineral which is frequently present in arkose is mica, both of the muscovite and biotite varieties (Table 4.2). Calcite is a common cementing material. Since arkoses are commonly derived from granitic sources, the proportion of K-feldspar always exceeds plagioclase. In chemical analysis this is reflected by a higher proportion of K_2O and a lower proportion of Na_2O (see Table 4.3).

The presence of large proportions of fresh grains of an unstable mineral such as feldspar in arkose is puzzling. For a long time sedimentologists maintained that lack of moisture in the environment is essential for preservation of freshness of the feldspar grains. A warm-arid or cold-dry condition was suggested as the essential prerequisite. A later discovery of fresh feldspars in regoliths produced under humid, tropical conditions led to the conclusion that the key factor in the preservation of feldspar is tectonics, not climate. Quick burial without much transportation, due to sudden upheaval of the source and rapid subsidence of the area of deposition, may lead to preservation of fresh feldspar grains. It should be remembered, however,

Fig. 4.6: Photomicrograph of an arkose under cross-polarisers. Note large grains of feldspar. Barakar Formation, Central Godavari Valley, India (thin section No. SSG-1, author's collection).

Table 4.3. Average chemical composition of major sandstone types (data from Pettijohn 1975, Table 7.3; averaged by Potter, Maynard and Pryor 1980)

	Orthoquartzite (wt.%)	Graywacke (wt.%)	Arkose (wt.%)
SiO_2	95.4	66.7	77.1
Al_2O_3	1.1	13.5	8.7
Fe_2O_3	0.4	1.6	1.5
FeO	0.2	3.5	0.7
MgO	0.1	2.1	0.5
CaO	1.6	2.5	2.7
Na_2O	0.1	2.9	1.5
K_2O	0.2	2.0	2.8
CO_2	1.1	1.2	2.8
C	—	0.1	—
H_2O	0.3	3.0	0.9

that both climate and tectonism could have played important roles in the preservation of fresh feldspars in some sandstones.

GRAYWACKE

Graywackes, which look very much like volcanic igneous rocks in hand specimens, belong to an exotic group of sandstones. Typical graywackes, as described by Krynine (1948), contain a medium- to fine-grained, angular framework of quartz, chert and rock fragments such as slate, schist, phyllite, set in a large proportion of fine-grained matrix. Silt-size grains of illite, chlorite, micas, and various alteration products of basic igneous rocks constitute the matrix. Feldspars, abundant in graywackes derived from high-rank metamorphic sources, are usually of the sodic plagioclase variety.

Graywackes, often found in association with a huge thickness of shales and radiolarian chert, are believed to be the product of sand carried into the deep sea by density (turbidity) currents although not all turbidites are graywackes (for example, see Okada 1966). Typical graywackes are commonly associated with altered basic intrusives ('greenstones') and often grade into volcanic wackes. Most of these rocks have been reported from orogenic belts of Precambrian to Cenozoic age.

The term graywacke was originally assigned by the 18th century miners in the Harz mountain in Germany to a rock with an exceptionally high matrix content (hence called grau + wacke, the German for gray, pasty matter; 'gray grit' or 'dark, dirty sand'; Krynine 1948). Since that time the term has been used for different types of rocks or even assemblages of rocks. The definition of graywackes became so ambiguous that any dark-coloured rock not positively known to be igneous came to be jestfully referred to as a graywacke in the Canadian Shield. For these reasons questions have been raised about the desirability of retaining the term graywacke. Tectonic or turbidite origin is implied in the classical usage of the term, but rocks similar to graywackes have been reported from fluvio-estuarine sediments also, while clean, non-graywacke sand has been found interbedded with argillaceous sediments of deep-water origin. Compositions of the type graywacke of the Harz mountain of Germany, and those of some arkose of France are known to overlap (Okada 1971). Considering all these, some authors recommend replacement of the term graywacke by a more general term, such as 'wacke' (Okada 1971; Friedman and Sanders 1978; Fritz and Moore 1988). In his three-component scheme of classification Okada (1971) recommended use of only two groups of sandstones: arenites and wackes. The feldspathic and lithic wackes of this scheme encompass all sandstones, including those that have been traditionally designated as graywackes.

The terms graywacke and arkose have long pervaded sedimentological literature and discontinuation of these terms at present would merely add to the existing confusion. Moreover, graywackes, with their abundant matrix, and angular framework grains, create a distinctive textural pattern (see Figs. 4.7 and 4.8). It is proposed here to include these sandstones in a new

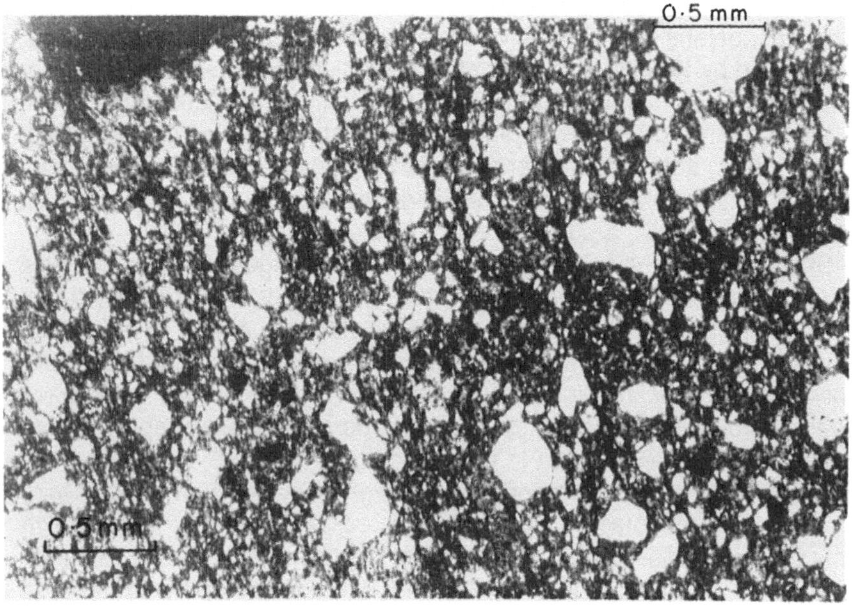

Fig. 4.7: Photomicrograph of a graywacke, Chandpur Formation (Cambro-Ordovician ?), northwest of Naini Tal, India. Plane polarised light. Note high proportion of matrix (thin section No. N-1, author's collection).

triangular field at the high matrix end of the revised Dott diagram (Fig. 4.3). Consequently, the term graywacke has been removed from its usual place in the original Dott diagram of 1964.

The problem of the occurrence of a large proportion of matrix in graywackes was studied experimentally (Kuenen 1966, Kuenen and Sengupta 1970). More recent experiments (Ghosh, Mazumder, Saha and Sengupta 1986) indicate that, contrary to common belief, a wide range of grain-sizes, from coarse to fine, may be simultaneously deposited from a suspension when the flow velocity of a stream carrying a large amount of suspension load is decreased. The fine-grained particles deposited in the intergranular pores of the coarser framework grains might, in such cases, produce a texture similar to that of a graywacke.

Chemical Compositions of Sandstones

Average chemical compositions of orthoquartzites, arkoses and graywackes are shown in Table 4.3. As expected, an orthoquartzite consists almost wholly of silica, whereas the proportion of alumina is quite high both in arkose and graywacke. The presence of aluminosilicates, like feldspars, accounts for the alumina in these rocks. In arkose potash exceeds soda

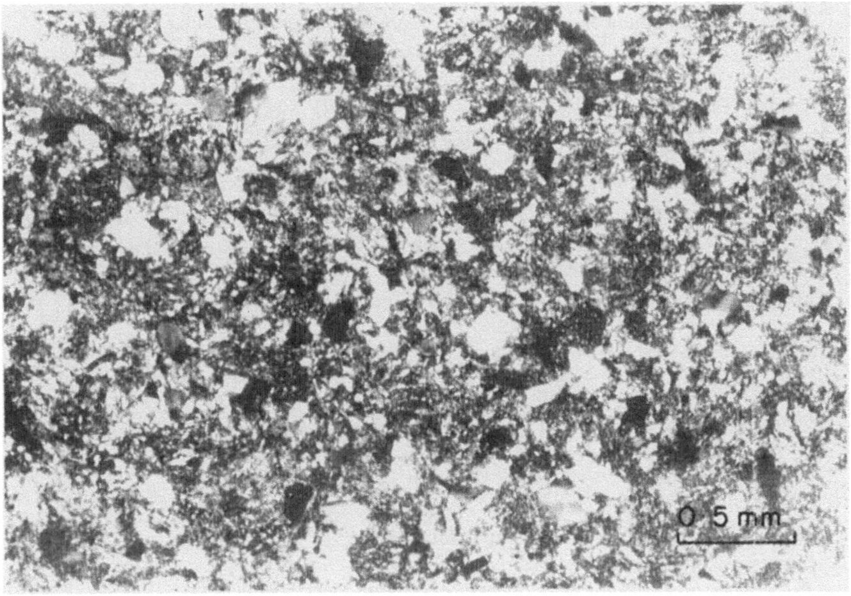

Fig. 4.8: Photomicrograph of graywacke from Mt. Marion Formation (Middle Devonian), northeast of Monroe, USA. Note angular framework grains 'floating' in a high proportion of matrix (Plane polarised light). Author's collection.

because of the presence of potash feldspars but the proportion is often reversed in graywackes due to the presence of sodic plagioclase.

Chemical analyses of Indian sedimentary rocks are rare. The average chemical composition of a few Gondwana sandstones from the Raniganj coalfield, West Bengal, are reproduced in Table 4.4 from Suttner and Dutta (1986). The average chemical composition of lithic arenites (LA) and arkoses (AR) is also included in this table. In each of these analyses potash exceeds soda, indicating the presence of K-feldspars. The calcium found in these rocks is apparently derived from the calcium carbonate cement.

Average modal analyses of the Gondwana sandstones from the Jharia and Raniganj basins have already been shown in Table 4.2. The feldspars of the sandstones belonging to the Ironstone Shale Formation of Raniganj coalfield account for the high Al_2O_3 content of these rocks. The presence of appreciable quantities of FeO and CaO is obviously due to the ferruginous and carbonate cements.

The chemical and mineralogical composition of graywackes may vary considerably, depending on their occurrence in relation to particular geotectonic settings. These are discussed in detail in Chapter 7.

Table 4.4. Average bulk chemical composition of Gondwana sandstones from Raniganj basin, India (after Suttner and Dutta 1986)

	North America		Gondwana, India				
	LA (Blatt et al. 1972)	AR	I (n = 5)	II (n = 7)	IV (n = 11)	V (n = 2)	VI (n = 4)
SiO_2	66.1	77.1	70.30	79.60	72.70	93.50	92.15
Al_2O_3	8.1	8.7	10.50	7.53	9.31	2.46	3.12
K_2O	1.3	2.8	1.99	1.71	2.25	0.78	0.35
Na_2O	0.9	1.5	1.29	0.41	1.20	0.04	0.02
CaO	6.2	2.7	2.34	1.23	3.68	0.08	0.05
FeO	1.4	0.7	3.42	1.01	2.90	0.60	1.31
MgO	2.4	0.5	0.68	0.49	0.77	0.07	0.06
MnO	-	-	0.58	0.03	0.07	0.01	0.02
TiO_2	-	-	0.27	0.29	0.47	0.11	0.48
P_2O_5	-	-	0.36	0.25	0.26	0.50	0.35
Ignition loss	-	-	5.50	4.51	4.18	-	-
SiO_2/Al_2O_3	8.3	8.7	6.7	10.9	8.2	38.4	29.5

LA — average lithic arenite; AR—average arkose.
VI — sandstones from upper part of Panchet Formation
IV — sandstones from upper part of Barren Measures ('Ironstone Shale') Formation + Lower Panchet Formation.
II— sandstones from Barakar and lower part of Barren Measures ('Ironstone shale') Formation.

Lutaceous (Argillaceous) Sedimentary Rocks

Lutaceous sedimentary rocks are composed mainly of grains smaller than sand size. Sediments finer than sand but coarser than clay are classified as silt. Silt and clay together constitute mud. Indurated silt is called siltstone. Indurated, non-laminated mud forms mudstone. When mudstone is laminated and fissile, the rock is designated as shale. About 1/3rd to 2/3rd of such rocks are composed of clay-size particles, the rest being silt (Table 4.5). The fissility of shales is attributed to parallel arrangement of the constituent flaky minerals. Black shales, rich in organic matter, are particularly fissile while calcareous shales have no fissility at all.

Shales constitute more than half of the stratigraphic record and hence are important. A study of shales is not as easy as that of sandstones, however, because the grain-size involved is very small. A study of the clay minerals which constitute a substantial part of all shales is possible only with advanced techniques. Shale chemistry is considered more important than shale petrology. SiO_2 and Al_2O_3 together constitute the dominant part of common shales. H_2O and CO_2 constitute about 10%. Five other oxides, Fe_2O_3, FeO, K_2O, MgO and CaO account for about 15–20% of common shales. Potash generally exceeds soda. In Table 4.6 the chemical compo-

Table 4.5. Classification of indurated argillaceous rocks (more than 50% of grains finer than sand). (Modified after Potter, Maynard and Pryor 1980)

Proportion of clay-size constituents (rest mostly silt)	0–32%	33–65%	66–100%
Beds > 10 mm thick	Bedded Siltstone	Mudstone	Claystone
Laminae < 10 mm thick	Laminated Siltstone	Mudstone (Shale)	Clay-shale

Table 4.6. Chemical composition of shales

	1	2	3	4	5	6
SiO_2	58.10	58.93	62.15	39.05	48.06	64.57
Al_2O_3	15.40	18.17	11.38	29.07	19.31	11.93
Fe_2O_3	4.02	9.55	6.09	0.51	1.56	13.47
FeO	2.45			24.97	26.05	
MgO	2.44	2.90	5.06	0.34	2.10	1.31
CaO	3.11	0.13	4.01	2.22	4.12	0.87
Na_2O	1.30	0.40	nd	1.36	nd	0.84
K_2O	3.24	5.78	4.07	0.95	nd	-
TiO_2	-	0.92	0.68	0.03	Tr	0.80
MnO	-	0.03	0.06	0.36	0.12	0.02
P_2O_5	-	0.12	0.09	0.06	nd	0.20
S	-	0.005	nd	nd	nd	2.27
Cr_2O_3	-	0.02	0.01	-	-	
NiO	-	0.004	0.001	-	-	
BaO	-	0.05	0.04	-	-	
CO_2	2.63					
C LOI	0.80	3.18	7.52	*	*	4.12
H_2O	5.00					

1. Average shale (Pettijohn 1975, Table 8.7, averaged by Potter et al. 1980).
2. & 3. Precambrian shales from Varikunta-Zangamrajupalle, Cuddapah, A.P., India (Geological Survey of India 1986, source: V. Subrahmanyam). /
4. Naturally fused Permian shale ('para lava'), Bokaro Gondwana basin, India (Fermor 1918).
5. Naturally fused Permian shale ('para lava'), Jharia Gondwana basin, India (Sengupta 1957).
*The volatiles have escaped during spontaneous combustion and fusion.
6. Bijaigarh Shale, Precambrian, Vindhyan 'System' (Awasthi 1964, cited by Singh 1980).
nd = not determined.

sition of some Indian shales of Precambrian and Permian age has been compared with that of the average composition of shales.

Composition and Classification

Besides clay minerals, which constitute the bulk, some grains of quartz, feldspar, carbonates and organic matter are found in all shales. The non-terrigenous constituents include glauconite, volcanic glass, biogenic silica and some phosphatic compounds. Shales may be classified by their colours, which reflect chemical composition. With increasing Fe^{3+}/Fe^{2+} ratio, the colour of a shale changes from grey to red, through green and yellow (McBride 1974). Red shales, found near the surface, indicate the presence of hydrated ferric oxide minerals, which are formed under oxidising conditions. Since an arid or semi-arid condition helps oxidation, red shales may be taken to be indicative of an arid palaeo-environment. A red colour is also common in deep-marine clays. This is partly due to oxidation of the ferruginous minerals carried into the sea by streams and partly due to the volcanic dusts that settle into the deep sea from the atmosphere. Clay-size ferruginous matter, settling through the oxygen-rich sea-water at an extremely slow rate, may become oxidised over the years. Similarity in the chemical composition of some deep-sea clays and an average igneous rock suggests their volcanic origin (Dunbar and Rodgers 1957, p. 57).

The proportion of organic carbon is another important constituent influencing the colour of shales. With an increasing amount of organic carbon, grey shales tend to be black. The presence of pyrite and siderite together with trace elements such as vanadium, nickel and uranium in black shales suggests that these shales were deposited under a reducing environment (see discussions under euxinic basins).

CLAY MINERALS IN SHALES

Clay minerals (hydrous aluminosilicates) such as kaolinite, illite and smectite are important constituents of shales. Gibbsite (aluminium hydroxide) is also common. The type of clay mineral present in a particular shale depends on the nature of the source rock, depositional environment, climate and also, to some extent, the degree of diagenesis undergone. Kaolinite is generally the product of warm, humid, tropical conditions of weathering. In areas of low rainfall the proportion of smectite increases compared to kaolinite. Illite, very common in shales, may be derived either from pre-existing shales in the source area or by diagenesis of clay minerals such as smectite. Chlorite can also be produced in this way. In fact, illite and chlorite are common in older shales while kaolinite and smectite are characteristic of younger shales.

Depositional Environment of Shales

A certain amount of low-energy condition is essential for deposition of all fine-grained sediments. The deposits of the deep sea include blue-grey mud of terrigenous origin, green mud of glauconitic origin, and red mud produced either out of volcanic dusts or from clay-size particles of ferruginous sediments washed into the sea. Planktonic fauna may be associated with these deposits when they are laid down at relatively shallow depths. At greater depths siliceous fauna predominates. Ocean-floor clay deposits may extend for hundreds of kilometres. Elongate bodies of silt may be deposited on the inner and outer shelf areas, if the energy condition there is sufficiently low.

Deposition of silt and clay is in no way restricted to a marine environment only. Finely suspended silt and clay spilling over river channels during high flood may produce extensive sheets of mud in the flood basins. Settling by flocculation in overbank areas also causes deposition of mud. In large lakes the coarser particles are trapped near the shore but the finer mud travels into the lake centre, to be dropped there from suspension. These form extensive, ovate muddy deposits at the lake bottom. Invertebrate shells of various types, spores and algae are common in these deposits. In closed basins, whether marine or non-marine, restricted circulation may lead to the deposition of black shales. Nearshore areas of fluctuating marine and continental conditions can also trap a large amount of mud. Deltas, lagoons and tidal flats are examples of mixed environments which can have large deposits of silt and clay.

Many shale sections in stratigraphic record demonstrate cyclicity. These may be either varves or non-varves, such as alteration of shale and limestone. Some of these features are discussed later (Chapter 5).

ROCKS OF CHEMICAL AND BIOCHEMICAL ORIGIN

Carbonates

Sedimentary rocks of chemical and biochemical origin are divided into two groups—evaporites and non-evaporites. Non-evaporites constitute the bulk of chemical and biochemical rocks. Among the non-evaporites the most important are the carbonate rocks. The carbonate minerals forming these rocks include calcium carbonates (calcite and aragonite), iron carbonate (siderite) and also calcium-magnesium carbonate (dolomite). The term limestone is generally restricted to the calcitic and aragonitic rocks. Limestones composed essentially of the mineral dolomite ($CaCO_3 \cdot MgCO_3$) are called dolostones. Approximately one-quarter of the stratigraphic record is constituted of limestones and dolostones. Limestones occur in all ages, from the Precambrian to the Recent, but are most common in the Late Precambrian (Proterozoic). They are also abundant in the early Palaeozoic.

Limestone is found today in almost all environments, both marine and non-marine (Fig. 4.9). It is very common in shallow (< 10 m) marine conditions of the tropics, the carbonate saturation in water being highest in tropical to warm, temperate seas. Under this condition a slight increase in temperature or alkalinity causes the carbonate to precipitate. These are the areas where extensive reefal growths are found today. The Great Barrier Reef off the Australian coast and the coral reefs of the Florida-Bahama banks of the USA are good examples of such reefal growth.

In relatively deeper water, up to a depth of about 3000 m, pteropod ooze, coccoliths and foraminifera predominate. The carbonate precipitation in this zone is essentially in the form of aragonite (orthorhombic $CaCO_3$). Between 3000 m and 5000 m globigerina ooze, consisting mostly of low-magnesian calcite (hexagonal $CaCO_3$), dominates the ocean floor. In the deeper parts of the ocean, where the water is cold, pH is low and pressure is high, carbonate remains mostly in solution. Increase of partial pressure of CO_2 (pCO_2) due to dissolved CO_2 at great depth also allows ocean-water to hold more carbonate in solution. The deposits of the abyssal ocean floor contain only about 20% carbonate, whereas the deposits in the shallower parts of the sea contain 90%. 'The depth at which the rate of dissolution of solid calcium carbonate equals the rate of supply is called the carbonate compensation depth' (Friedman and Sanders 1978). Deep-water carbonate, wherever found, has been either transported from the shallower parts of the ocean by suboceanic density currents, or dropped onto the ocean floor from the pelagic zone. Coccolith ooze, formed by the latter process, is an important constituent of fine-grained limestone (chalks).

Limestones of non-marine origin include algal mounds formed in fresh-water lakes. These are mixtures of carbonates and argillaceous sediments. The corresponding rocks are called marlstones. Travertines in hot springs, spongy, porous carbonate deposits (tufa) associated with natural springs and dense, elongate, vertical bodies (stalactites and stalagmites) formed within caves, are other examples of freshwater limestone deposits. All these are mostly calcitic. In dry and semi-arid regions a thin layer of carbonate deposit occurs on the surface soil together with some silica, alumina and iron oxide. This is formed by evaporation of lime-rich water drawn to the surface by capillary action. This deposit is termed caliche and its indurated counterpart is the calcrete. Similar calcareous crusts are called Kankar in India, Nari in the Middle East and Kalkkruste in Germany. In Chile and Peru the term caliche is used for deposits of impure nitrate of soda together with sulphates, iodates and phosphates.

Chemical Composition of Limestones

All limestones contain appreciable quantities of Si, Al and Fe. These are the acid insoluble residues (Table 4.7). A number of trace elements,

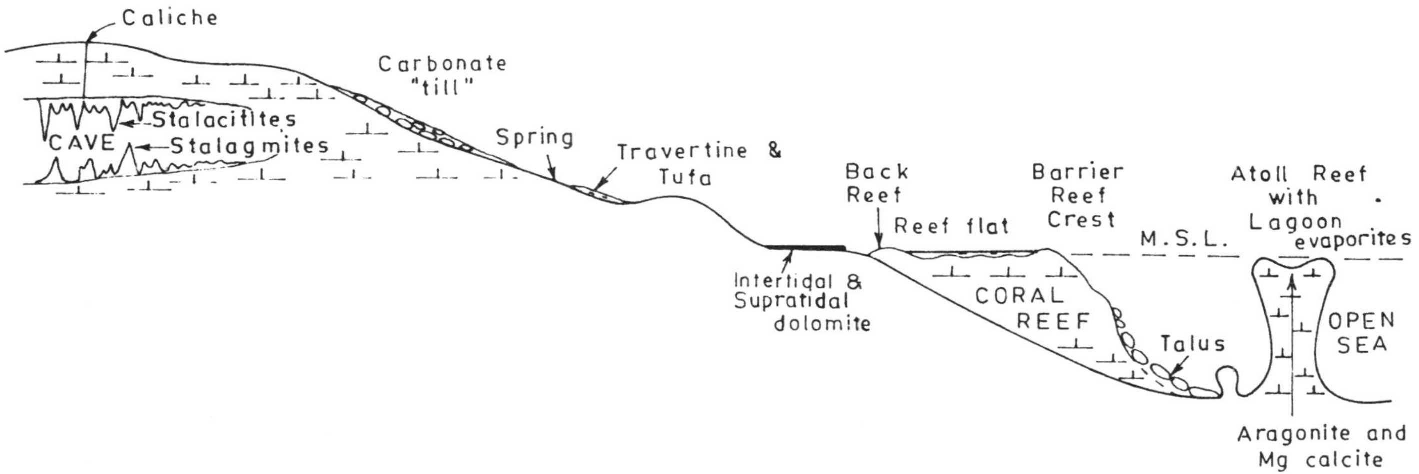

Fig. 4.9: Schematic representation of depositional environments of carbonates.

Table 4.7. Chemical composition of limestones and dolomites (percentage of major oxides)

	1	2	3	4	5	6	7	8
SiO_2	5.19	1.15	0.09	8.774	15.88	10.51	11.44	4.12
Al_2O_3	0.81	0.45		0.762	6.79	3.28	2.85	1.24
Fe_2O_3	0.54	-	0.11	0.720	8.69	3.36	1.15	0.92
FeO		0.26		0.745			0.78	0.16
MgO	7.90	0.56	0.35	15.363	1.08	15.49	17.50	2.03
CaO	42.61	53.80	55.37	30.319	34.06	25.79	24.24	50.42
Na_2O	0.05	0.07	-	0.994	nd	nd	1.26	0.84
K_2O	0.33		0.04	0.784	1.43	1.74	0.86	0.67
TiO_2	0.06	-	-	0.141	0.26	0.14	0.33	0.22
MnO	0.05	-	-	0.105	1.71	0.62	0.39	0.30
P_2O_5	0.04	-	-	0.109	0.03	0.02	0.45	0.36
S	-	-	-	-	0.005	nd	-	-
Cr_2O_3	-	-	-	-	0.01	nd	-	-
BaO	-	-	-	-	0.19	0.03	-	-
H_2O^+	0.56	0.69	0.32				0.55	0.77
H_2O^-	0.21	0.23		LOI 41.624	29.86	38.94		
CO_2	41.58	42.69	43.11				38.26	38.65

1. Composite analyses of 345 limestones (Clarke 1924)*
2. Lithographic limestone, Solenhofen, Bavaria (Clarke 1924)*
3. Travertine Hotsprings, Yellowstone, Wyoming (Clarke 1924)*
4. Average composition of 38 limestone samples, Krol Formation, Simla Himalayas (Bagati 1979).
5. Limestone, 6. Dolomite, both from Varikunta-Zangamrajupalle, Cuddapah, India (Geological Survey of India 1986, source: V. Subrahmanyam).
7. Fawn dolomitic limestone (microdolsparite), Son Valley, India (Mehrotra et al. 1975).
8. Rhotas limestone (micrite), Son Valley, India (Mehrotra et al. 1975).

Note: nd = not determined. *Cited by Pettijohn 1975, Table 10.4.

constituting 10^{-2} to 10^{-3} wt% of the rock, are also found in limestones (see Table 4.8). These trace elements attach themselves to the silicates, oxides or carbonates. Intensive studies on these trace elements have been carried out in recent years with the hope that they will provide clues to the microenvironment of deposition and diagenesis of the carbonates. Whether this contention can be statistically substantiated, however, remains to be seen. For example, Sr, commonly believed to be characteristic of lagoons and central reefs, has been found in evaporites and inland lake deposits also (Flügel 1982). Non-reefal carbonate sediments are characterised by trace elements such as Ni, Co, Cr, V, Ba and also, occasionally, Sr. The high concentration of P in the carbonates from the Krol Formation of the Simla Himalayas has been taken to be indicative of nearshore lagoonal environment, and the high concentration of Mn and P as indicative of a warm, humid climate (Bagati 1979).

Table 4.8. Average concentration of trace elements (in ppm) in relation to carbonate microfacies, Krol Formation, Simla Himalaya (Bagati 1979)

Trace elements	Dolomicrite #1	Chert-bearing dolomicrite	Dolomicrite #2	Dolosparite	Siliceous oolite-bearing oodolomicrudite	Intradolosparudite	Micrite	Sandy micrite	Total no. of samples
	4 samples	4 samples	2 samples	5 samples	1 sample	4 samples	2 samples	1 sample	23
Pb	60	50	25	50	50	70	58	50	
Ga	68	145	-	43	55	113	97	88	
V	59	-	75	84	140	14	-	-	
Bi	27	187	30	92	15	129	162	62	
Sn	30	60	-	37	-	57	28	20	
Cu	69	44	30	43	45	42	19	50	
Ni	50	45	25	50	50	50	50	50	
Cr	-	35	-	-	-	21	-	25	
Ba	19	200	15	26	75	99	82	90	
Sr	91	160	135	126	140	131	175	290	
Zn	112	25	13	25	25	112	25	40	

- Sought but not detected.

Note: For definition of rock names such as dolomicrite see text under dolomite.

Carbonate sediments, it must be remembered, form close to the area of deposition either by inorganic precipitation or by organic activity. The chemical conditions prevailing at the site of deposition therefore play important roles in carbonate sedimentation. Oxygen isotopes in limestones have been used extensively for interpreting palaeotemperature and palaeosalinity. The problem in palaeotemperature interpretation is that the oxygen isotope is affected, not only by temperature variation but also by physiology and life habitat of organisms. Interpretation of palaeosalinity from oxygen isotopes gives more dependable and consistent results. Isotope studies can also be used in the interpretation of diagenetic and cementation history of limestones.

The chemical composition of some Indian limestones of different geological ages is compared to the composition of some well-known limestones from Europe and USA in Table 4.7. An average carbonate rock, like the one tabulated in the first column of this table, contains a fairly large quantity of MgO together with large proportions of CaO and CO_2. In fact, calcite ($CaCO_3$) and magnesite ($MgCO_3$) form an isomorphous series within which two varieties of calcite occur. One of the variety, designated as high-magnesian calcite, contains more than 10 mole per cent of $MgCO_3$. The other, designated as low-magnesian calcite, contains less than 8 mole per cent $MgCO_3$. This division is arbitrary. A calcite containing 9 mole per cent $MgCO_3$ may be designated either high- or low-magnesian calcite (Friedman and Sanders 1978, p. 49).

Carbonate Petrography and Classification

Limestones can form by precipitation within a basin or may be redeposited elsewhere, after erosion and transportation from their place of origin by water or air. The *in situ* limestones have been termed autochthonous by Pettijohn (Folk prefers the term orthochemical for such rocks) while the term allochthonous is used for the transported variety (allochemical of Folk). The two types of limestones can be distinguished by their texture and structure. While the grains of allochthonous limestones can be well sorted as to size and have their interstices filled with clear calcite cement, the grains of autochthonous limestones are usually unsorted. The intergranular spaces of such limestones are filled with lime mud. The allochthonous limestones show evidence of transportation, such as cross-bedding structures, while the autochthonous limestones have articulated organic structures, such as undisturbed coral colonies. Carbonate rocks made up of materials eroded from older, lithified carbonate sediments have termed calclithite by Folk (1959) to distinguish them from calcarenite, a term coined by Grabau (1904) for rocks composed of carbonate sand produced by chemical or biochemical precipitation.

The transported and redeposited allochemical carbonates are divided into the following classes by Folk (1959).

Intraclasts. Reworked fragments of penecontemporaneous carbonate sediments. Most of these fragments are made up of microcrystalline or fine-grained carbonates.

Fossils. Broken or unbroken skeletons and tests of carbonate-secreting organisms. These include foraminifera, corals, echinoderms, bryozoans, molluscs, brachiopods and other shells.

Oolites. Spherical carbonate bodies, 0.1 to 1.0 mm in diameter, having concentric or radial internal structures.

Pellets. Extremely small (0.03–0.15 mm) carbonate bodies of spherical or elliptical shape, similar to fecal pellets. They lack the internal structures which characterise oolites.

The orthochemical components are divided into two groups:

Micrite. Microcrystalline calcitic deposits of a size comparable to terrigenous clay (0.001 to 0.004 mm). The constitutent grains can hardly be resolved, except under high magnification. White or grey (with waxy to dull lustre) to nearly black in hand specimens, micrite is believed to be deposited as biochemical ooze or produced by attrition of organic shells. Some of the ancient micrites are possibly of diagenetic origin.

Sparite ('spar' or sparry calcite). Calcite crystals deposited within veins, pore spaces or fissures are called sparite. Individual spar crystals generally range from 0.01 to 0.10 mm in size but as an exception may exceed 1.0 mm in diameter. Some of the ancient spars are believed to be the products of recrystallisation of micrite. Sparry calcite is deposited only as a secondary vein- or pore-filling material, but micrites may be consolidated to form carbonate rocks.

Folk's scheme of classification of limestones uses various combinations of the allochemical and orthochemical components (see Figs. 4.10–4.13). Each term used in Folk's scheme is divided into two parts: a prefix and a stem (Table 4.9). The prefix indicates the nature of the allochemical component (intraclast, fossil, oolite or pellet), the stem indicates the nature of the orthochemical component (micrite or sparite). The speciality of Folk's scheme is that the nomenclature gives a complete petrographic description of the limestone. Thus Folk's scheme of limestone classification is extremely popular among petrographers.

A quick method for on-the-spot classification of limestones was developed by Dunham (1962) with a view to helping well-site and field geologists. This classification scheme is based on easily identifiable characters—whether the limestone is grain-supported or mud-supported. This scheme has been popular, particularly among petroleum geologists, because it provides a quick method of identification of rock properties useful for reservoir evaluation. A comparative study of the schemes proposed by Folk (1959), Dunham (1962) and Pettijohn (1975) is given in Table 4.10. The

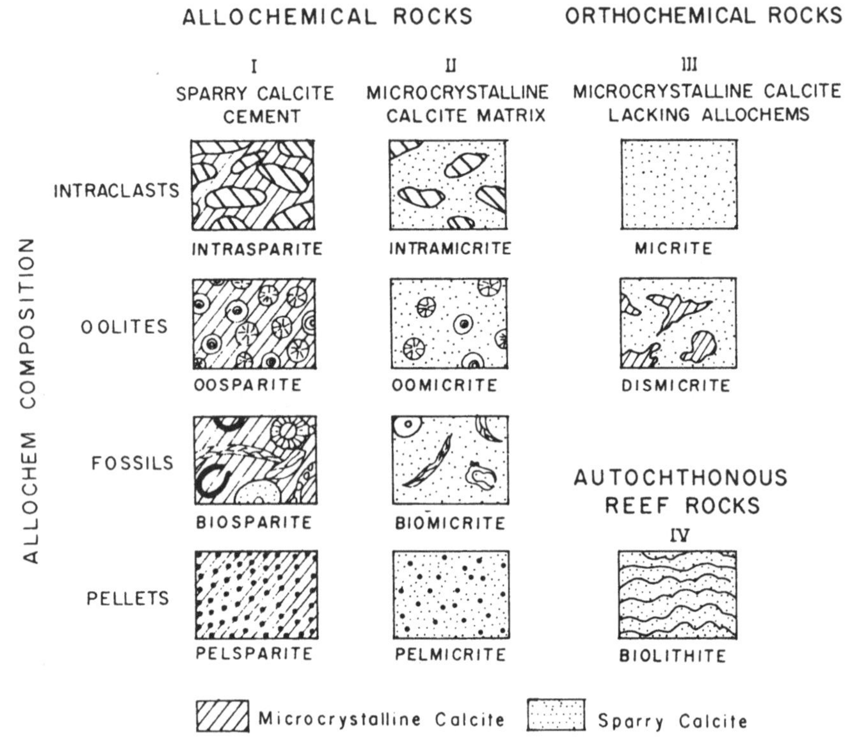

Fig. 4.10: Graphic representation of Folk's scheme for classification of limestones (redrawn after Folk 1959 with permission of the author and American Association of Petroleum Geologists).

major groups of limestones identified by Dunham (1962) are described below (see Fig. 4.14).

Grainstones of Dunham (1962) are rocks having a grain-supported framework. The intergranular space is filled with clear calcite cement. The carbonate detritus may consist of fragments of pre-existing limestones (intrasparite of Folk), fossil fragments (biosparite of Folk) or oolites (oosparite of Folk). Depending on the grain-size of the constituting carbonate grains, a grainstone may be grouped either as calcarenite (sand-size grains) or calcirudite (larger than sand size) of Grabau (1904).

The sandbanks of the Bahamas are similar to the ancient calcarenites. This has led many researchers to presume that the ancient calcarenites were deposited either in shallow seas under the influence of strong tidal action, or as subaerial carbonate dunes. Calcarenites, in fact, can be produced in marine conditions when oolites and skeletal debris of sand size are bound together by calcareous mud. Sparry calcite may also serve as the cementing

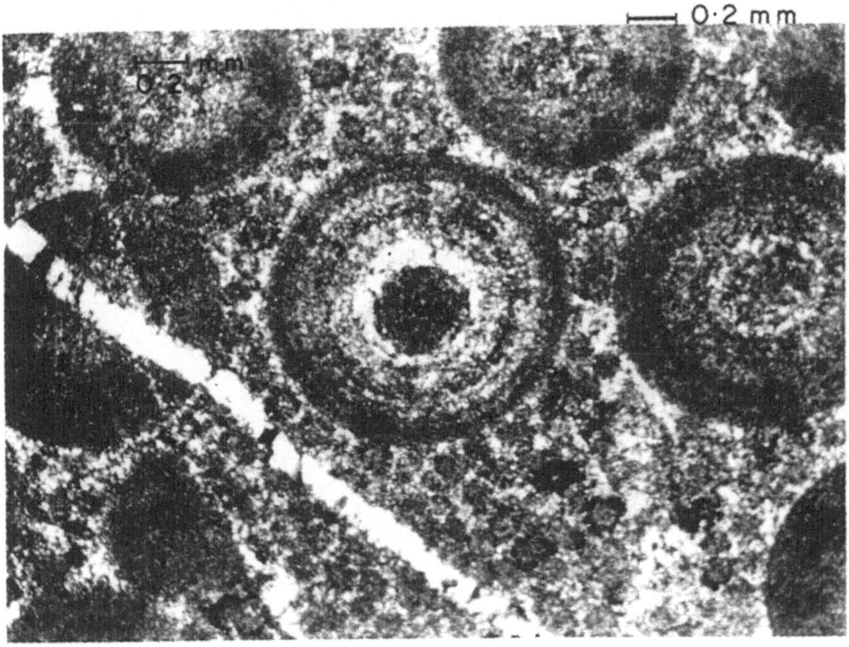

Fig. 4.11: Photomicrograph of an oosparite from Vempalle Formation, Papaghani Group (Upper Proterozoic), Cuddapah, Vempalle, A.P., India (thin section No. A11/3, courtesy of D. Dasgupta and C.N. Rao).

material for calcarenites (Pettijohn 1975, p. 353).

Boundstone is the same as reefoidal carbonate. This represents a loose network of carbonate with large voids. The framework is constituted of organisms such as corals, algae and bryozoa.

Wackestones are micrites having more than 10% coarse grains. These may be composed of skeletons of organisms or other materials.

Packstone has a grain-supported framework but contains a matrix of mud.

Mudstone consists of extremely fine-grained carbonate, comparable to the calcilutite of Grabau (1904) or the micrite of Folk (1959). The fine-grained, perfectly homogeneous variety of 'mudstone' used for lithographic purposes is called a lithographic stone. Mudstones may be produced by attrition, or may be chemically or biochemically precipitated. The choice of the term 'mudstone' to designate a particular type of limestone is not a happy one because argillaceous rocks of terrigenous origin and composed of silt-clay are also called mudstones (see Table 4.5).

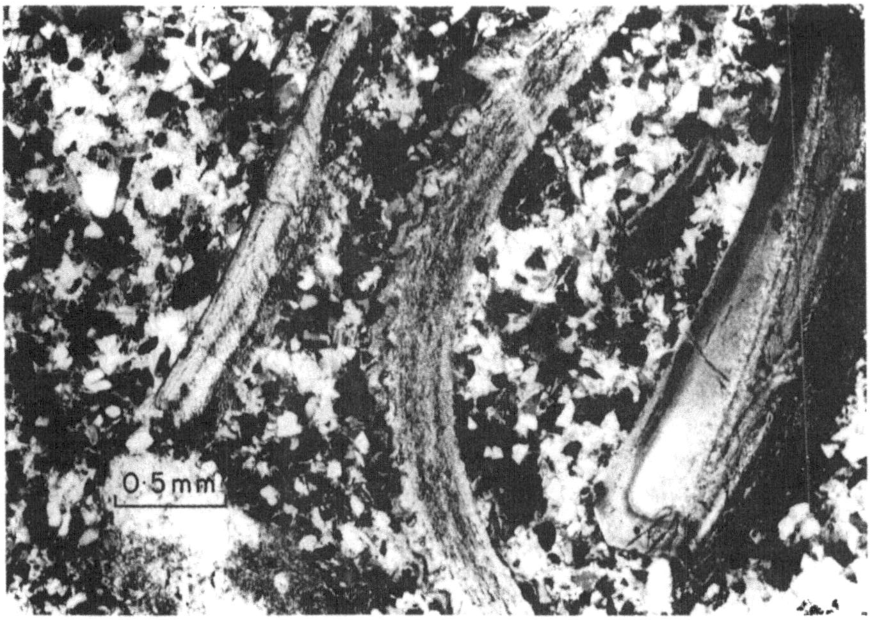

Fig. 4.12: Photomicrograph of a biosparite from 'Shell Limestone', Trichinopoly (thin section No. 6, author's collection). Cross polarised light.

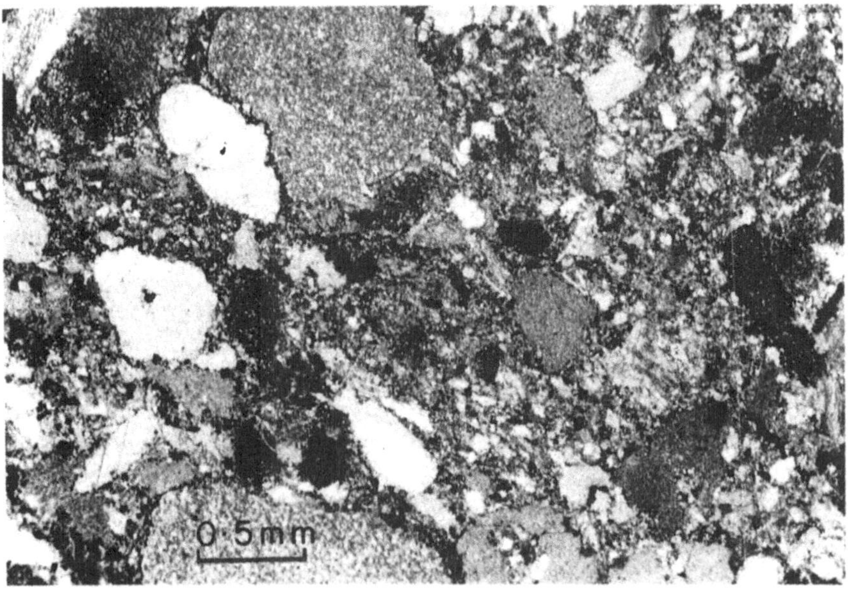

Fig. 4.13: Photomicrograph of an intrasparite. Locality unknown (thin section supplied by Hindusthan Minerals & N.H.S.S. Co., Calcutta, author's collection).

Table 4.9. Folk's scheme of classification of limestone (1959). Terms proposed by Grabau (1904) and Pettijohn (1975) have been inserted for comparison

Dominant Allochemical component	Orthochemical component			
	Grain size	Allochems > 10%		Allochems <10%
		sparite > micrite	micrite > sparite	Microcrystalline rocks
Intraclasts	>sand	INTRASPARUDITE *Calcirudite* (intraformational conglomerate)	INTRAMICRUDITE	Micrite and Dolomicrite *(Calcilutite)*
	sand	INTRASPARITE *Calcarenite* (lithic calcarenite)	INTRAMICRITE	
Oolites	>sand	OOSPARUDITE *Calcirudite* (pisolite)	OOMICRUDITE	
	sand	OOSPARITE *Calcarenite* (oolitic calcarenite)	OOMICRITE	
Fossils	>sand	BIOSPARUDITE *Calcirudite* (coquina)	BIOMICRUDITE (coquinoid limestone)	
	sand	BIOSPARITE *Calcarenite* (biocalcarenite)	BIOMICRITE (fossiliferous calcilutite)	
Pellets	<sand	PELSPARITE *Calcilutite* (pellet calcarenite)	PELMICRITE (pelletiferous calcilutite)	

Note: Terms proposed by Folk (1959) are given in block letters, those proposed by Grabau (1904) in italics. Those proposed by Pettijohn (1975) in parentheses.

Dolomite and Dolomitisation

When more than 50% of a carbonate rock is composed of the mineral dolomite ($CaCO_3$, $MgCO_3$), the rock is called a dolostone (Shrock 1948) or simply a dolomite. Pettijohn would name a rock dolomite only if the proportion of dolomite in it exceeds 90%. When calcite occupies more than 10% in such a rock (between 10 and 50%), the resultant rock is named a 'calcitic dolomite'. A dolomite with an appreciable proportion of ferrous carbonate is called a ferrodolomite ('ferrandolomite', Pettijohn 1975, pp. 359–360). Ferrous carbonate entering into a solid solution with dolomite imparts a buff

Table 4.10. A comparative study of the major schemes of classification of limestones

Criterion	Authority	Classification	Name
Grain-size	Grabau (1904)		
		1. >Sand size	Calcirudite
		2. Sand size	Calcarenite
		3. <Sand size	Calcilutite
Mode of deposition	Pettijohn (1975)		
		1. Precipitated *in situ*	Autochthonous
		2. Transported from outside	Allochthonous
Mode of deposition	Folk (1959)	1. Precipitated *in situ*	Orthochemical
			Sparite
			Micrite
		2. Transported from outside or locally deposited	Allochemical
			Intraclasts
			Oolites
			Fossils
			Pellets
Textural features			
Nature of binding of original components	Dunham (1962)	1. Components bound together during deposition	Boundstone
		2. Components not bound together	
		2(A.) Rock is mud-supported	
		grains < 10%	Mudstone
		grains > 10%	Wackestone
		2(B.) Rock is grain-supported	Packstone
			Grainstone

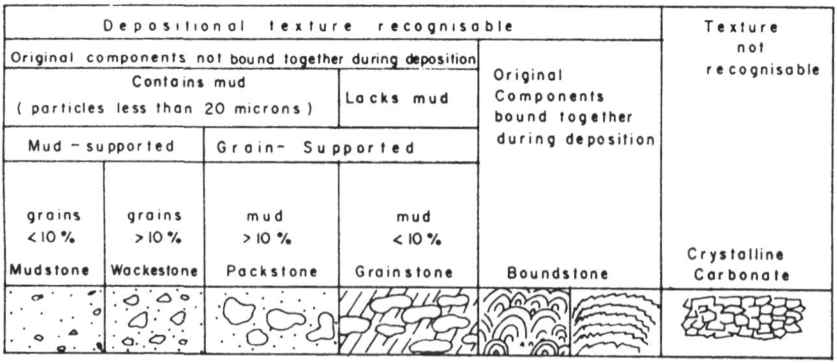

Fig. 4.14: Dunham's (1962) scheme of classification of carbonate rocks according to depositional texture (modified by changing per cent mud from 1 to 10).

colour to the rock upon oxidation. In hand specimens dolomite is often recognised by this colour, which is uncommon in calcitic limestones. The other common impurities in natural dolomites are anhydrite and gypsum.

In ancient geological records dolomites are more common than calcitic limestones although, strangely enough, dolomite-forming environments are extremely rare in the world today. Only some of the sabkha* flats of the Persian Gulf, playa lakes of California, the hypersaline lagoons of South Australia and the supratidal areas of Florida and the Bahamas are known to be congenial to the development of dolomite. In rare cases, dolomite may form under deep-sea conditions also (see Folk and Land 1975 and Fig. 4.15). No living organism is known to secrete dolomite today although dolomitic reefs are known from ancient geological records. Much of the ancient dolomite

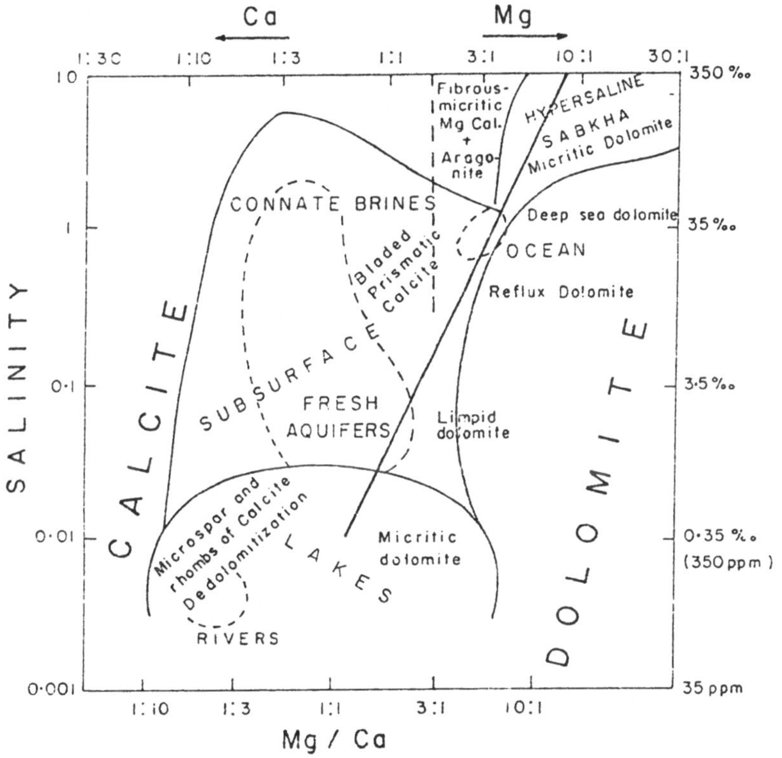

Fig. 4.15: Influence of salinity and Mg/Ca ratio on precipitation of calcite, dolomite and aragonite in different natural environments (compiled from diagrams by Folk and Land 1975 with permission of R.L. Folk).

* See footnote on P. 79

is thus believed to have originated by replacement of the original calcitic limestone.

Two types of dolomites are recognised: (i) primary dolomite and (ii) secondary dolomite. The primary are extremely fine-grained (maximum size 0.020 mm, generally much smaller) dolomite ('dolomicrite' of Folk). These contain unmistakable signs of primary deposition, such as lamination, ripple marks and mud cracks. Dolostones laterally passing into calcitic limestones, microbeddings of dolomite and limestone and limestone intraclasts within dolostones also indicate the primary nature of dolostone. The secondary dolomites are the coarsely crystalline, saccharoidal rocks bearing evidence of replacement. The evidence of replacement includes faint outlines ('ghosts') of former grains in rocks composed wholly of dolomite, preservation of original shapes of dolomitised fossils, presence of grains of detrital particles within dolomites and other similar features. Dolomite veins cutting into calcitic limestones, dolomite rhombs crossing the boundary of $CaCO_3$ fossils, also indicate the secondary nature of the rock.

Observations in nature today indicate that hypersaline brines, richer in magnesium than ordinary sea-water, are needed for precipitation of dolomite. Alkalinity and temperature of the magnesium-rich fluid must also be sufficiently high to cause replacement of calcite by dolomite. This led to the formulation of the 'reflux hypothesis' by King in 1947. This hypothesis envisages dolomitisation of porous calcite by a magnesium-rich heavy brine moving through the pore space of the rock. Such a brine may be produced by inflow of normal sea-water into isolated pools under arid conditions, leading to the concentration of a variety of salts (see Fig. 6.17 and also discussion under evaporites). In the course of more modern studies the reflux sabkha process is believed to be restricted to the surface. The 'mixed-water aquifer model' and the 'hydrothermal convection model' (Morrow 1990) are believed to be more effective at depth. The 'burial compaction model' operative at depth, not only accounts for the occurrence of a vast amount of dolomite but also explains modification of pre-existing porosity.

Localised replacement of calcite by magnesium-rich solutions along post-depositional structures such as faults or fissures, might have produced epigenetic dolostone (Friedman and Sanders 1978, p. 179), whereas syngenetic dolostones could have developed penecontemporaneously by the reaction of lime carbonate with magnesium-bearing sea-water. The observation that the older dolomites are richer in magnesium content led some authors to imagine that the primordial sea-water was richer in magnesium.

The physico-chemical conditions leading to the precipitation of calcite and dolomite were studied in detail by Folk and Land (1975). It emerged from their study that salinity and Mg/Ca content of the medium are the two most important factors controlling precipitation of calcite and dolomite. When salinity is low, dolomite can precipitate at a Mg/Ca ratio of 1 : 1. With an increase in salinity, a much higher proportion of Mg is needed for precipitation of dolomite (Fig. 4.15).

In the hypersaline and Mg-rich sabkha* environment the crystallisation rate of dolomite is high. Only very fine-grained, micritic dolomite precipitates in this environment. In the reflux (schizohaline) condition, dilution of a hypersaline sea or sabkha takes place by encroachment of fresh water. This decreases salinity while the Mg/Ca ratio still remains high. The condition is ideal for the precipitation of prefectly rhomb-shaped dolomite crystals. In freshwater lakes and caves some micritic dolomite may be precipitated even when the Mg/Ca ratio is around 1 : 1. Under such a condition, with a further decrease of Mg content in the solution, sparry calcite precipitates. This is also the condition under which dedolomitisation may take place and calcite starts replacing dolomite (Fig. 4.15).

The crystal habit of calcite is also influenced by the Mg/Ca concentration of the medium of deposition. Low Mg content of the solution favours growth of large calcite crystals but an increase in Mg content in the solution retards sidewise growth of crystals, thereby producing only elongate rhombs of calcite (Folk 1974).

Dolomitisation involves large-scale recrystallisation. The process often leads to the generation of a mozaic of euhedral dolomite crystals, producing a texture called the idiotopic fabric. When the crystal faces are not so well developed a mozaic of anhedral dolomite crystals result (Fig. 4.16). This is called the xenotopic fabric (Friedman 1965).

As early as 1836 it was suggested that a replacement of limestone by dolomite would result in a volume shrinkage of 12–13%. This idea, subsequently discarded, became popular again, following the studies by Hohlt in 1948. It is now believed that shrinkage of the c-axis of crystals during transformation of calcite to dolomite creates voids. The packing being comparatively loose, the dolomite crystals orient themselves randomly. Many dolomite bodies act as good reservoirs of petroleum because of the secondary porosity acquired in this way (Levorsen 1958, p. 121). However, more modern workers have pointed out that a volume change can take place only when the replacement occurs within a 'closed system', a situation rarely to be found in nature (Longman 1981).

Chemical Sediments Other than Carbonates

The chemical sediments other than carbonates belong to two groups of precipitates, primary and secondary. The primary precipitates include evap-

* Sabkhas are depressions formed in extremely arid conditions, for example deserts, by removal of loose particles by wind. Sabkhas go down to the ground-water level or to the zone where concentration of salts drawn by capillary action takes place.

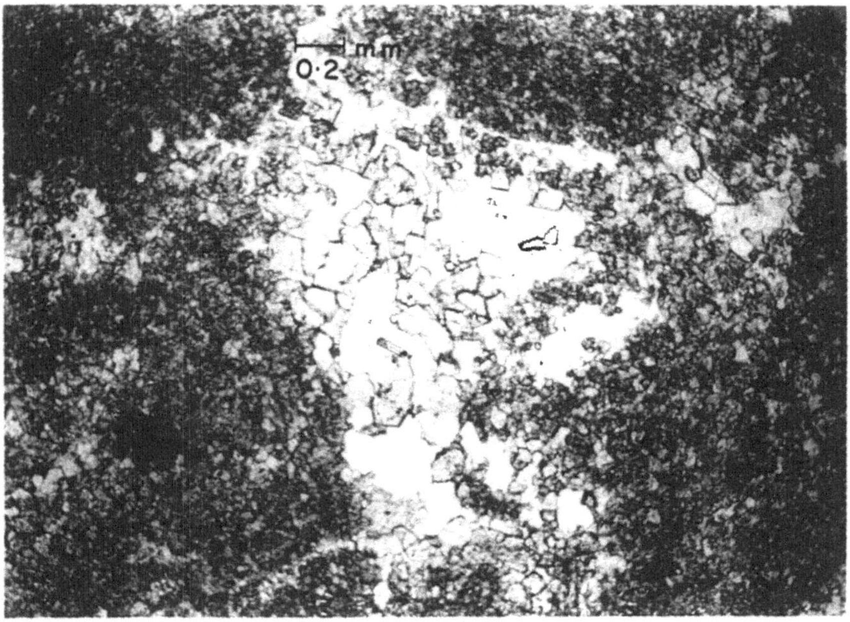

Fig. 4.16: Sparry dolomite crystals filling a cavity ('bird's eye') in a limestone. Note progressive increase in crystal size towards the centre. Location: Pulivendla, Andhra Pradesh, India (thin section no. H 13(11), courtesy of D.K. Dasgupta and C.N. Rao).

orites such as chlorides, nitrates and sulphates. The secondary precipitates include non-evaporites such as the siliceous, phosphatic and ferruginous deposits.

Evaporites

Evaporites are precipitated from solution and concentrated by evaporation. Sodium chloride (halite), and calcium sulphates (gypsum and anhydrite) are common examples of evaporites. Potassium chloride (sylvite), hydrated magnesium chlorides (bischovite and epsomite) are relatively rare varieties of evaporite. All these rock-forming salts are deposited by the process of evaporation of sea-water. The sequence of deposition followed by the common rock-forming evaporites is as follows (beginning with the least soluble mineral, which is always the first to be deposited): gypsum or anhydrite, sylvite and halite.

Areas of extreme aridity and local areas of internal drainage with a high rate of evaporation, as in continental or coastal sabkhas, are ideal for precipitation of evaporites. Another ideal situation for evaporite deposition is a barred (silled) basin formed as a bay or cut-off arm of a sea (see discussions on euxinic or silled basins). Circulation in such a closed basin is maintained by the influx of salt-laden sea-water flowing above the sill and return (reflux) of this water into the sea as a subsurface current. The amount of evaporite deposited into the closed basin is governed by the amount of salt flowing into the basin and the amount carried back to the sea during reflux. In this process the least soluble salts (anhydrite and gypsum) are deposited nearer to the sill and the highly soluble salts (halite and sylvite) are deposited away from the sill, in the distal part of the basin (see Chapter 6, Fig. 6.17).

Due to precipitation of the calcium ions in the proximal part of the basin in the form of gypsum and anhydrite, the sea-water becomes enriched in magnesium. The magnesium-rich sea-water flowing back into the sea causes dolomitisation of the Ca-rich carbonates precipitated on the sill. This is what is commonly called the 'reflux theory' of dolomitisation (see discussions under dolomitisation). The physical environments of evaporite deposition and the main facies found in these environments are summarised in Fig. 4.17.

Evaporites are found in all geological ages, from the Precambrian to the Recent, but they are possibly most common in the Permian. The best examples are the Permian salt deposits of Texas, Mexico and New Mexico. The well-known evaporite deposits of north-west Europe, Germany, and the North Sea are also of Permian age. Also well known are the salt deposits of the Persian Gulf belonging to a wide range of geological ages, from the Permian to the Tertiary. The salt deposits of the Rann of Kutch of Western India, the coastal sabkha of the Persian Gulf, the Great Bitter Lake of Suez, the Death Valley of the USA and the Dead Sea are some of the well-known evaporite deposits of the present day.

Siliceous Sediments

The common siliceous minerals of sedimentary origin include chert, flint, opal, chalcedony and a variety of other microcrystalline silica. Under the microscope they show the aggregate polarisation effect of silica although the individual crystal boundaries cannot be recognised. They occur either in association with the shallow-water orthoquartzite-carbonates, or with the deep-water geosynclinal sediments. High temperature and a high degree of alkalinity (pH > 8.5) are known to increase the solubility of silica in water. A decrease in temperature and alkalinity allows precipitation of silica, either near the surface or in deep water.

The origin of siliceous sediments can be either inorganic or organic. Silica may precipitate from sodium silicate gels in alkaline lakes. Silica of deep-water geosynclinal association is ascribed to silica-secreting organisms such as the oozes of radiolaria and diatoms. Some of the bedded cherts are believed to be lithified ooze of organic origin. Some of the chert can also be of secondary origin—the product of diagenetic differentiation and replacement (see discussions under diagenesis).

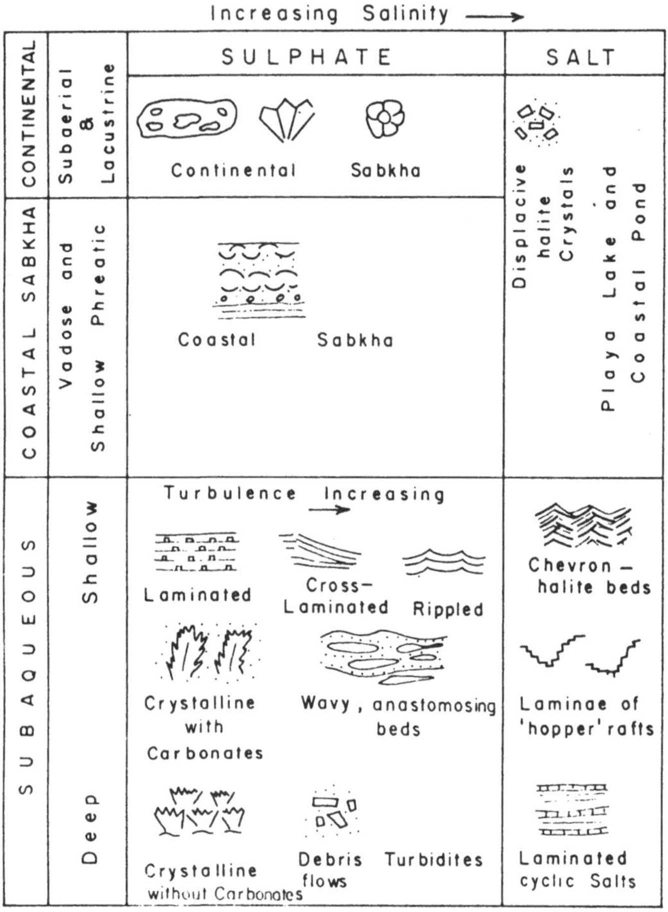

Fig. 4.17: Physical environments of evaporite deposition (modified after Schreiber et al. 1976 and Kendall 1984).

Phosphatic Sediments

All sediments containing more than 7.8% P_2O_5 (equivalent to an apatite content of about 20%) are classified as phosphatic sediments. Those containing more than 19.5% of P_2O_5 (apatite content of 50%) are called phosphorites.

The phosphate in sediments is mostly of organic origin, derived from the hard parts of organisms which extract phosphorus from sea-water. Guano (organic excerta) deposits are some of the best-known organic phosphate deposits found on land. Nodular phosphatic conglomerates and metasomatic phosphatic deposits are of inorganic origin.

Like all other chemical sediments, the deposition of phosphatic sediments is controlled by the acidity-alkalinity and oxidation-reduction potential of the medium of deposition. A slightly reducing, alkaline condition provides the environment most congenial for deposition of phosphate (see Fig. 4.18). Phosphates can occur either as a platform facies, as in the glauconite-bearing marine sediments, or as a basinal (deep-sea) facies produced under anaerobic conditions.

In general, phosphate-bearing sediments are more common in the Phanerozoic than the Precambrian because of the abundance of organisms in the former. The phosphorites in India, however, are mostly of Precambrian or Cambrian age. These have been assigned to three groups according to their association (Banerjee 1986): 1) bedded cryptocrystalline phosphorites associated with dolomitic stromatolites; 2) intraclastic and microcrystalline phosphorites associated with carbonates, black shales and dark cherts; 3) phosphorites associated with ferruginous sediments and residual weathering products.

All the above-mentioned phosphorites are of marine origin. Silled (restricted) basin conditions, controlled by basement topography, were responsible for the high concentration (10–35%) of P_2O_5 in most cases. The specific conditions of deposition of these phosphates have been discussed in detail by Banerjee (1986). In Rajasthan, a restricted condition was provided by the presence of a shallow carbonate embankment along a Precambrian basement ridge. Similar conditions also existed in the Himalayas where the phosphorites were deposited over the Upper Proterozoic Krol Formation. In Peninsular India, dolomites of Proterozoic age provided the silled basin conditions for the Vindhyan phosphorites. Association with stromatolites in many cases indicates that micro-organisms played important roles in concentrating phosphorus from sea-water. In some cases, as in the Bijawar rocks of Central India, processes such as diagenetic mobilisation, weathering, leaching, precipitation in voids and fissures also helped in concentrating phosphorites. Phosphorites found in association with ironstones in the Raniganj coalfield, India, are an example of a shallow-water phosphate deposit, perhaps formed under some marine influence.

Iron-bearing Sediments

Iron-bearing sediments include banded (bedded) iron formations (BIF), ironstones, ferruginous ooids and fossil fragments. The bedded iron formations consist of alternating thin layers of silica and ferruginous minerals of chemical origin. The iron-bearing minerals are mostly oxides (magnetite and hematite). Silica occurs as quartz, chert or jasper. The bedded iron rocks, found in all the Precambrian cratons of the world, are often modified by diagenesis and metamorphism. The ironstone formations, which are of younger age, are composed mostly of iron carbonate (siderite) with some iron sili-

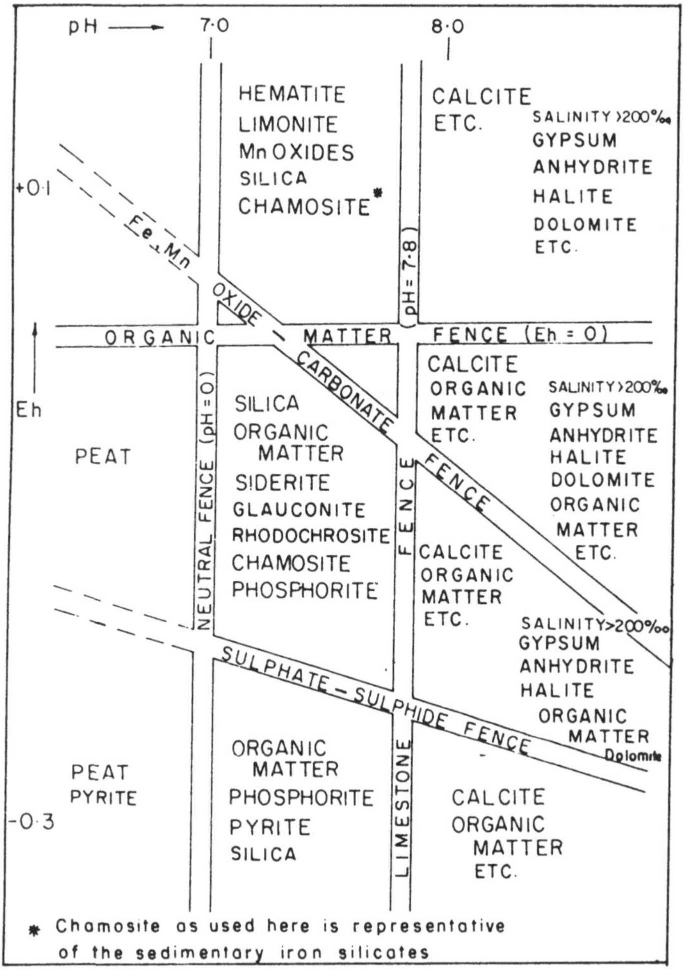

Fig. 4.18: Fence diagram illustrating influence of Eh and pH on generation of mineral assemblages in sedimentary rocks (after Krumbein and Garrels 1952, fig. 8, with permission of the University of Chicago Press).

cates (chamosite). Ooids and fossil fragments, often replaced by iron oxide minerals, are also known to occur within ironstones. The siderites are often altered to limonite near the surface.

As in the case of other chemical sediments, acidity-alkalinity (pH) and oxidation-reduction potential (Eh) play important roles in the precipitation of iron-bearing minerals. As indicated by the Krumbein-Garrels diagram

(Fig. 4.18), an oxidising and alkaline condition is ideal for precipitation of the iron oxide minerals. The production of near-surface carbonate or bicarbonate of iron could be wholly inorganic (James 1966). The inorganic iron may be transported together with clay minerals as colloidal iron oxide, Organic matters can also play an important role in the transportation of iron because large proportions of ferric oxide can be kept in colloidal suspension by small quantities of organic matter. Following a more modern approach to the problem, Berner (1981) concludes that hematite and goethite are precipitated under oxic ($C_{O2} \geq 10^{-6}$) conditions, while pyrite and marcasite are precipitated in anoxic ($C_{O2} < 10^{-6}$) conditions. Siderite is formed under anoxic conditions which are nonsulphidic ($C_{H_2S} < 10^{-6}$). These conclusions are based on the measurement of dissolved oxygen and dissolved sulphide in modern sediments at the time of authigenic mineral formation. C in the notations refers to concentration of oxygen or total sulphide in terms of moles per litre.

The conditions leading to simultaneous deposition of silica and iron oxides are not fully understood but a marine or restricted marine environment of sedimentation, somewhat isolated from the open sea, is believed to be responsible for deposition of the banded iron formations of Precambrian age. A Precambrian ocean with an upper oxic layer and a bottom anoxic layer containing Fe and SiO_2 is the commonly assumed model. Large-scale precipitation of BIF followed the upwelling of deep, Fe–SiO_2-rich waters to the newly developed continental margins and shelves.

In some cases, as in the Precambrian rocks of the Lake Superior District, silica and iron, produced from weathering of land masses, were deposited in alternate layers under freshwater lacustrine conditions. The model proposed for such a deposition by Hough (1958) envisages a lacustrine condition wherein settling of the dead, oxygen-consuming organisms from the surface creates a reducing condition in the lake bottom which is cold as well as acidic. The condition is ideal for precipitation of silica, while iron brought in suspension by streams during the summer remains in solution to be precipitated as iron oxide only when the lake 'turns over' (see lacustrine depositional environment in Chapter 6) during the autumn. Repetition of the process leads to the production of alternating bands of iron oxide and silica.

Volcanic activity, sea-bottom reactions on clastic materials and deep-seated weathering processes on land are some of the commonly suggested sources of iron and silica. These conditions, it has been suggested, were repeated periodically but occurred on a unique scale in the early Proterozoic (2.0–2.5 b.y.), when an extraordinary concentration of BIF (92% of the total) was formed in almost all continents round the globe (Goodwin 1982).

Erosion under climatic conditions similar to what prevails today will not allow transportation of large quantities of iron in suspension without a substantial amount of accompanying detritus, but surprisingly most of the thick sequences of the Precambrian iron deposits are devoid of clastics. As an explanation, it has been proposed that the Precambrian atmosphere had a large content of CO_2, resulting in acidic surface waters which allowed easy leaching and transportation of iron with no accompanying detritus. It is also possible that near-absence of oxygen in the Precambrian atmosphere did not allow oxides of iron, the chief inhibitor in the process of iron transportation, to form. Local concentrations of iron oxides in this model were explained as the result of local biological photosynthesis. As an alternative to these suggestions, it has also been proposed that the sluggish surface run-off over a senile, deeply weathered Precambrian land surface did not allow transportation of particles coarser than particulate iron oxide together with fine clay. Many workers are of the opinion, however, that the contributions of volcanic, hydrothermal or biological processes in the generation of Precambrian iron deposits cannot be ruled out altogether (Stanton 1972).

The Precambrian Banded Hematite Jasper (BHJ) Formation of Noamundi, Bihar, India contains numerous syndepositional structures suggesting their deposition in a shallow-water environment. The possible biological origin for at least part of the iron and silica is suggested by the presence of microstructures resembling organic forms (Rai, Sarkar and Paul 1980; Sarkar 1984). According to one view, the banded iron formations represent exhalative phases of volcanism along a fracture zone. Iron and silica, discharged into an inner sea, were precipitated out in alternating bands through this fracture (Banerji 1977). A source of pyroclastics and exhalatives (volcanic and/or hydrothermal) adjacent to the existing BHJ occurrence together with a rapid rate of precipitation of the detrital-pyroclastic/chemical sediments close to their place of generation has been suggested as an alternative hypothesis for the iron and manganese deposits of the Noamundi basin (Dasgupta and Batabyal 1988).

Ironstones are also known from the Ironstone Shale Formation of the Raniganj coalfield in West Bengal, and its stratigraphic equivalent in other Gondwana basins of India. Apart from iron and silica, these rocks contain small quantities of phosphorus and manganese (Walker 1914, cited by Krishnan, 1964). The assemblage of minerals in these ironstones suggest their precipitation in environments which were slightly alkaline as well as reducing. This and other lines of evidence also have led to the suggestion of a marine invasion into some of the Gondwana basins of India during the Permian (Bose and Sengupta 1993).

VOLCANICLASTIC ROCKS

Sedimentary rocks constituted of fragments of volcanic origin are called volcaniclastic rocks. These rocks contain the product of weathering and erosion of pre-existing volcanic rocks or pyroclastic debris. Pyroclastic is the name given to rocks composed of the ejecta from volcanic vents.

Volcaniclastic sediments are broadly divided into two groups according to grain-size—the coarser volcanic breccia (angular fragments) and agglomerates (rounded fragments), and the finer tuffs. Breccia and agglomerates consist of volcanic bombs and rock fragments. Tuffs are composed of fine crystals, rock fragments and volcanic glass. These are named according to their principal constituent—crystal tuffs, lithic tuffs and vitric tuffs (Fig. 4.19).

The origin of most volcaniclastic deposits is ascribed to eruption of rhyolitic lava which, being highly viscous, with a high gas content, is particularly explosive. The finer ash is carried to the upper atmosphere and transported for long distances. Deposition of ash may lead to extensive pyroclastic beds (the 'ash fall'). Clouds of hot gas ('glowing cloud' or *nuée ardente*), produced out of some acidic volcanic eruptions, may also be responsible for sedimentation of an unusual nature. The basal flows resulting from these eruptions contain ash, lapilli and blocks which are kept in suspension by escaping gases. The mass may flow downstream as a turbulent mixture. The deposits formed from this mixture are called ash-flow tuffs, welded tuffs, or ignimbrites. These poorly sorted, poorly bedded units have compositions

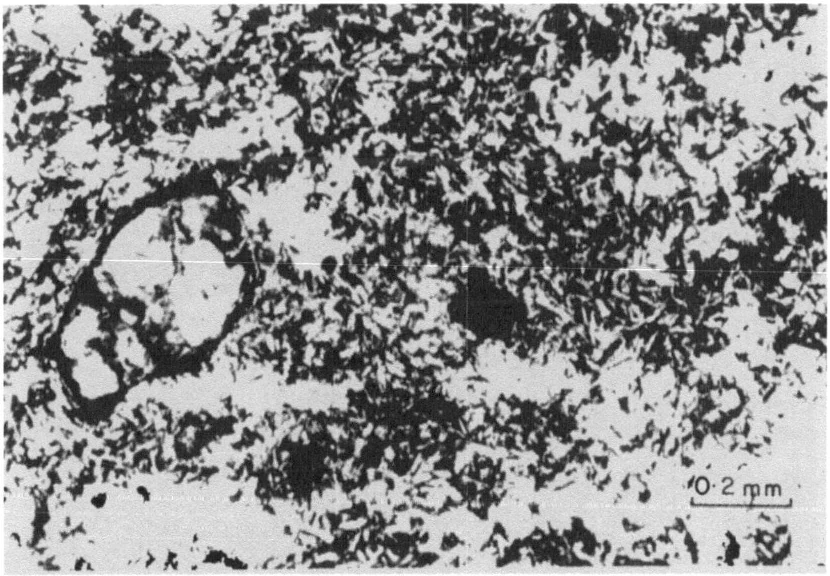

Fig. 4.19: Photomicrograph of a tuff (plane polarised light). Locality near Padberg, Germany (author's collection).

comparable to rhyolitic lavas. A single deposit of ignimbrite may be as much as 100 m thick. The ash-fall deposits, on the other hand, are well sorted according to fall velocity of the grains, with mean thickness and grain-size decreasing away from the source. In some cases they can be used as palaeocurrent indicators (Scheidegger and Potter 1968).

Silicified tuffs, called porcellanites, reported from the Proterozoic Vindhyan sediments by early workers of the Geological Survey of India, contain fragments of quartz and feldspar of vitroclastic shapes. Wedge-shaped fragments of feldspar indicate their pyroclastic origin (Auden 1933, p. 154). These volcanigenic deposits contain bombs, lapillis and huge ash falls (Ghosh 1972). Lava flows, tuffs and pyroclastic breccia around 'crater-like structures' have been reported from the Proterozoic sediments of the Godavari valley, India (Saha and Ghosh 1987). These tuffs show distinct colour lamination, parallely arranged lapilli, flat-ended and stretched pumice fragments. These have been assigned to syntectonic volcanism in Proterozoic island arcs (Fig. 4.20). Ash-beds, interbedded with late Quaternary sediments, have been reported from the Son and Narmada valleys of India. This ash contains angular glass shards and some accessory minerals. In the absence of any evidence of Quaternary igneous activity in

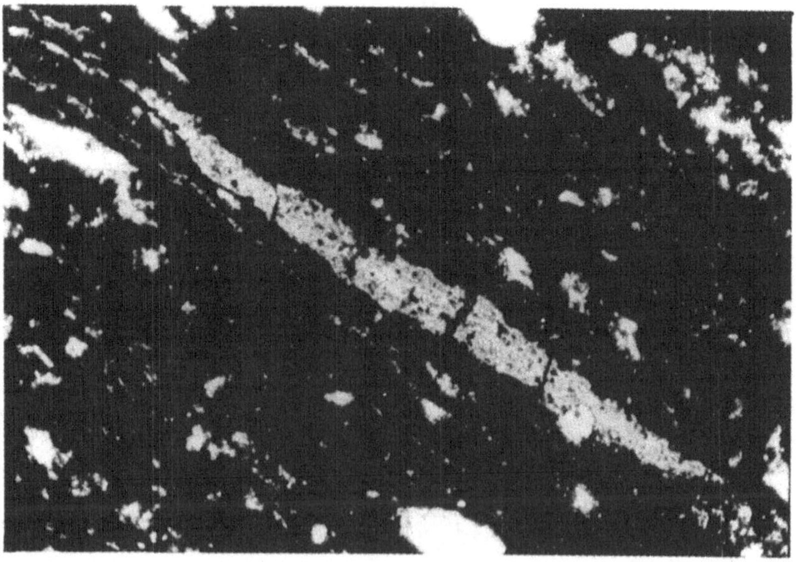

Fig. 4.20: Photomicrograph of a deformed tuff-lava, Sommanpalli, Godavari valley, India. Length of the stretched pumice fragment is about 0.76 mm (courtesy of D.K. Saha).

the areas, this ash-bed has been tentatively assigned to neotectonic activity along the Son-Narmada lineament, known to be an area of high heat flow (Basu, Biswas and Acharyya 1987).

DIAGENESIS

The changes which sediments undergo between deposition and lithification under normal pressure-temperature conditions are called diagenesis. More recently, the parameters of diagenesis have been extended to depth (4–7 km) and temperature (120°–220°C) conditions where incipient metamorphism sets in. Near-surface (telegenetic), shallow depth (eogenetic) and deeper level (mesogenetic) changes are included in this definition (Burley, Kantorowicz and Waugh 1985).

Diagenesis of Siliceous Sediments

Near-surface diagenetic changes are caused by the interaction of minerals with interstitial pore fluids. This may lead to dissolution of the existing minerals and production of new (authigenic) minerals. Authigenic growths ('overgrowths') of quartz and feldspar are common features at this stage. The newly formed quartz is added to the detrital quartz grains maintaining the same crystallographic orientation (Fig. 4.21). Authigenic precipitation of feldspar takes place at a higher order of pressure and temperature than that provided by near-surface conditions. Excess potassium in the pore fluid is an essential condition for development of authigenic feldspar. Dissolution features in feldspars and production of kaolinite as pore-filling material are common at this stage.

At a shallow depth, in a warm, humid, non-marine environment, bacterial activity may impart a distinctly black colour to fine-grained sediments. Lateritisation takes place in warm, humid climate whereas calcrete forms in warm arid to semi-arid conditions. In a warm humid climate, when the pore fluids are alkaline and rich in calcium, sodium and silica, but deficient in carbonates, a variety of zeolites may be produced authigenically. When the climate is hot and dry ferric oxides are produced by oxidation of ferruginous sediments. These oxides impart a red colouration to sediments at shallow depths. In marine conditions a variety of clay minerals, illite, smectite, interbedded illite-smectite, may be authigenically produced. Quartz, feldspar and clay authigenesis is often followed by carbonate precipitation under marine conditions.

At greater depth mechanical processes play important roles in diagenesis. Fracturing, crushing and squashing of detrital grains and wrapping of mica flakes around quartz grains are common mesogenetic changes. Reduction of porosity, partly due to compaction with increasing depth of burial

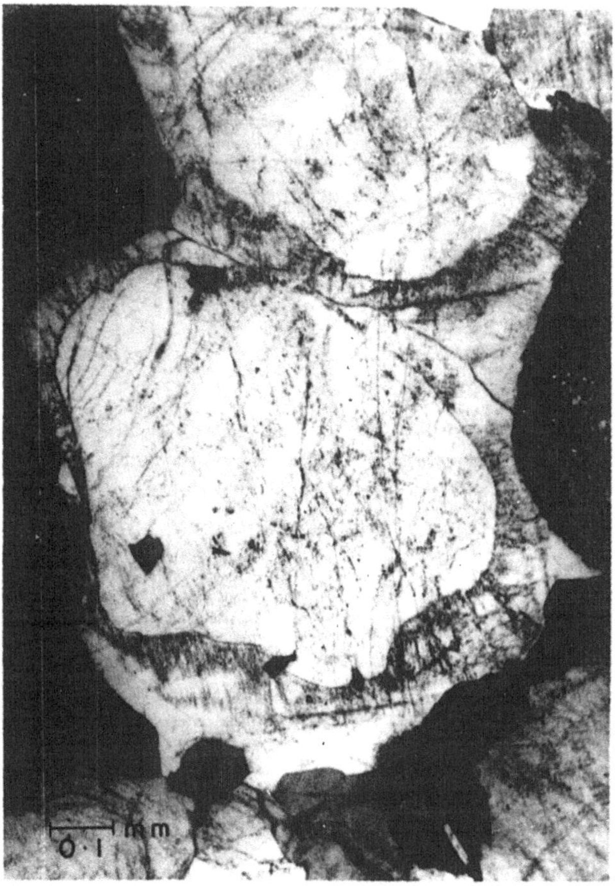

Fig. 4.21: Photomicrograph showing authigenic overgrowth on a quartz grain in quartz arenite from the Upper Proterozoic Chandarpur Formation, Raipur, India (thin section courtesy of S.S. Das and C.N. Rao).

but largely due to chemical processes, takes place. With increasing depth the porosity of sand changes only marginally while that of shale decreases logarithmically. Reduction of intergranular pore space is caused by increase of solubility at the contact of solid clastic grains under pressure. This phenomenon is called 'pressure solution'. Pressure solution may change the nature of grain contact from tangential to sutured through concavo-convex with increasing depth of burial.

A minimum depth of 1 to 1.5 km is necessary for pressure solution to be effective. Limestones and calcareous sandstones need lesser temperature and pressure for welding of grains than arenites. Finer grained rocks are affected more because they provide larger surface areas for grain contact. Thin films of clay around quartz grains promote pressure solution, but larger quantities of clay deter the process by 'cushioning' (Tada and Siever 1989). Optical microscopes are of little use in distinguishing between overgrowths and pressure solution effects such as concavo-convex contacts. The cathodo-luminescence technique often helps in such cases because the nature of the luminescence differs for detrital quartz and overgrowths (Sippel 1968).

With increasing (mesogenetic) depth of burial, clay minerals such as smectite and kaolinite are replaced by more stable products, illite and chlorite. With increasing depth of burial in the Neogene sediments of the Bengal basin, Bangladesh, a progressive decrease in the proportion of kaolinite and calcite and an increase in the proportion of illite by illitisation of illite-smectite were noted. Within the same sediments the following order of diagenesis was reported: chlorite authigenesis, siderite crystallisation, development of illite rims, quartz overgrowths and production of late Fe-calcite cement (Imam and Shaw 1984).

Source of Silica Cement

On land silica may be taken into solution during weathering of rock-forming silicates (see equations in Chapter 2). Kaolinitisation of feldspar and diagenetic alteration of clay minerals may also release some silica. In the early stage of diagenesis the pores of quartz sand may be filled up by silica produced from weathering. Opal and chalcedony cements in a desert environment represent a transition between weathering and diagenesis.

Solubility of both amorphous silica and quartz increases with an increase in temperature and alkalinity of the medium. A sharp increase in silica solubility is noted when pH exceeds 8.5. Hot springs and hydrothermal waters which satisfy these conditions might serve as sources of silica in sediments. Precipitation of secondary silica by hydrothermal solution is a common feature. Solutions saturated with silica at high pressure-temperature conditions at depth become further concentrated when they are pushed up to cooler zones by compaction.

In deeply buried sandstones some silica may be produced by pressure solution. The silica thus produced migrates into clay-free, fine-grained layers where authigenic overgrowth can take place readily (von Engelhardt 1977). Silica released by clay minerals during diagenesis of shales might migrate into adjacent sandstones to be deposited in voids.

In a marine environment the silica seems to be of primary origin, generated at the sea-bottom due to dissolution of diatom tests. This leads to concentration of silica in the pores of deep-sea clay and silt. Escape of silica-saturated water from the pores of clay or silt to the intergranular space of sand might account for much of the silica saturation in sandstones of marine origin. In some cases, as in the Buntsandstein sandstones of Germany, silica cementation has actually been found to increase towards shale (Füchtbauer 1974).

Diagenesis of Carbonates

Replacement of aragonite and high-magnesian calcite by calcite and consequent reduction of pore space are the two most important effects of diagenesis on carbonate sediments of marine origin. These processes are to be studied with reference to the physicochemical conditions prevailing at sea. In a coralline island for example, four different environments prevail. These are: (i) a marine phreatic zone, where the pore spaces are filled with saline water; (ii) a freshwater phreatic zone immediately below the exposed part of the island; (iii) a mixed zone with brackish water in the pore spaces and (iv) a vadose zone occupying the space between the surface and the water table. The pore space here contains water as well as air (Fig. 4.22).

In the marine phreatic zone, where salinity and Mg/Ca content of the seawater are fairly high, aragonite is secreted by corals (Fig. 4.15). The porosity of freshly deposited aragonite may be as high as 55%, but micritisation and submarine cementation inside the grains may reduce this porosity to about 40%. Intergranular cementation may further reduce the porosity to 5–10% only. The binding material here is provided by coralline algae which secretes magnesian calcite.

A lowering of sea level following this primary stage of cementation translocates the carbonate to the freshwater phreatic zone. Decrease of salinity and of Mg/Ca ratio in this zone causes sparry calcite to precipitate within the shells and tests, thereby reducing the porosity to about 10%. Interestingly, the original shapes of the shells are not altered during this conversion. This is due to the presence of a micritic envelope which allows the calcitic pore fluid to pass through it, without destroying the shape of the outer shell. The envelope itself is not involved in this reaction (Friedman and Sanders, 1978). In the case of cavity filling larger crystals are formed at the centre and the smaller ones remain confined to the inner rim of the cavity (see Fig. 4.16). In some cases cementation takes place round the outer boundary of the pre-existing grains in the form of syntaxial rims. The process has been termed aggrading neomorphism by Folk. In this case the boundary between the nucleus and the syntaxial rim is sharp. Irregular boundaries are also produced when the larger grains are converted to smaller ones during diagenesis. This process of grain diminution has been termed micritisation by Bathurst (1975).

When the carbonate is left in the freshwater zone for a long time, leaching of aragonite and Mg-calcite takes place by meteoric water since these

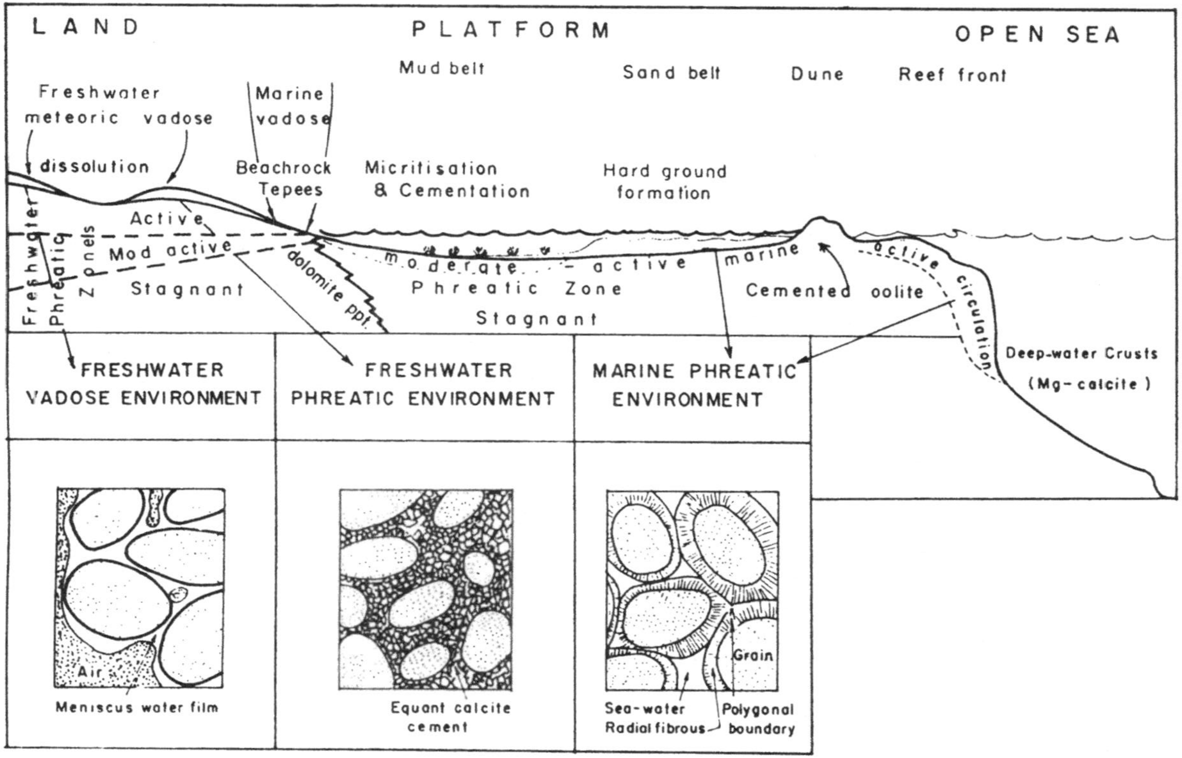

Fig. 4.22: Diagenesis of carbonates in freshwater and marine environments (modified and redrawn after a compilation by Sellwood 1986, with permission of the author).

minerals are unstable under freshwater conditions. Although porosity may increase temporarily at this stage, neomorphism of calcite which follows, fills up the moulds, thereby reducing the porosity to about 5–10%. A vuggy porosity may develop, however, if the limestone is brought up to the vadose zone and retained there for a sufficient length of time. The secondary porosity developed in this way may provide space for holding hydrocarbons. Much of the reservoir porosity of the oil-bearing 'Bombay High' structures located off the western coast of India, for example, is due to the vuggy porosity developed in the vadose zone (Roychoudhury and Deshpande 1982). Dissolution of carbonate and feldspar in the post-cementation phase, leading to generation of secondary porosity, has also been reported from the Neogene sediments of the Bengal basin, Bangladesh (Imam and Shaw 1984).

Although vadose zone leaching makes an important contribution to the development of porosity, prolonged exposure of a carbonate rock in the vadose zone may cause a reduction in porosity. This is effected by precipitation of sparry calcite within the vugs. When the aragonite is of marine origin, bicarbonate with a positive value of δC* is produced because the mean δC of marine carbonate is positive. On the other hand, the humus-derived bicarbonate occurring in the near-surface vadose zone has a highly negative δC. The reaction between these two bicarbonates produces a pore fluid with δC ranging between −6 and −10, values not as highly negative as the vadose zone bicarbonate. The condition is ideally suited for precipitation of sparry calcite in the vadose zone. The process may drastically reduce the porosity.

The occurrence of both silica and carbonate cement in the same sandstone has been a mystery. Precipitation of carbonate is possible only when the medium is alkaline, but solubility of silica is known to increase with an increase in alkalinity. When an algal matte growing over a carbonate sediment is periodically exposed to atmosphere, the interstitial water attains exceptionally high alkalinity, leading to supersaturation, and eventually to precipitation, of silica on carbonate. This is one of the special conditions wherein simultaneous precipitation of silica and carbonate may take place (Baas-Becking, Kaplan and Moore 1960). In India, a case of replacement of early diagenetic rhombic carbonates by silica was reported by Mukhopadhyay and Chanda (1972) from the Iron Ore Group of rocks of Orissa. Another interesting post-depositional sequence of replacement of calcite by secondary silica in the Lameta Beds of India, followed by reprecipitation of calcite from percolating waters, has also been reported by Chanda (1963).

* δC represents deviation, measured in parts per thousand (per mil or ‰), from the standard. The standard is taken to be a belemnite from the Cretaceous, Pedee Formation in North Carolina, USA.

APPENDIX

Techniques of Sedimentary Petrography

1. Scheme for Description of Sandstones under the Microscope

A. FRAMEWORK GRAINS

1. Mineralogy

Quartz	a) Non-undulatory b) Undulatory c) Polycrystalline (2–3 crystals/grain) d) Polycrystalline (>3 crystals/grain)	The proportions of quartz, feldspars and rock fragments should add up to 100. Use Fig. 4.23 for visual estimation of proportions. Also determine proportions of quartz grains of different varieties separately, so that they add up to 100.

Feldspars
Rock Fragments—specify
Micas—muscovite/biotite
Heavy Minerals—specify
Other Minerals—specify

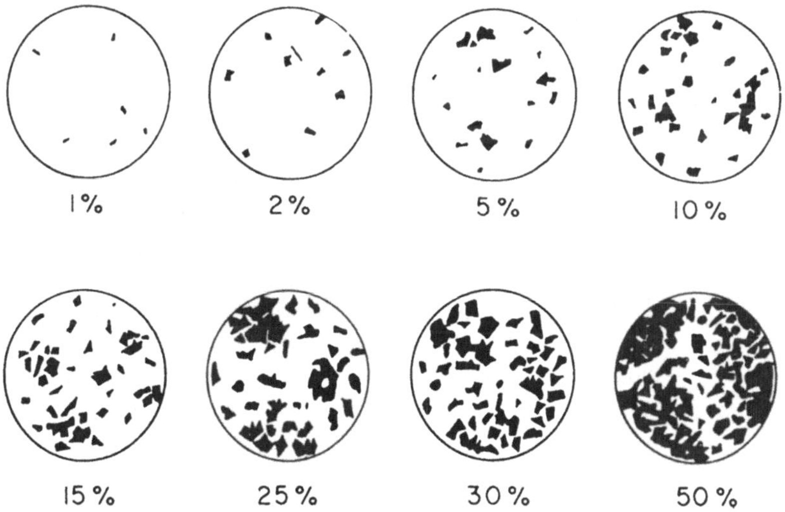

Fig. 4.23: Charts for estimating percentage composition of minerals under the microscope. Prepared by M.S. Shvetsov (modified and redrawn from Terry and Chilingar 1955, with permission of SEPM, Society for Sedimentary Geology).

2. Texture

i) Grain-size—use pre-calibrated eyepiece micrometer for size determination. See Table 3.1 for Wentworth nomenclature of size classes. Designate proportions also in qualitative terms: abundant, frequent, few, rare, e.g., medium sand-size particles abundant.

ii) Sorting—use the following scheme of classification:

Sorting	Matrix content	Grain-size variation
Well sorted	Low	Most grains lie within one or two Wentworth size classes
Moderately sorted	Moderate (~ 10%)	Moderate
Poorly sorted	High	Large

Range

iii) Roundness & Sphericity—use Powers' chart (Fig. 3.9) for visual estimation. For computation of the quantitative value of roundness follow the procedure discussed in Chapter 3 (see Figs. 3.8 & 3.9).

iv) Nature of grain contact—see Fig. 3.13 for comparison.

B. CEMENT—identify type—siliceous/calcareous/ferruginous/any other.

C. MATRIX—determine proportion (greater or less than 12.5%). Name the rock wacke or arenite accordingly.

D. CLASSIFICATION AND NOMENCLATURE—use revised Dott's (1964) scheme (see Fig. 4.3), e.g., Arkosic Wacke.

E. COMMENT ON THE POSSIBLE NATURE OF THE SOURCE ROCK
Use proportions and types of quartz, feldspar and rock fragments (see discussions on provenance in Chapter 2). Plot 'a', 'b', 'c', and 'd' types of quartz on Fig. 2.3 to draw conclusions regarding igneous or metamorphic nature of source rock.

Note: An easy way to classify a sandstone is to determine whether the total matrix content is greater or less than 12.5%, i.e., whether the matrix fills one-half of one quadrant of the field of the microscope. The rock may be named arenite or wacke depending on whether the matrix content is smaller or greater than 12.5%. Proportions are to be determined in three stages: (i) proportion of matrix (ii) proportion of the three important constituents—quartz, feldspars and rock fragments (iii) proportions of four varieties of quartz, undulatory, non-undulatory etc.

2. Scheme for Description of Limestones under the Microscope

A. ALLOCHEMICAL COMPONENTS

1. Intraclasts 2. Fossils 3. Oolites 4. Pellets	Determine proportion and size (sand size or larger) of each of these and tabulate in Folk's scheme (Table 4.9).

B. ORTHOCHEMICAL COMPONENTS

Micrite (microcrystalline ooze, 0.001-0.004 mm)
Sparite (sparry calcitic or dolomitic mozaic, 0.02-0.1 mm).

Decide name of the limestone by using appropriate combination of A and B. Use Folk's scheme of classification (see Table 4.9). Add 'rudite' as suffix if the majority of the allochemical grains are larger than sand size.

5

HYDRAULICS, SEDIMENT TRANSPORTATION AND STRUCTURES OF MECHANICAL ORIGIN

HYDRAULICS OF OPEN CHANNEL FLOWS

Much of the sediment in nature is transported by running water. To follow the process of sediment transportation therefore, one must have a clear idea of the dynamics of flowing water. The mechanism of sediment transportation by wind is also comparable in many respects to that of running water. The basic principles of hydraulics are discussed in this section to provide an insight into the mechanism of sediment transportation under aqueous and eolian conditions.

The most fundamental relationship between water discharge and flow velocity in an open channel is given by the continuity equation:

$$Q = w \cdot d \cdot v = A \cdot v,$$

where Q = total flow discharge, w = channel width, d = water depth, v = flow velocity, A = cross-sectional area of channel. It follows from this equation that discharge and water level remaining constant, the velocity of flow through an open channel varies inversely with the channel width. All open channel flows, including natural streams, are guided by this fundamental relationship.

The physical state of flow through an open channel is controlled by several factors, of which the flow velocity and fluid viscosity are more important. When the resistance offered by viscosity of the medium is high compared to the flow velocity, the fluid particles move in straight, parallel paths in the form of laminae. This flow is called laminar flow. The resistance offered by fluid viscosity is overcome when the flow velocity increases. At this stage the particles of fluid move in a haphazard fashion, multidirectional secondary motions are superimposed on the primary flows and eddies are generated (Fig. 5.1). The flow at this stage is called turbulent flow. The conditions governing the physical state of flow were expressed by Osborne Reynolds in 1883 by the following equation:

$$R = v \cdot l / \nu,$$

where v = mean flow velocity; l = a linear measure such as depth of water, grain-size, or hydraulic radius obtained by dividing the cross-sectional area of the channel by the wetted perimeter; ν = kinematic viscosity of the fluid obtained by dividing the coefficient of viscosity (η) by density (ρ). R, the Reynolds Number, is a dimensionless parameter. The changeover from laminar to turbulent flow, in the case of water at 20°C, takes place at a critical Reynolds Number of 500. This changeover is temperature-dependent, the viscosity of a fluid being inversely proportional to its temperature. It should be realised, however, that even for a given temperature of water there is no sharp demarcation between a laminar and a turbulent flow. A zone of transition marks the conversion even when the temperature is constant (Fig. 5.2; also see Chow 1959).

While the relationship between viscous and inertial forces determines whether a flow will be laminar or turbulent, the ratio of inertial forces within the flow to gravity forces (given by $\sqrt{g \cdot d}$) determines whether a particular flow will be tranquil (subcritical) or rapid (supercritical). These conditions are governed by the following equation after Froude:

$$F = v/\sqrt{g \cdot d},$$

where v = flow velocity, d = water depth, g = acceleration due to gravity. F, a dimensionless parameter, is called the Froude Number. When flow velocity equals the velocity of propagation of the gravity wave, that is, $F = 1$, the flow condition is said to be critical. At $F < 1$, the flow is tranquil or subcritical. At $F > 1$, the flow is rapid or supercritical. In the subcritical stage the water level is high but the flow velocity is low. This is the Lower Flow Regime of the

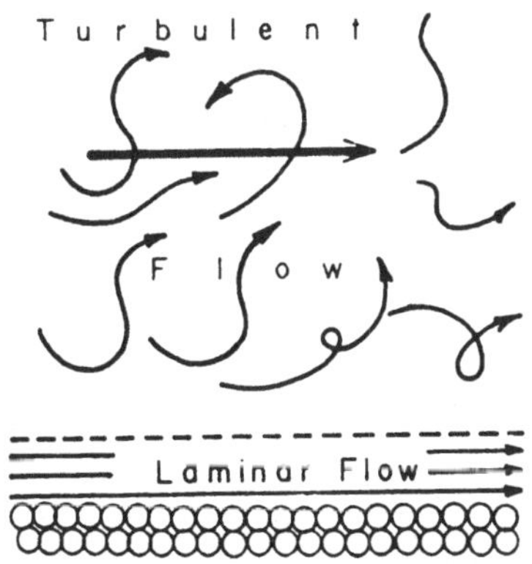

Fig. 5.1: Laminar and turbulent flows over a fine-grained bed (modified and redrawn after Fritz and Moore 1988, fig. 4.12 a).

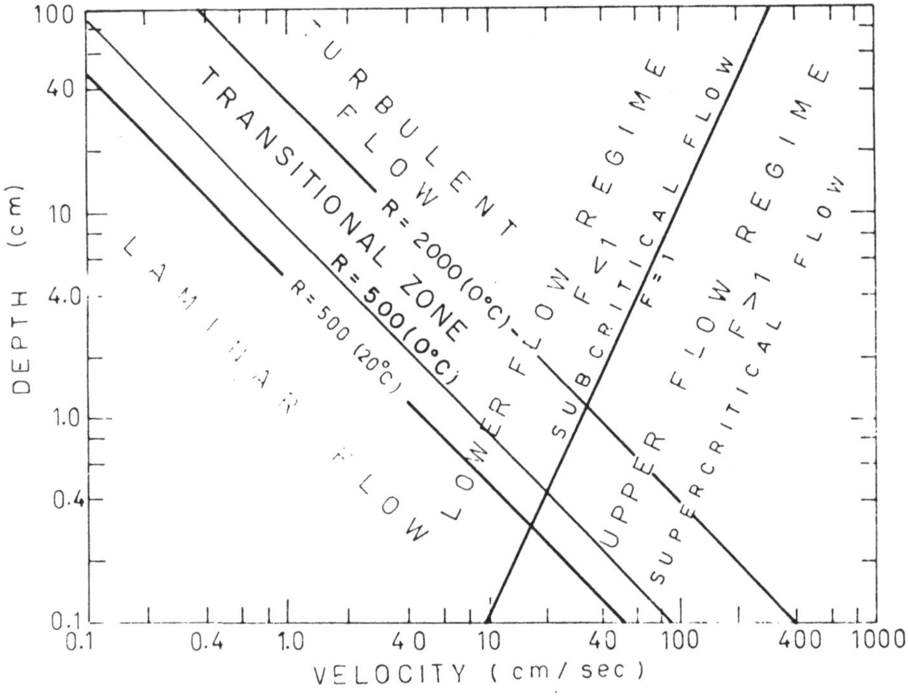

Fig. 5.2: Hydraulic conditions at various combinations of water depth, flow velocity and temperature (modified after Sundborg 1956, with permission of the author).

stream. In the supercritical stage the water level is low but the flow velocity is high. This is the Upper Flow Regime of the stream. The changeover from the higher to the lower flow regime is often abrupt. At this stage some of the kinetic energy associated with supercritical flow is suddenly transformed into turbulent energy and surface rollers appear. The phenomenon is known as hydraulic jump. It should be noted that the fields of critical Reynolds Number and Froude Number may overlap in nature. Thus a supercritical flow may be either laminar or turbulent (see Fig. 5.2). Moreover, it has been suggested that for small depths (< 1 cm) the Froude criterion is not reliable.

SEDIMENT TRANSPORTATION BY RUNNING WATER

When the velocity of water flowing over a bed of non-cohesive sediment exceeds a certain threshold value, the grains are put into motion (entrained). At this stage the flow overcomes the resistance offered by the grain due to its mass, cohesion (attachment to the neighbouring grains) and other forces which are determined by its shape and position of its centre of gravity. The relationship between critical erosion velocity and grain-size was revealed by a diagram compiled by Hjulström (1935; see Fig. 5.3). The influence of the drag force of the fluid on sedimentary particles (τ_*), and the resistance offered by the grain in terms of turbulence at the boundary (R_*) on the critical erosion velocity was experimentally determined by Shields in 1936 (see Fig. 5.4). The field above the curve shown in Fig. 5.4 indicates the condition suitable for entrainment of grains. It is clear from these diagrams that the particles most easily eroded from a bed are of fine sand size (0.3 to 0.6 mm).

Particles coarser than fine sand are difficult to erode because of their weight. Those of silt and clay size resist erosion due to their smoothness and cohesion. The velocity needed for eroding a particle from a bed is not exactly the same as that at which the same grain, if already in motion, comes to rest. The eroding velocity must be decreased by about 30% to cause deposition of the coarse grains. Clay and fine silt, once entrained, may remain in suspension even at a low velocity. Curve B in Hjulström's diagram (Fig. 5.3) corresponds to the lowest transporting velocity. Hjulström's studies, although of fundamental importance, have their limitations. For example, later workers have demonstrated that for very coarse sediments the Hjulström curve underpredicts the entrainment velocity (Novak 1973).

The flow velocity immediately above a sediment bed is retarded due to friction with the bed surface. The thin layer of fluid adjacent to the bed surface

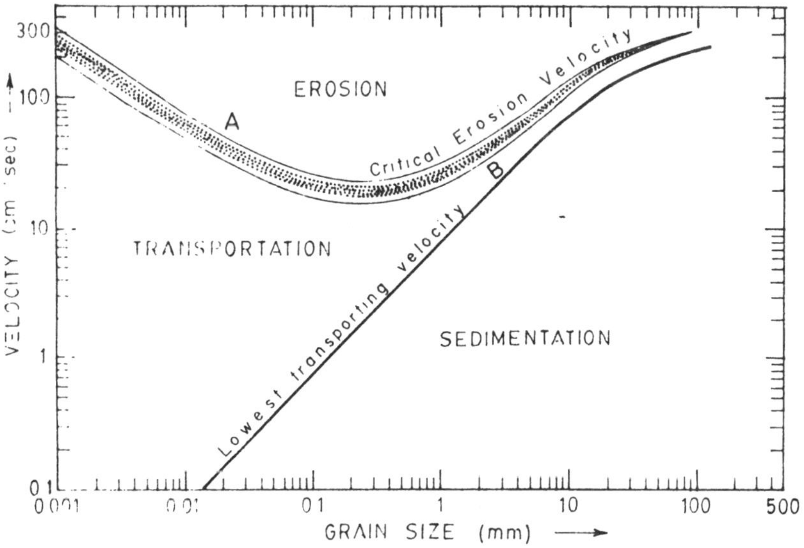

Fig. 5.3: Relationship between flow velocity (at a height of 1 m above bed), grain-size and erosion-transportation-deposition of sediments (after Hjulström 1935).

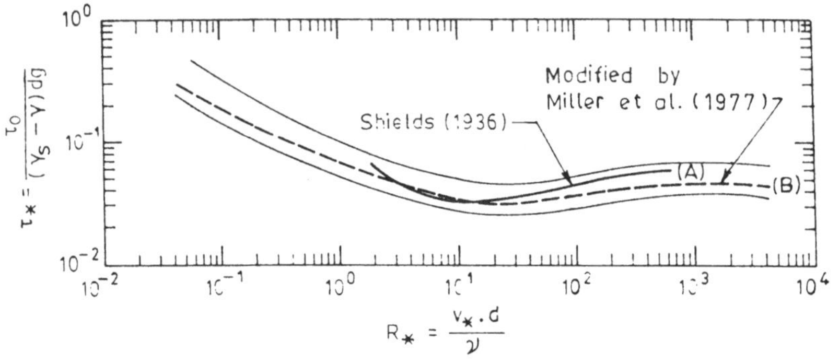

Fig. 5.4: Curves for sediment transportation (A) as determined by Shields 1936, and (B) modified and extended by Miller et al. 1977. v_* = shear velocity, d = particle diameter, ν = kinematic viscosity, γ = specific gravity of the fluid, γ_s = specific gravity of the particle. Redrawn with permission of International Association of Sedimentologists.

where this retardation takes place is called the boundary layer. Within this layer the flow can be either laminar or turbulent, but even when the flow is turbulent a very thin film, immediately in contact with the bed, shows laminar motion. This laminar sublayer generates over extremely smooth beds which offer no resistance to flow (see Fig. 5.1). Shields' experiments showed that the ratio of the force exerted by the flowing water along the bottom to the resistance offered by a layer of sand grains is a universal constant of the ratio of the grain-size to the thickness of the laminar sublayer. This significant conclusion led to many important studies on sediment transportation in later years (see Sundborg 1956).

Shields' curve is valid for grains transported by water as well as wind. All one needs to do is to change the parameters of the equations used by Shields. Experiments conducted by Bagnold (1954) showed that in the case of wind also fine sand is most easily moved. The surface provided by silt and clay is aerodynamically smooth and hence difficult to erode. Experiments conducted with wind and water have led to the conclusion that for initiation of grain movement a certain amount of form resistance is essential. This may be in the form of extraneous bodies lying on the bed surface or some irregularities projecting above the laminar sublayer, so that an otherwise smooth bed is converted into a rough one.

The nature of form resistance provided by a bed is given by a number of well established equations:

The Chézy equation, $V = C\sqrt{R_h \cdot S}$.

The Manning equation, $V = \left[1.49(R_h)^{2/3} \cdot S^{1/2}\right] /n$.

The Darcy equation, $\sqrt{\frac{8}{f}} = R_h^{1/6}/3.8n$,

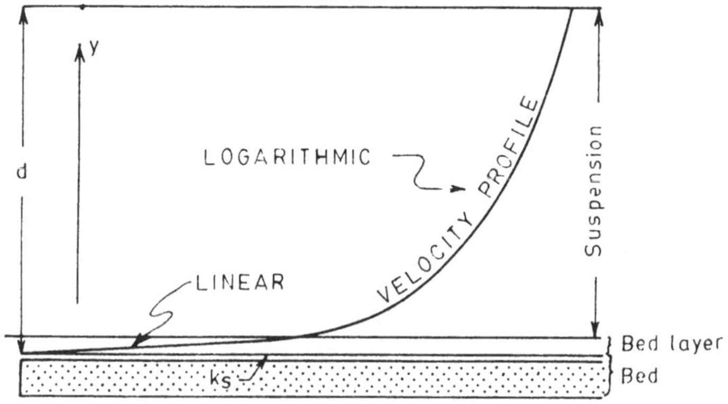

Fig. 5.5: Velocity distribution over a non-cohesive sand bed.

where V = uniform flow velocity; R_h = hydraulic radius, S = channel gradient; C = Chézy friction coefficient, n = Manning roughness coefficient; f = Darcy's friction factor (Chow, 1954). The expression for Manning equation given above is in f.p.s. units. In metric units the constant 1.49 is replaced by unity.

The velocity profile of the body of water flowing through an open channel is logarithmic with respect to height except at the very bottom where it is linear (Fig. 5.5). This velocity profile, deduced from the similarity theorem of von Kármán varies, depending on whether the bed concerned is smooth or rough. The constants in the Kármán equations were experimentally determined by later workers. After inserting these constants the well-known Prandtl-Kármán velocity distribution equations read as follows (Sundborg 1956):

$$u/u_* = 5.75 \log_{10}(zu_*/\nu) + 5.5 \text{ for smooth boundary;}$$

$$u/u_* = 5.75 \log_{10}(z/k) + 8.5 \text{ for rough boundary,}$$

where u is the average velocity at a distance z from the boundary surface, u_* is the shear-stress velocity, ν is the kinematic viscosity, and k is the diameter of sand grains on a sandy surface.

MODES OF SEDIMENT TRANSPORTATION BY WATER AND WIND

Once entrained, sediments are transported by running water in four different ways: sliding, rolling, saltation and suspension. The first two, confined to the stream bed, contribute to what is commonly called the bed load. In natural streams rolling is more important than sliding. A grain rolls forward in response to the impact of fluid against it (called the 'velocity pressure'). Larger particles offering greater surface areas move more easily than smaller grains. This movement is opposed by the viscosity pressure whose magnitude depends on the fluid's coefficient of viscosity, density and velocity. The total drag (F_c) is due to the pressure difference between the two forces. As a grain moves forward a negative pressure, called the static or internal pressure, acts on its upper part due to the Bernoulli effect. Whenever the fluid velocity is accelerated, the Bernoulli law demands a reduction of pressure transverse to the direction of fluid motion.

A decrease in static pressure on top of the grain compared to the pressure at the bottom causes the grain to be lifted above (hydraulic lift) the bed in the direction L_c passing through the centre of gravity of the grain (Fig. 5.6). The forces required to roll or lift a grain are dependent on its size, shape, immersed density and also on the angle of repose of the grain with respect to the mean drag level of water or wind flow. Two other factors on which the lift depends are the degree of grain packing on the sediment surface and the impulse of wind turbulence associated with drag and lift. In the course of his experiments with soil grains transported by wind, Chepil (1959) observed that L_c is about 0.85 F_c for spherical grains, and the angle of repose (ϕ) of the topmost grains with respect to the mean level of drag is about 24 degrees.

When the force required to push a particle up is greater than that required to roll it forward, the particle tends to leap forward into a state of tem-

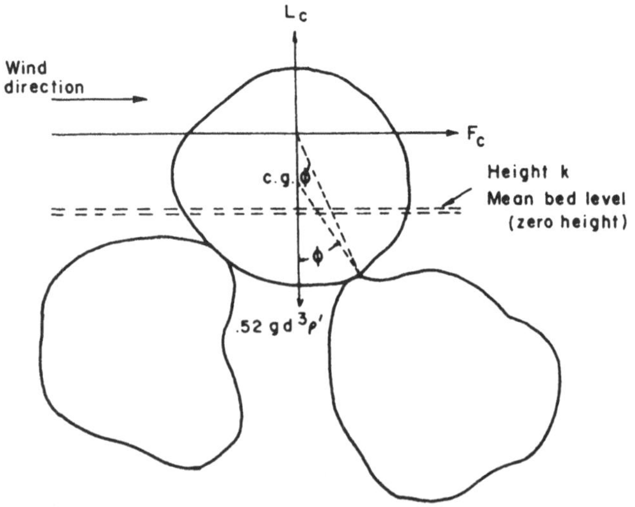

Fig. 5.6: Forces of lift, drag and gravity acting on a soil grain in a windstream at the threshold of movement of the grain. Net movement opposing F_c is (0.52 g $d^3\rho'$ - L_c) tan ϕ. Redrawn from Chepil 1959.

porary suspension until the force of gravity brings it down to the bed. The whole process is then repeated and the particle moves forward in a series of jumps, constituting what is called the saltation load (German: *Satz*—jump). Saltating grains follow low, smooth trajectories and are governed largely by ballastic forces. The velocity of a particle during saltation is less than that of water (Francis 1973).

Much of the sand-size particles is transported by wind in saltation while the coarser grains are left behind as the lag deposits. The saltating sand grains impinging on the sand surface at low angles put into motion the grains which are normally too large to be entrained. Only the upper few millimetres of a sand bed are affected by this surface creep but the saltating grains may themselves be raised up to a height of one metre or so. At higher levels finer dust, constituted of silt and clay sized particles, moves in suspension, being supported by the upward component of turbulence. This is true for wind as well as water in motion. Windblown dust extends to many kilometres above desert sand; the turbulence in the case of wind is particularly high because of the low viscosity of air (Fig. 5.7). On the contrary, the suspension load in water rarely extends to more than a few metres above the bed.

Estimation of Bed Load

An early attempt at estimation of bed load was made by H.A. Einstein in 1942. In the modified version of his equation, Einstein (1950) assumed that particle movement is initiated when the hydrodynamic lift force exceeds the submerged weight of the particle. He further assumed that the probability of this particle being redeposited is equal at all points of the bed and the average distance travelled by the particle as bed load is constant between consecutive points of deposition. Einstein's assumption that this distance (about 100 times the diameter of the particle) is independent of the

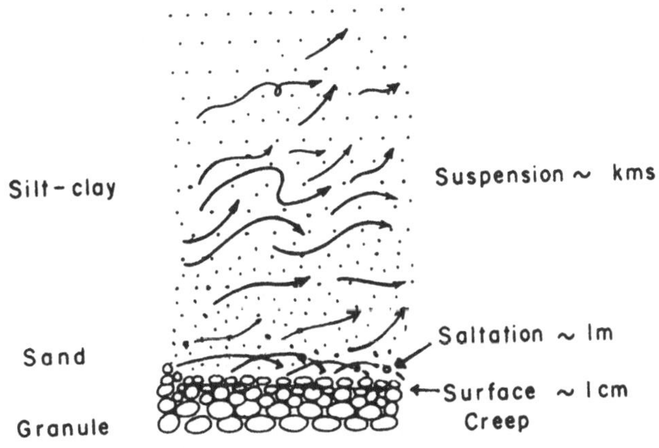

Fig. 5.7: Eolian sediment transportation (based on Fritz and Moore 1988, fig. 6.5).

rate of transport, bed- and flow conditions, was questioned by later workers. Einstein's bed-load equation relates the rate of bed-load transport (ϕ_*) to the properties of the grain and the flow (ψ_*). A correction factor (θ) was introduced in 1953 by Einstein and Chien for materials which are fine or non-uniform (see Garde and Ranga Raju 1977, p. 145 for detailed discussion).

Bed load equations were later developed by many authors (for example, Meyer-Peter and Müller 1948, Laursen 1958, Shen and Hung 1972, Yang 1973; see Shen 1978). Shen (1978) reviewed these equations and recommended the use of Einstein's method as modified by Colby and Hembree (1955) for estimation of bed and suspended loads relative to some measured data when these are available. In India, Goswami (1988) applied as many as eight different formulae to estimate the bed-load transportation rate in the Brahmaputra River and obtained widely varying results, suggesting that no one of these equations can fully accommodate all the natural variables. Laursen's (1958) equation, according to Goswami, provides the most acceptable result in the case of the Brahmaputra.

Estimation of Suspension Load

A suspension (suspended) load is defined as one in which the weight of the suspended particles is continuously supported by the fluid. The force necessary to keep the particles heavier than water or wind in suspension is provided by the vertical component of turbulence. As this is a fluctuating phenomenon, attempts to compute in absolute terms the amount of suspension load in a turbulent flow have led to extraordinarily complex exercises. On the other hand, equations have been developed for estimation of *relative* concentration of suspended loads in open channel flows (Rouse 1938, p. 341):

$$C_y/C_a = [\{(d-y)/y\} \cdot \{a/(d-a)\}]^z ,$$

where $z = v_s/\chi u_*$, C_a and C_y are sediment concentrations at distances a and y above the bed respectively, d is total depth of water, v_s is the settling velocity of the grain, u_* is the shear velocity. χ, the von Kármán constant, is 0.40. A simpler expression of the exponent z has been given by Sundborg (1956, p. 214):

$$z = 6.25c \ln(d/z_0)/U_{\text{max}},$$

where U_{max} is maximum velocity in profile; c is settling velocity of the grain; z_0 is a measure of the roughness of the bed given by $k_s/30$, where k_s, the representative grain diameter of the bed material, is given by that sieve size of which 65% of the mixture (by weight) is finer. The equation provides a simple method for estimation of suspension concentration at any level (C_y), provided the concentration at the reference level (C_a) is known.

Several modified curves for critical erosion velocity, cessation of grain movement and relative concentration in suspension were constructed by Sundborg (1956). Ljunggren and Sundborg (1968) also combined the curves for transportation of minerals of different densities in a composite diagram (Fig. 5.8). It must be noted that all these diagrams are based on experiments with (grain-size wise) homogeneous material. In natural conditions of sediment transportation, where a large range of grain-size is involved, the mechanism of sediment movement is far more complex than that shown in these diagrams. Only a few attempts have been made so far to study the mechanism of grain sorting in heterogeneous sediment loads (Sengupta 1975, 1979; Sengupta et al. 1991; Ghosh et al. 1981, 1986).

BED CONFIGURATION DURING SEDIMENT TRANSPORTATION

Water flowing over a plane, non-cohesive sediment surface throws it into a series of rhythmic patterns called bedforms. Observations in streams reveal that the nature and magnitude of bedforms are dependent on parameters such as the flow velocity and sediment grain-size. The precise manner in which these variables influence the shape and magnitude of bedforms, however, remained unclear until the natural processes could be simulated to some extent under controlled conditions in the laboratory. The results of one such pioneering study, carried out in the United States (Simons, Richardson and Nordin 1965) are reproduced in Fig. 5.9. The nature of the bedform generated at various combinations of stream power and fall velocity of sedimentary particles is shown in this diagram. Stream power ($\tau_0 \cdot \bar{v}$) was used by the experimenters as a measure of the erosive power of the stream and fall velocities were used as an index of the hydraulic behaviour of the grains. The flows responsible for generation of bedforms were divided into three regimes according to Froude number: a lower flow regime ($F < 1$), a transitional flow regime ($F \approx 1$) and an upper flow regime ($F > 1$).

The first to develop in the lower flow regime were the small sand ripples. With increasing stream power and fall diameter of the bed material, larger bedforms (dunes) appeared. Long, flat sand-waves marked the transitional zone ($F \approx 1$). With increasing stream power these were replaced by plane beds of the upper flow regime and eventually by antidunes. At this stage the water waves broke in a direction opposite to the flow and the sediment transport direction was also reversed. The antidunes were in phase with the water waves.

Repetition of these studies in different laboratories during the last thirty years has led to modification and refinement of the original Simons' curve (see Harms, Southard and Walker 1982). One of the latest available results (Southard and Boguchwal 1990) is reproduced in Fig. 5.10. It is clear from all

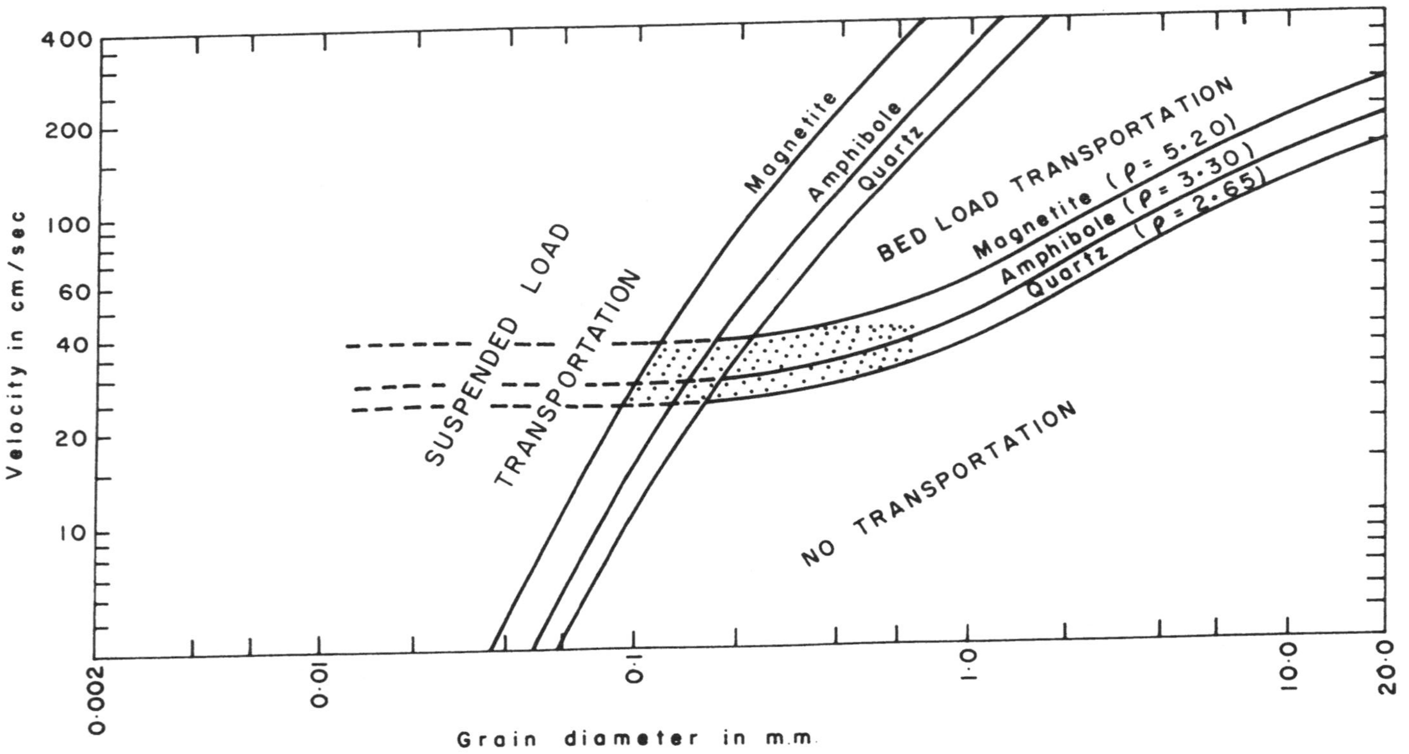

Fig. 5.8: Relationship between grain-size, flow velocity and transportation of minerals of different specific gravities. Stippled area indicates environment suitable for heavy mineral enrichment (after Ljunggren and Sundborg 1968, with permission of Å. Sundborg).

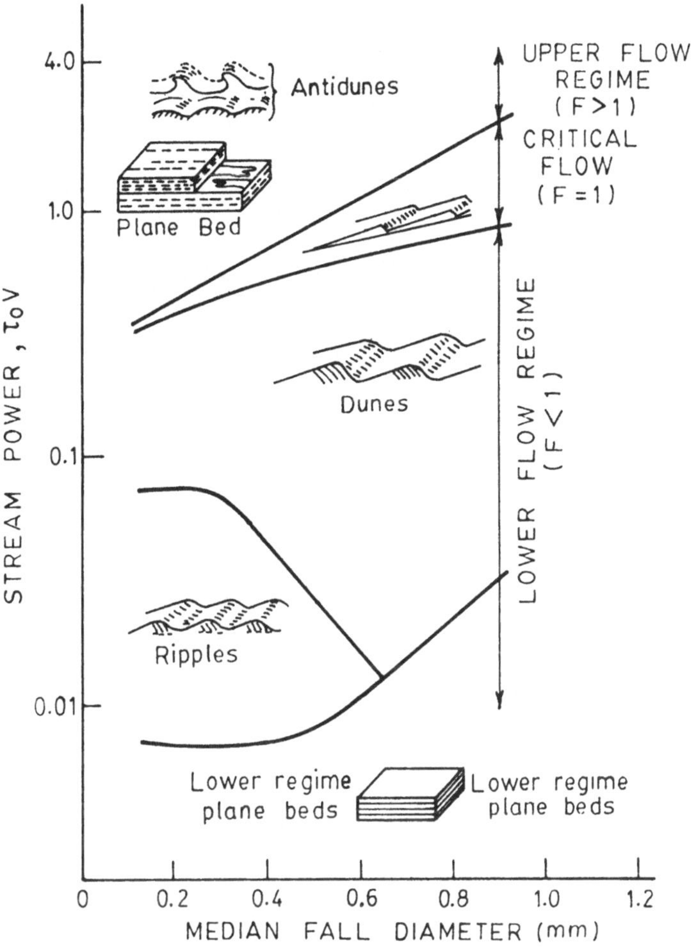

Fig. 5.9: Bedforms generated in experimental channels at various combinations of fall diameter and stream power (modified and redrawn after Simons et al. 1965).

these diagrams that grain-size has an important role to play in the passage of one bedform to the other. Only when the grains are of medium sand size do all the four types of bedforms appear at increasing velocities. When the grain-size is very small (< 0.15 mm), small ripples grade directly into upper plane beds without passing through an intervening dune stage. Similarly, small ripples do not generate at all in coarse sand. Another important factor in the development of bedforms is the depth of water. A minimum depth of about 4 cm of water is essential for the appearance of all the four different types of bedforms. The influence of the three independent variables—water

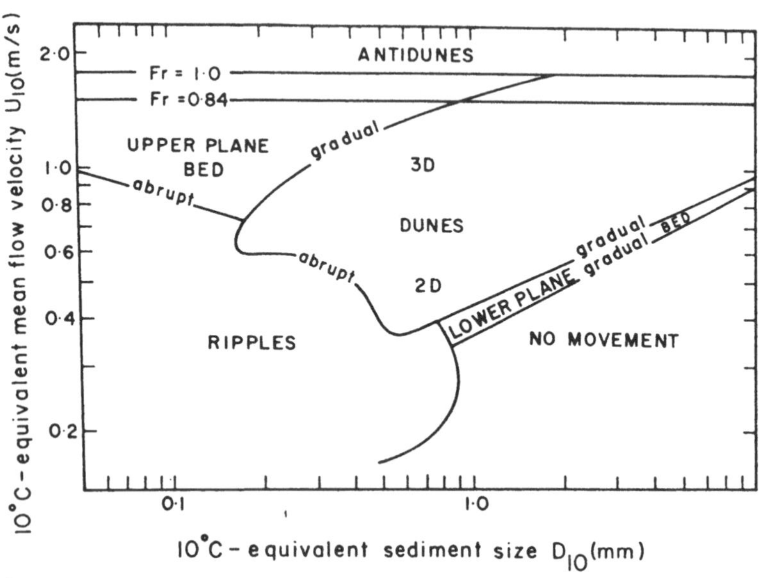

Fig. 5.10: Schematic representation of stability fields for bedforms of different types at various combinations of flow velocity and grain-size for flow depths = 0.25–0.40 m. Fr indicates Froude Number (after Southard and Boguchwal 1990, fig. 8, with permission of SEPM, Society for Sedimentary Geology).

depth, flow velocity and grain-size—on the nature and magnitude of bedforms can be presented only in three-dimensional diagrams of the type shown in Fig. 5.11 (see also Rubin and McCulloch 1980).

Introduction of the flow regime concept in bedform studies was prompted by the idea that an assemblage of bedforms implies a state of hydrodynamic equilibrium controlled by a number of variables, all of which cannot be properly defined. Not only stream power and fall velocity of particles, but even parameters such as temperature and the amount of clay suspension, both of which influence fluid viscosity, play important roles in the process of generation of bedforms. The concept of flow regime in a closed system, as in a laboratory channel again, is not the same as that applicable to nature. In the latter case variations in the rates of sediment supply and deposition, and also non-availability of the time needed for attaining hydraulic equilibrium, may cause complications which are not faced in laboratory experiments (Blatt, Middleton and Murray 1972, pp. 123–124).

Morphology of Bedforms

The simplest type of repetitive bedform, produced transverse to unidirectional flows, is a sand ripple. The steeper face of a ripple dipping in the downcurrent direction is called the lee face. The gently sloping face dipping in the opposite direction is the stoss. The crest of the ripple is called the

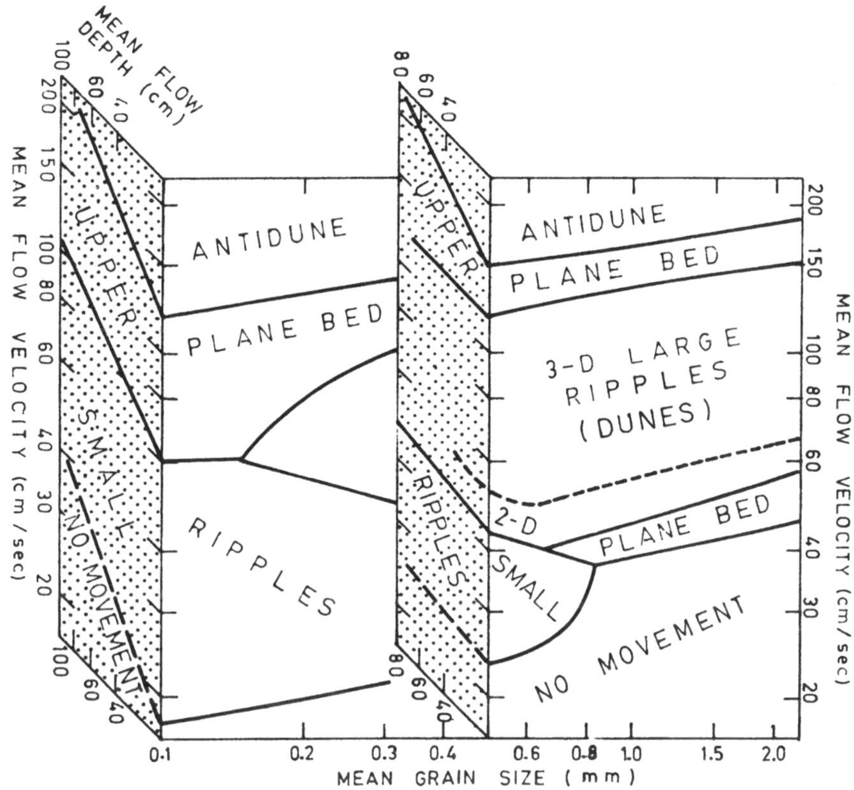

Fig. 5.11: A three-dimensional representation of the stability fields of bedforms at various combinations of water depth, flow velocity and grain-size.

summit point. The vertical height of the crest above the base is the height (H) of the ripple. The horizontal distance from one crest to the next is the length or wavelength of the ripple (L or λ). The ripple index, also called the vertical index, is given by the ratio of length to height (L/H).

The stream flow lines associated with an asymmetric sand ripple are shown in Fig. 5.12. Erosion of sediments from the stoss face of the ripple and deposition of these sediments on the lee side cause the ripple to 'migrate' downstream. The internal stratification developed out of this process produces cross-stratification on the lee face. At the same time flow separation at the crest of the ripple produces a backflow vortex in front of the

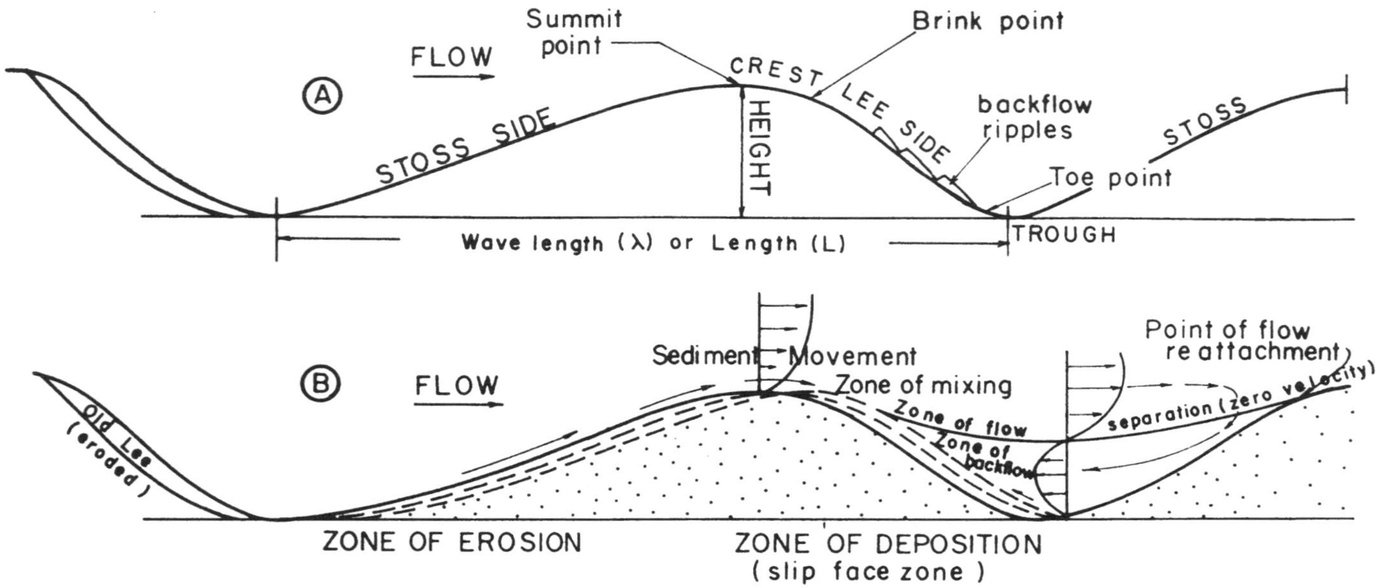

Fig. 5.12: (A) Morphology of an asymmetric sand ripple. (B) Flow lines related to an asymmetric sand ripple migrating by stoss-side erosion and lee-side deposition.

advancing lee. Backflow ripples are generated in front of the advancing lee by this vortex. The point of flow separation corresponds to the brink point. The point of flow reattachment is the toe-point of the ripple.

Ripples and dunes which make sharp angles with the horizontal are called angular bedforms. Flume experiments have shown that the stability of angular bedforms is controlled to a large extent by the lee side sediment fallout pattern. When sediment grain-size and flow velocity conditions cause the sediment to avalanche immediately near the ripple crest, so as to maintain a concave-up lee profile, an angular bedform pattern is maintained. Upper flow regime flat beds generate when the coarser grains, at high flow velocity condition, are deposited at a greater distance from the toe of the lee. At this stage the lee face becomes convex and the brink point vanishes. Coarse sands do not ripple at or near the threshold condition because they cannot develop or maintain a slip face (Hand and Bartberger 1988).

The criteria commonly used for classification of sand ripples are shape and magnitude. The former is defined by crestline, and the latter, by wavelength and height. Four different shapes of ripples—straight, catenary, linguoid and lunate—were recognised by Allen (1968a; see Fig. 5.13). Linguoid ripples are arranged in *en echelon* (out-of-phase) patterns. When the ripples are in phase, the pattern is called cuspate. These ripple patterns were correlated with increasing velocity and decreasing water depth conditions (see Fig. 5.14).

Magnitude-wise ripples have been traditionally classified into ripples, megaripples, dunes and sandwaves, but lack of uniformity in the definition of these terms led to a good deal of confusion in the past. Attempts were made from time to time to review the bedform nomenclature to resolve this confusion (see, for example, the review by the American Society of Civil Engineers Task Force 1966). According to current thinking, bedforms should be classified according to descriptive parameters which are objectively definable, easily measurable and have genetic implications. The classes should be mutually exclusive and their names should unambiguously define the nature of the bedform (Bridge in Ashley, 1990).

The SEPM, Society for Sedimentary Geology Committee, who reviewed the bedform classification problem in 1987, recommended use of the term 'dune' for eolian as well as subaqueous ripples of large ($L > 0.6$ m) dimension. This Committee also recommended classification of sand-dunes into four groups: small ($L = 0.6$–5 m), medium (5–10 m), large (10–100 m) and very large (> 100 m). Only two groups of shape terms were accepted by the Committee. When the dune geometry can be adequately described by a single transect parallel to the flow, that is, when the dune is straight crested, it is designated as a 'two-dimensional' dune. Dune forms characterised by scour pits and curved lee faces are called 'three-dimensional'.

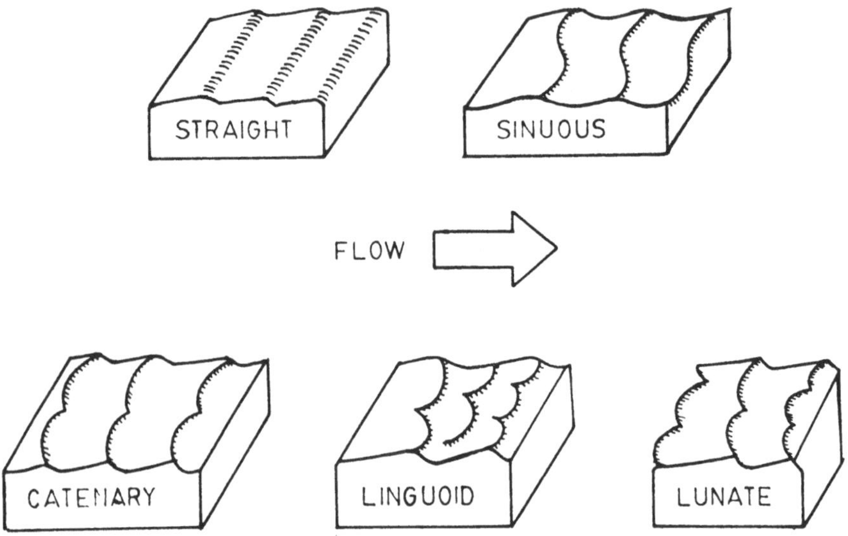

Fig. 5.13: Classification of ripple forms according to crest line (modified and redrawn after Allen 1968).

The other, secondary criteria recommended by the SEPM Committee are listed in Table 5.1.

Table 5.1. Classification scheme for subaqueous ripples and dunes (modified after SEPM recommendations in Ashley 1990)

I.	*Essential Characters*	Ripples	Dunes			
			Small	Medium	Large	Very large
	Size: Spacing (L or λ)	= 0.1–0.6 m	0.6–5 m	5–10 m	10–100 m	> 100 m
	Height (H)	≈ 0.04 m	0.075–0.4 m	0.4–0.75 m	0.75–5 m	> 5 m
	Shape: 2-dimensional (2-D)—Straight crest line					
	3-dimensional (3-D)—Curved crest line					
II.	*Important Characters*					
	Superposition: simple or compound (sizes and relative orientation)					
	Sediment characteristics (size, sorting)					
III.	*Useful Characters*					
	Bedform profile (stoss and lee slope lengths and angles)					
	Full beddedness (fraction of bed covered by bedforms)					
	Flow structure (time-velocity characteristics)					
	Relative strengths of opposing flows					
	Dune behaviour-migration history (vertical and horizontal accretion)					

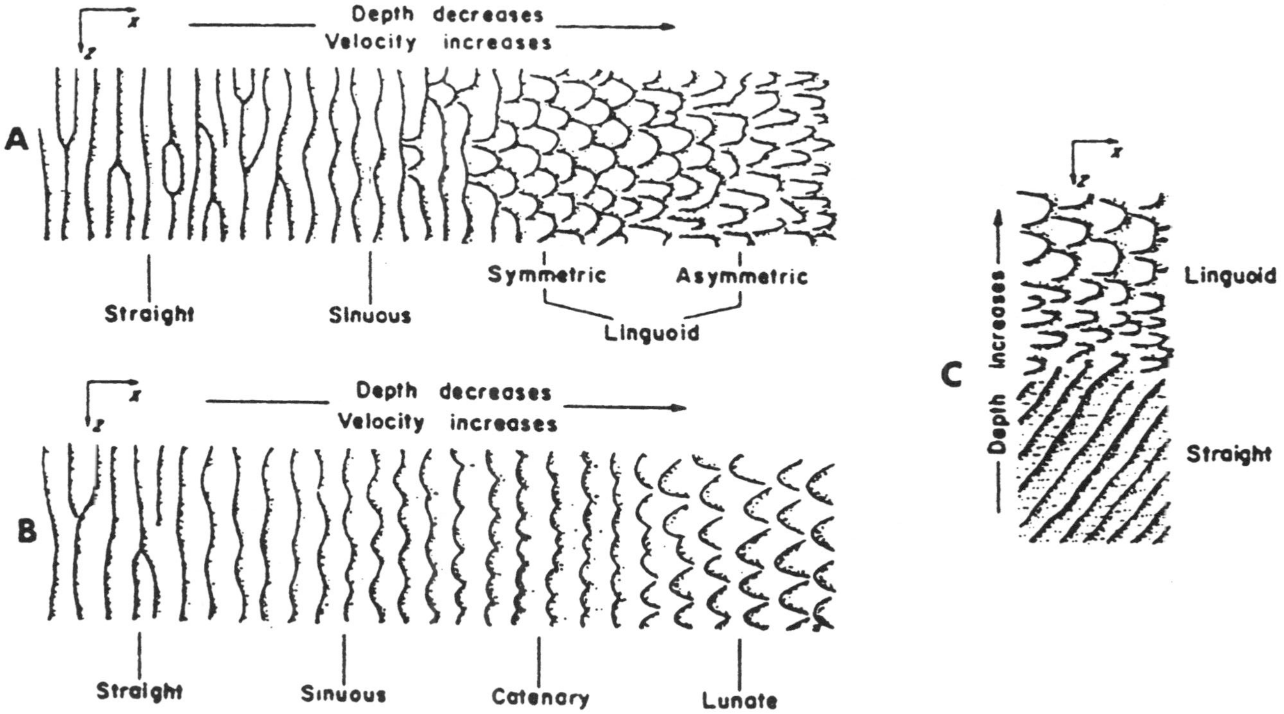

Fig. 5.14: Influence of flow velocity and/or water depth on ripple form (after Allen 1968a, fig. 4-61). A = small scale ripples, B = large scale ripples, C = small-scale ripples transverse to flow. Leesides are stippled. Reprinted with permission of J.R.L. Allen.

STRUCTURES GENERATED BY BEDFORM MIGRATION

Current Ripple Structures

As current flows the stoss side of a ripple is progressively eroded and the eroded materials continuously accumulate at the brink of the ripple. When the lee slope exceeds the angle of repose, which is about 35° for subaqueous ripples, the eroded materials avalanche down the lee slope. If new sediments are added to the system at this stage, the ripple not only migrates forward but also builds upward. The net direction of ripple movement now is described by an angle of climb, θ (Fig. 5.15A). The lamination produced by this process of accumulation is called ripple-drift cross-lamination. Jopling and Walker (1968) recognised three different types of ripple-drift cross-laminations. When the grain movement is restricted to the bed, the ripple stoss is eroded almost completely and climbing sets develop on the lee side (type-A ripple-drift, Fig. 5.15). When fallouts from suspension are also involved, sinusoidal ripple lamination, consisting of a continuous series of sine-wave profiles (type-B) is generated (Fig. 5.15B). Intermediate between these two is the 'type C' of Jopling and Walker (1968). The type-C ripple laminae consist of climbing sets of lee-side faces together with stoss-side laminae. Alignment of ripple crests in type-B climbing ripples may sometimes produce the illusion of cross-stratification ('pseudostratification'; also see Fig. 5.16B).

When the depositional system contains mud as well as sand, intermittent breaks in the current flow may lead to preservation of streaks of mud within the ripple troughs. The resultant structures have been termed flaser bedding (Reineck and Wunderlich 1968). With increasing proportions of mud, wavy beddings are produced. When the proportion of mud is very high compared to that of sand, only isolated lenticles of sand may be preserved within thick layers of mud, leading to what has been termed lenticular bedding (Fig. 5.17).

Contemporaneous cutting and filling by a migrating ripple produces one of the very common sedimentary structures called cross-stratification. These structures consist of beds having a higher inclination than the principal surface of accumulation. Cross-stratifications may also generate in two stages—an initial stage of scouring followed by a stage of deposition (scour-and-fill). Emptying of sediments carried by streams into a relatively quiet pool of water may also produce cross-bedded sets. These cross-beds have three components: a topset lying parallel to the principal surface of accumulation, a middle part (foreset) having stratification inclined at a much higher angle than the topset, and a lower bounding surface parallel to the depositional floor (bottomset). An ideal delta depicting all the three cross-bedding components is often called a Gilbert-type delta, after G.K. Gilbert who described such features from the delta of Lake Bonneville about a century ago.

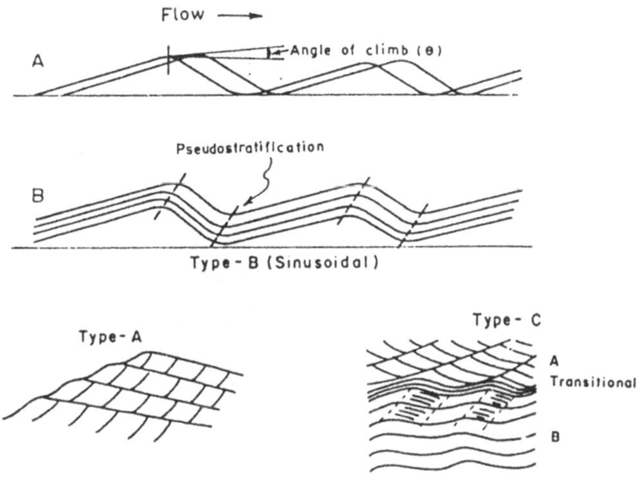

Fig. 5.15: A) The mechanism of generation of ripple drift lamination (modified after Harms et al. 1982). B) Different types of ripple drift laminations (modified after Jopling and Walker 1968).

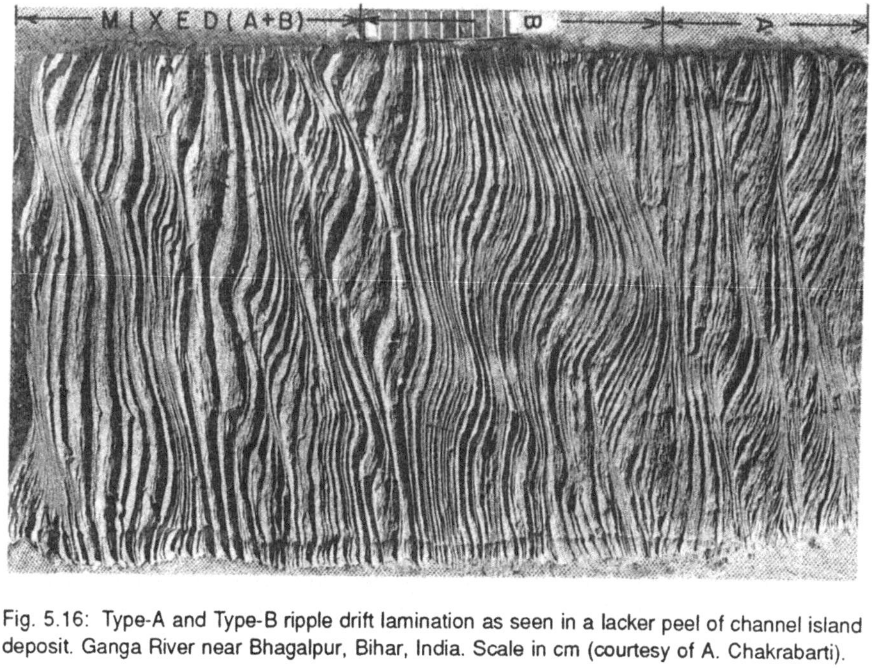

Fig. 5.16: Type-A and Type-B ripple drift lamination as seen in a lacker peel of channel island deposit. Ganga River near Bhagalpur, Bihar, India. Scale in cm (courtesy of A. Chakrabarti).

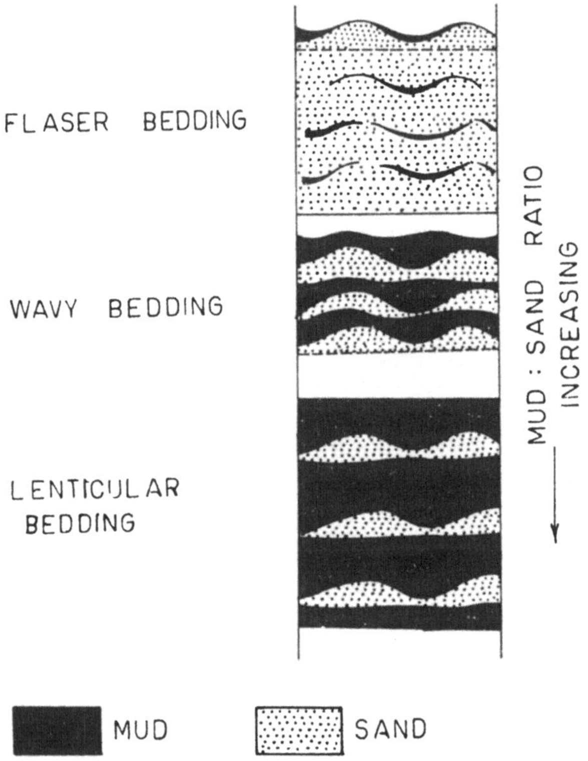

Fig. 5.17: Influence of mud and sand proportions on flaser, wavy and lenticular bedding (simplified and redrawn after Reineck and Wunderlich 1968).

Cross-strata may be classified according to their magnitudes and shapes. Those measured in ranges of centimetres are called cross-beds (> 1 cm), and those measured in millimetres are called cross-laminations (< 1 cm). The shape classification uses the character of the lower bounding surface (simple, planar and trough) and the shape of sets of cross-strata (tabular, lenticular and wedge) as the main criteria (McKee and Weir 1953). Several other less important criteria are also used (see Fig. 5.18 and Table 5.2). Allen (1963) proposed a very comprehensive scheme of classification based on the character of the lower bounding surface and five other morphological features (Fig. 5.19 and Table 5.3).

Tabular and trough types of cross-stratifications may be produced by dune migration under aqueous and eolian conditions. In tabular cross-strata the bedding surfaces are planar or parallel and the foreset laminae meet the lower bounding surface at sharp angles. In plan the traces of

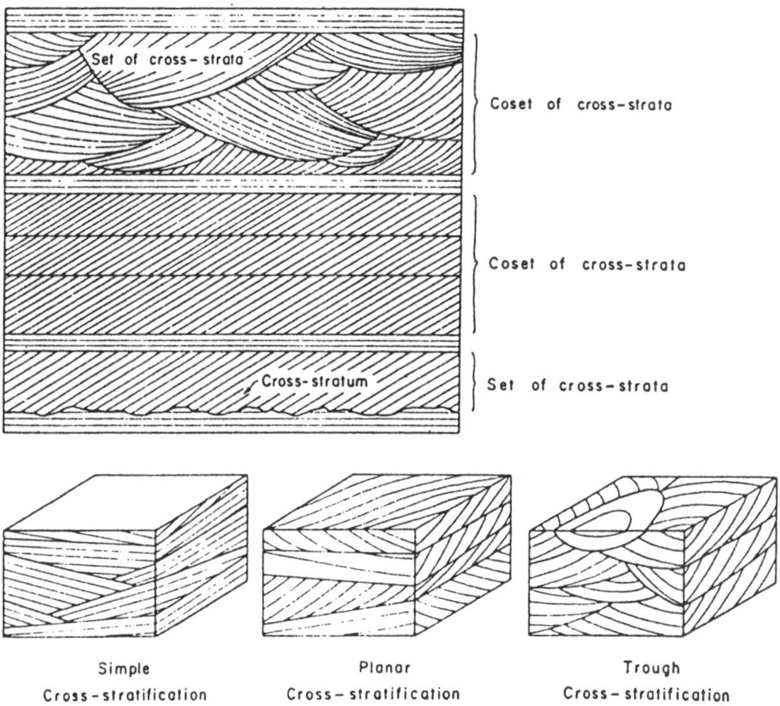

Fig. 5.18: Classification of cross-bedding (after McKee and Weir 1953, figs. 1 & 2).

Table 5.2. Classification of cross-stratification (abridged from McKee and Weir 1953)

			Terminology
I.	NATURE OF LOWER BOUNDING SURFACE (LBS)		
	A. LBS - non-erosional	—	Simple
	B. LBS - planar surface of erosion	—	Planar
	C. LBS - curved surface of erosion	—	Trough
II.	SHAPE OF EACH SET OF CROSS-STRATA		
	A. Sets bounded by converging surfaces	—	Lenticular
	B. Sets bounded by planar, parallel surfaces	—	Tabular
	C. Sets bounded by planar, converging surfaces	—	Wedge

Note: Based on criteria I and II cross-bedding sets are termed planar-tabular, planar-wedge or trough-lenticular.

the laminae are straight or nearly so. These are produced by migration of two-dimensional, straight-crested dunes (Fig. 5.20B). A trough type of cross-bedding is produced by migrating, three-dimensional, sinuous-crested dunes. The set boundaries are curved in longitudinal section. The foreset

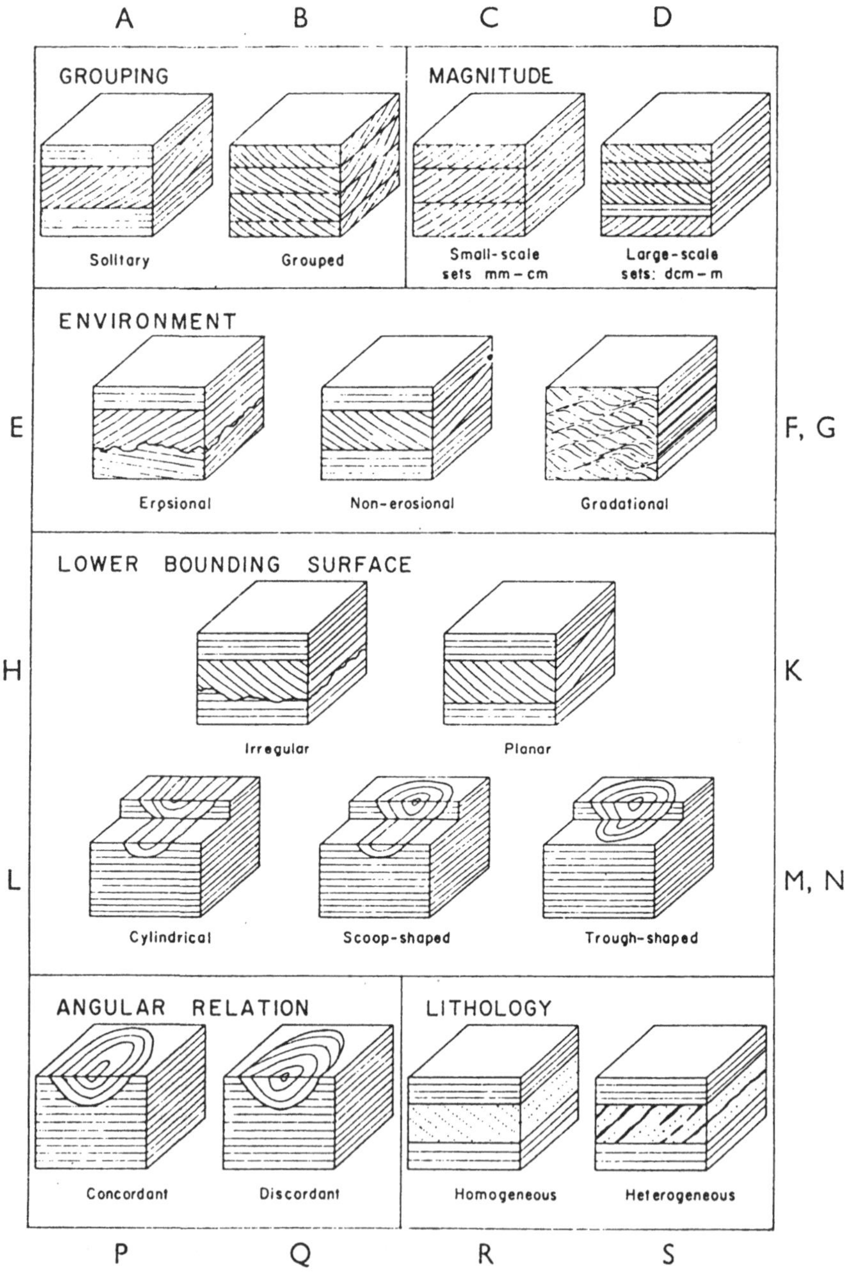

Fig. 5.19: Classification of cross-bedding (after Allen 1963, fig. 2, with permission of the author and Elsevier Science Publishers B.V.).

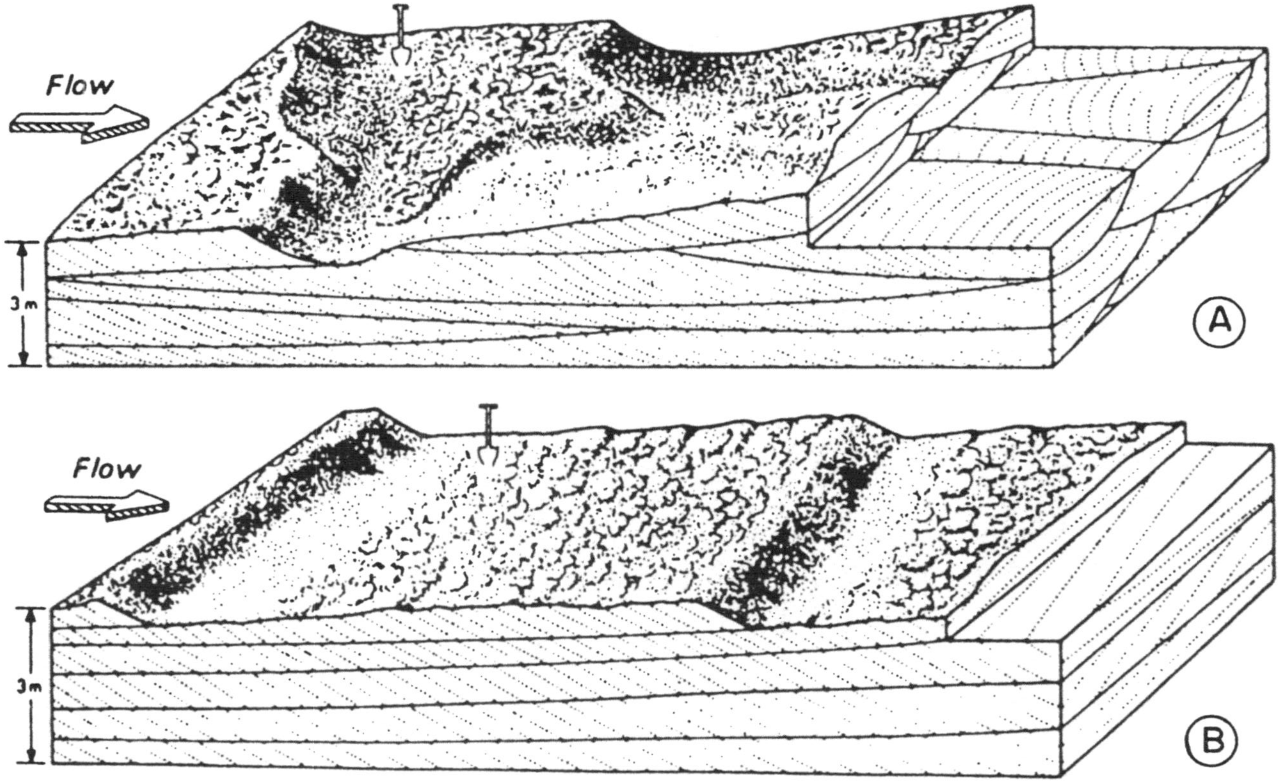

Fig. 5.20: (A) Large-scale trough-lenticular cross-bedding produced by migrating sand dunes, (B) Planar-tabular cross-bedding produced by migrating sand waves (after Harms et al. 1975, figs. 3-1 & 3-2, with permission of SEPM, Society for Sedimentary Geology).

Table 5.3. Criteria of classification and descriptive terms applicable to cross-stratified units (Allen 1963)

Criteria	Grouping	Magnitude	Character of lower boundary	Shape of lower boundary	Relation of cross-strata to lower boundary	Lithology of cross-strata
				Irregular		
				Planar		
Descriptive terms	Solitary	Small-scale	Erosional	Cylindrical	Concordant	Homogeneous
			Non-erosional	Scoop-shaped		
	Grouped	Large-scale	Gradational	Trough-shaped	Discordant	Heterogeneous

laminae of trough cross-beds are also curved in plan and have a tangential relationship to the underlying erosional surface (Fig. 5.20A).

The type of cross-bedding generated by the process of dune migration is largely controlled by flow velocity and the amount of suspension load available in the system. With increasing flow velocity and suspension load, a tabular cross-bedding is likely to be modified into a lenticular one with a curved, tangential contact at the base. The experiments conducted by Jopling (1965) provided insight into the mechanism of this change (see Fig. 5.21). Flow separation at the brink of an advancing dune causes the sediment load to split into two fractions: a bed load and a suspended load. The former, accumulating at the brink of the dune, avalanches down the lee slope, adding foreset laminae to the lee face. The latter, caught up in the zone of turbulent mixing, is transported further downstream, to be deposited at a distance from the lee, below the zone of mixing. When the flow velocity is high, these deposits are pushed up the lee slope by the backflow vortex, thereby modifying an angular contact into a tangential one. Backflow ripples of this type have been reported by Boersma (1967) from ancient as well as modern sediments.

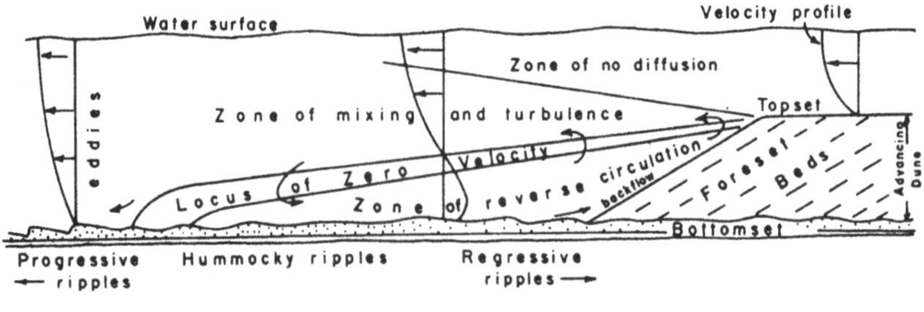

Fig. 5.21: Flow over a delta (dune) shaped bedform in an experimental channel indicating the origin of trough cross-bedding (modified and redrawn after Jopling 1961).

A trough-lenticular type of cross-bedding may also be generated from contemporaneous cutting and filling during the course of ripple migration. Impingement of the flow jet in front of an advancing dune may cause the bed to erode at the point of flow reattachment. When the dune crest is sinuous, the process of erosion may produce a series of spoon-shaped scours at different distances from the advancing dune front. These are soon filled by lenticular, lee-face laminae of the migrating dune, producing a trough-lenticular type of cross-bedding. Large-scale, trough-shaped scours may also be scooped out of stream-beds by eddies (called 'kolks') swirling with semi-vertical axes. Such eddies are commonly generated at the confluence of stream channels meeting at sharp angles. The filling of these trough-shaped scours by advancing dune laminae may produce trough-lenticular cross-beds of large dimensions. Such features have been described from ancient and modern sediments among others, by Harms and Fahnestock (1965), Coleman (1969) and Sengupta (1974) (see Fig. 5.22B).

The internal stratifications of eolian dunes are not exactly the same as those of subaqueous dunes. Four different eolian stratifications have been identified: plane bed lamination, translation ripple lamination, grainfall lamination and sand-flow lamination (Hunter 1977). Thin, parallel laminae (parallel bed lamination) develop on the stoss of eolian dunes when the wind velocity is too high to generate ripples. Accretion on the stoss side of dunes produces translation (climbing) ripple lamination. When the angle of climb exceeds the dip of the stoss face, wavy lamination and internal cross-lamination may develop. This is the supercritical stage. When the angle of climb is less than the slope of the stoss side due to a slow rate (subcritical) of accretion, tabular lamination, a few millimetres thick, may develop, and the laminae may also show inverse grading. Grainfall laminations develop when flow separation by strong winds causes deposition of saltating grains below the dune crest. These may show poorly defined internal structures. When the angle of repose at the dune crest is exceeded, *en masse* avalanching may produce linguoid flow lobes (sandflow lamination) on the lee face of small dunes and cover the pre-existing sedimentary structures (see Chapter 6 and Fig. 6.15).

Wave Ripple Structures

The simplest type of wave motion is caused by movement of wind on the free surface of water. At the water-air interface globules of water move in circles of the same diameter as the wave. At depth the diameters decrease exponentially. Wave ripples are generated on a non-cohesive sediment surface when the oscillatory motion of the water globules is strong enough to move the grains. The wave ripples thus produced are usually straight crested in plan. Only in exceptional cases are they sinuous or three dimensional. The crests of wave ripples are sharp and the round troughs are often

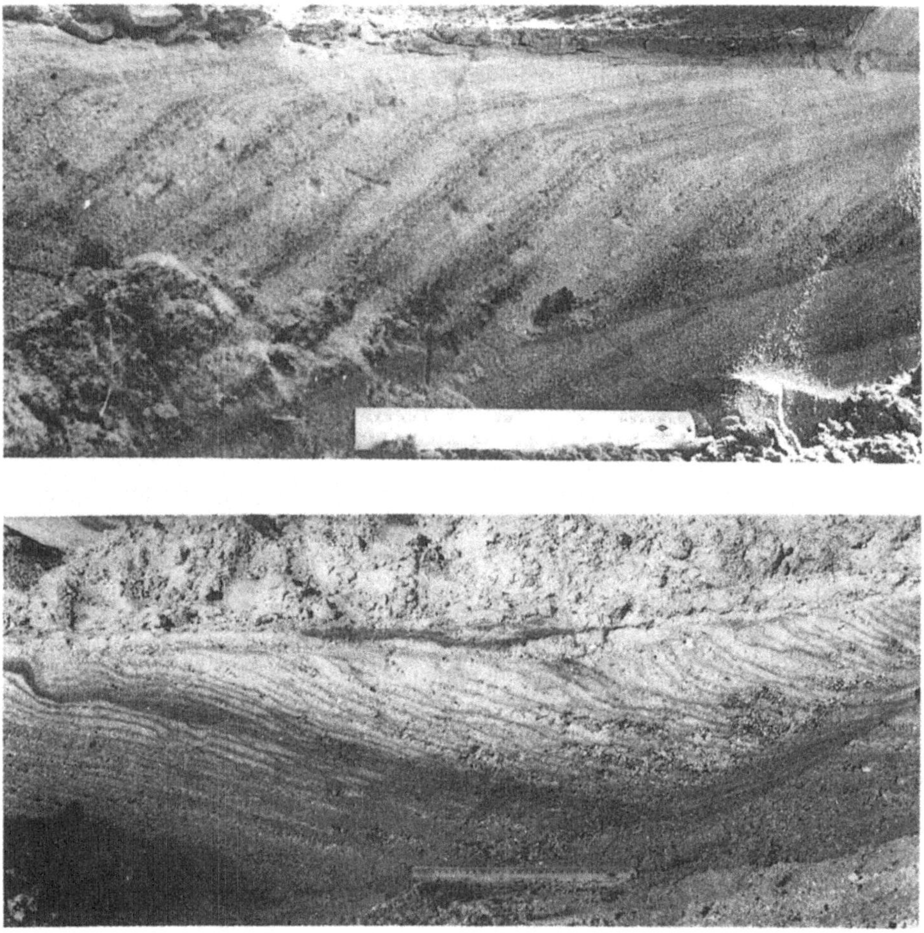

Fig. 5.22: Longitudinal sections of (A) planar-tabular and (B) trough-lenticular cross-stratifications Trench wall sections in a modern point bar, the Usri River, Bihar, India. Length of scale is 30 cm (from Sengupta 1974).

indented. These ripples are symmetrical in longitudinal section except when the forward and backward speeds of water waves are unequal.

It may be difficult to distinguish between wave- and current-generated ripples on a purely morphological basis. Current ripples generally have longer lengths than heights. At a ripple index (R.I. = ripple length/ripple height) exceeding 15 only current ripples occur (Reineck and Wunderlich 1968, in Reineck and Singh 1980, p. 35). Symmetry in the outward pattern of a current ripple may be generated due to erosional processes also, but the true nature of a ripple in such a case will be indicated by the internal stratifications.

The type of oscillation ripple generated on a bed surface is a function of the orbital speed (U_m) and time period (T) of oscillation. Density, mean grain-size and shape of the sedimentary particles, and density and viscosity of the fluid are some of the other factors controlling the nature of wave ripples. A sequence of oscillation ripples of varying shapes and dimensions may appear as the orbital speed and time period of oscillation wave change (see Fig. 5.23A). The first to generate on a non-cohesive sediment bed at low orbital speed are small, straight-crested ripples with gentle slopes. These have been termed 'rolling-grain ripples'. With increasing orbital speed these ripples change over to steeper-sided 'vortex ripples' in which lee-side vortices erode and carry sand each time the flow direction reverses. The vortex ripples are stable over a wide range of orbital speed and time period. At higher orbital velocity and at a small to moderate oscillation period the vortex ripples change over to flatter, post-vortex ripples. During each half of the flow cycle these post-vortex ripples show strongly unidirectional flow. With an increasing time period and at a larger duration of one-way flow, the post-vortex ripples are converted to 'reversing-crest ripples'. At this stage the grain transport in each direction levels off the stoss sides of the ripples. When the oscillation periods are very long, both post-vortex and reversing-crest ripples change over to flat beds (Harms, Southard and Walker 1982).

The internal stratification of an oscillation ripple is controlled by the rates of migration and aggradation (Fig. 5.23B). When the former is high compared to the latter, ripple troughs erode parallel to the already deposited laminae and the internal strata dip in one direction only. On the other hand, when the aggradation rate is much higher, minor shifts in crest positions cause the laminae to be dovetailed in a chevron-like pattern. At intermediate stages, when some migration takes place in an environment which allows rapid vertical build-up, the climb angle is steep and the entire ripple profile is preserved (Fig. 5.23B). Box cores taken from shallow marine sand (Newton 1968) show complex internal structures within features produced under oscillatory conditions. A large number of laminae dip randomly within such structures.

Wave ripples have been reported from various environments but they are most common in nearshore areas and in tidal flats. Symmetric and asymmetric wave ripples of small magnitude, often with tuning fork-like bifurcation in plan, are found on margins of water bodies not more than a few centimetres deep. Symmetrical wave ripples with unidirectional internal stratification have been reported from shallow-water, nearshore areas. With sufficient sediment supply from suspension, these may develop into climbing ripples or ripple laminae in phase.

Features described in sedimentological literature as hummocky cross-stratification have been assigned to oscillatory flows of moderate to long periods (5–10 seconds) and high orbital velocity (> 0.5 m/sec) correspond-

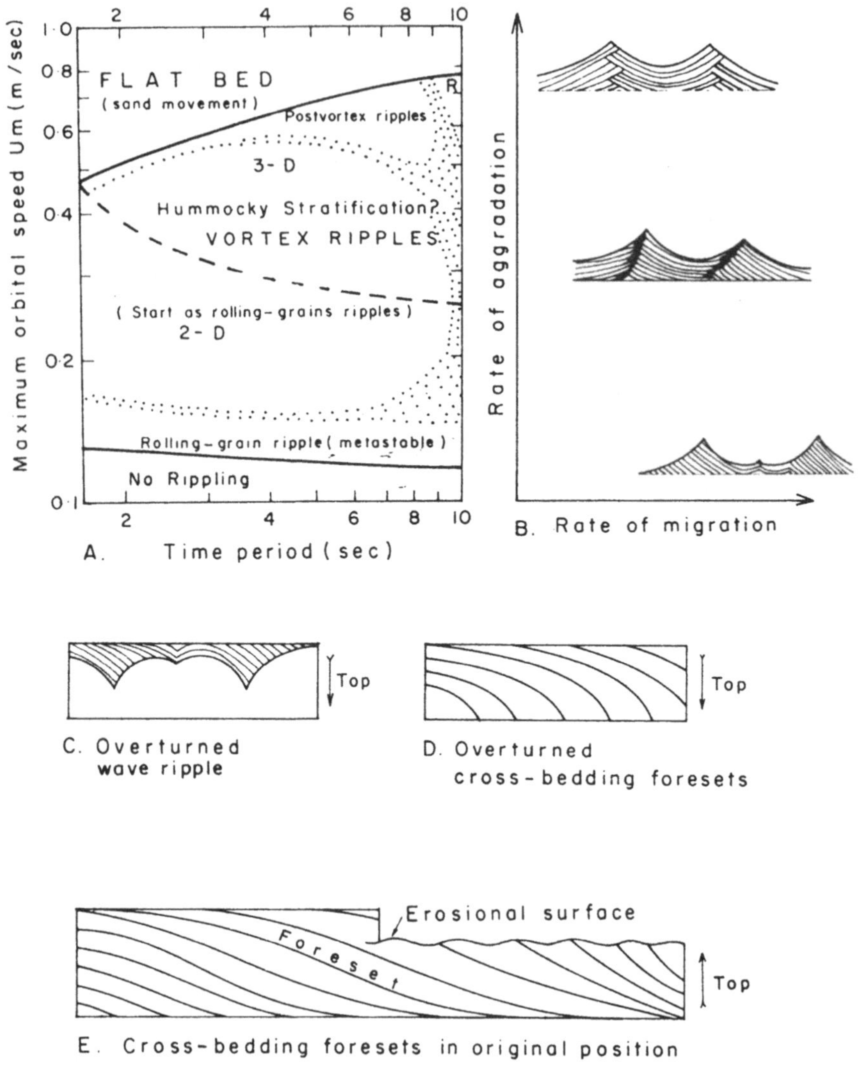

Fig. 5.23: (A) Physical conditions governing generation of wave ripples of various types (R indicates the field of reversing-crest ripples). (B) Internal structures of wave ripples as functions of rates of migration and aggradation. (C-D) Internal structures of wave ripple and cross-bedding as indicators of depositional top and bottom of beds. (E) Cross-bedding foresets in original position. A and B modified after Harms et al. 1982, figs. 2-14, and 3-13, with permission of SEPM, Society for Sedimentary Geology).

ing to the field of three-dimensional vortex ripples (Harms et al. 1982). The surface features of hummocky cross-stratifications are those of three-

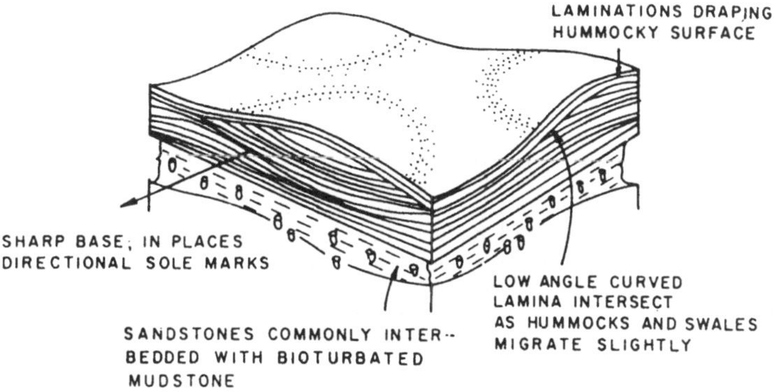

Fig. 5.24: Hummocky cross-stratification (HCS) commonly found in sandstones interbedded with bioturbated mudstone (modified after Walker 1984).

dimensional, circular to elliptical hummocks and swales of large magnitude (L = 1-5 m, H = 10s cm). In cross-section these features show sets of low-angle, concave-up laminations and generally low-angle (<10°) curved laminae intersections (Fig. 5.24). Found originally in the Cretaceous shoreface deposits of the Rocky Mountains, USA (Harms et al. 1975), hummocky cross-stratifications have now been reported from stratigraphic horizons of many different ages and localities (Dott and Bourgeois 1982).

The mechanism of origin of these features, which could neither be produced in the laboratory nor observed in any present-day depositional environment, remained an enigma for a long time. Marine fauna and bioturbated mudstone associated with some hummocks suggest marine origin, at depths varying between fair-weather and storm wave bases (30–50 m). Inner shelf storm surges were considered to be mainly responsible for their origin.

Hummocky megaripples (L = 2-5 m) with gently inclined internal laminations discovered in the inner Atlantic shelf of North America are believed to have been produced by storm-generated 'combined-flow currents'. It is inferred that the arrays of scour pits and mounds excavated on the shelf floor by intense storms were later modified by wind-driven flows. The flows were somewhat similar to those responsible for generation of megaripples in tidal environment (Swift et al. 1983).

STRUCTURES RELATED TO FLAT BEDS

Bedding

Bedding is defined as a sequence of parallel layering within a body of sedimentary rock wherein individual layers differ from the overlying and underlying ones in physical properties such as colour, composition or tex-

Table 5.4. Classification of bed thickness and splitting properties (modified after McKee and Weir 1953; and Potter, Maynard and Pryor 1980)

Thickness (mm)	Terminology for: Bedding	Terminology for: Splitting property
	Very thick-bedded	Massive
1200.0		
	Thick-bedded	Blocky
600.00		
	Thin-bedded	Slabby
50.0		
	Very thin-bedded	Flaggy
10.0		
	Thickly laminated	(Flaggy)
5.0		
	Medium-laminated	Platy
1.0		
	Thinly laminated	Fissile
0.5		
	Very thinly laminated	Papery

ture. Beddings are classified according to their thickness. Sedimentary rocks often display a tendency to split along bedding planes. The commonly used terminologies for bedding and splitting properties are given in Table 5.4.

Subaqueous horizontal bedding may be produced either during settling from suspension (without traction) or during transportation of sediments by running water. In the lower flow regime, plane beds may generate when the grain-size is large but the flow velocity is small. The upper flow regime plane beds are generated when the grain-size is small but the flow velocity is high. The upper flow regime plane beds may be identified by parallel arrangement of elongate mineral grains producing parting lineation (Fig. 5.25).

Gradational, fining-upward sequences are common features in suspension deposits. Each layer of sediment within a graded bed contains an assortment of grains, both coarse and fine, but the mean size of the grains gradually decreases upward (Fig. 5.26). In the laboratory graded beds could be generated by allowing sediment laden flows to deposit their load from suspension under decelerating velocity condition (Kuenen and Migliorini 1950). Some of the graded beds in ancient marine and lacustrine sediments could have been produced by turbid flows which progressively deposited finer grains of sediments as flow velocity and carrying capacity of the current steadily decreased. Graded gravel beds are produced when slow moving unsorted debris flows (mud flows) deposit their sediment loads.

Grading may also be produced on the foreset of sand-dunes as a result of sorting during bed load transportation. As stoss faces of dunes erode, and the materials transported to the dune crest avalanche down the lee slopes,

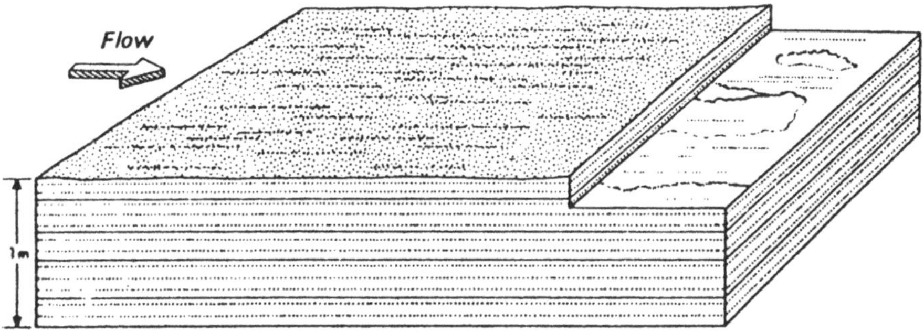

Fig. 5.25: Current lineation and parting lineation produced in the upper flow regime (after Harms et al. 1975, fig. 3-3, with permission of SEPM, Society for Sedimentary Geology).

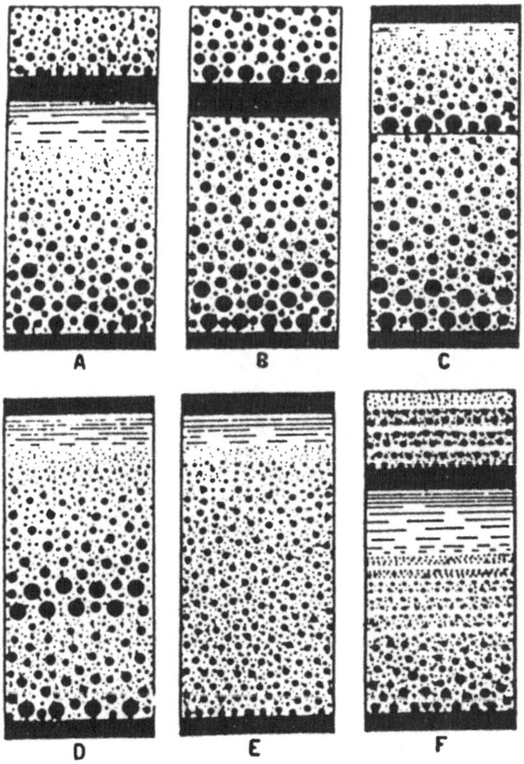

Fig. 5.26: Various types of graded bedding (after Kuenen 1953). Deep water lutites are shown in black.

a vertical grading is produced due to accumulation of coarser grains towards the bottom. On rare occasions dispersive pressure caused by grain collision may force larger particles towards the surface of the slip face leading to what is termed inverse grading (Bagnold 1954; Jopling 1965). Some thinly bedded layers of marine origin are characterised by basal shell lags and a crude grading of carbonate clasts. These have been termed 'tempesites' because of their inferred origin by storm surges. The term is also used for storm sand layers developed as graded rhythmites (Reineck and Singh 1980).

Structures Preserved on Bed Surface

Shrinkage cracks, raindrop imprints, water-level (recession) marks, wrinkle marks, swash and rill marks are some of the common features preserved on top of a bed surface. Shrinkage cracks may form either at the sediment-air interface by a desiccation process (desiccation cracks) or at the sediment-water interface by a synaeresis process. In plan desiccation cracks may look polygonal or non-polygonal (Fig. 5.27), whereas synaeresis cracks are generally spindle shaped or sinuous. In cross-sectional view either may be U- or V-shaped. Desiccation cracks originate by algal or mud shrinkage (mud cracks) due to atmospheric drying. These are frequently used as evidence

Fig. 5.27: Mud cracks produced by subaerial exposure of modern flood-basin deposits in the Usri River, Bihar, India.

of subaerial exposure. Synaeresis cracks originate by surface or substratal dewatering of submerged mud. Distinction between these two is difficult. Therefore, a crack should be taken to be diagnostic of subaerial exposure only when it is associated with other shallow-water, periodically exposed features such as raindrops, foam impressions or animal tracks.

Raindrop imprints are preserved as circular, small, crater-like depressions with raised rims. Other evidence of intermittent subaerial exposure include parallel marking on sloping banks (water recession marks) and wrinkle marks (*Runzelmarken* in German) produced by the action of wind on a thin film of water covering cohesive sediments. These are minute, ripple-like features, a few millimetres in length and only a fraction of a millimetre in thickness.

Swash marks are indicative of the farthest line of encroachment of waves on land. These marks consist of lobate ridges of fine-grained sand which are convex towards the land. The edges of these marks are often fringed by broken shells, wood fragments or similar light materials. Dendritic, bifurcating erosional features (single or multiple), called rill marks, are produced by erosion when water drains out of soft sediment surface at the falling water level (Fig. 5.28). These also indicate a thin water cover.

Crescentic marks ('current crescents') are produced on an unconsolidated sediment surface around obstacles (such as shells, pebbles, wood fragments) in the way of water flow. These are generated by continued excavation of sand due to vortex action at the upstream end of an obstacle and deposition of the eroded material at the downstream end in the form of a flare (Sengupta 1966; Johansson 1976; see Fig. 5.29). The flare, when found in an ancient rock, may serve as a palaeocurrent indicator.

Structures Preserved on Undersurface of Beds

Scour marks of various shapes and sizes are produced when unconsolidated sediment surfaces are eroded by current. These, as well as the so-called tool marks produced by objects dragged on the sediment surface, are best preserved as moulds on the undersurface of the overlying beds. Of these, the most common are flute marks. These are discrete elongate hollows that flare and shoal out in the downcurrent direction to eventually merge to the sediment surface. Flute marks may be symmetrical or asymmetrical in plan view and may range from a few millimetres to a few decimetres in length. Scour marks of various shapes and sizes have been attributed by Dzulynski and Walton (1962) to the erosive action of sediment-laden density (turbidity) currents, but flutes may also be formed by currents which are not sediment-laden. The most important prerequisite for flutes is the generation of a vortex by a strongly turbulent flow. Experiments have shown that flow separation accompanied by a strongly eddying current rotating in a near horizontal axis, transverse to the flow, is essential for generation of flutes. Allen

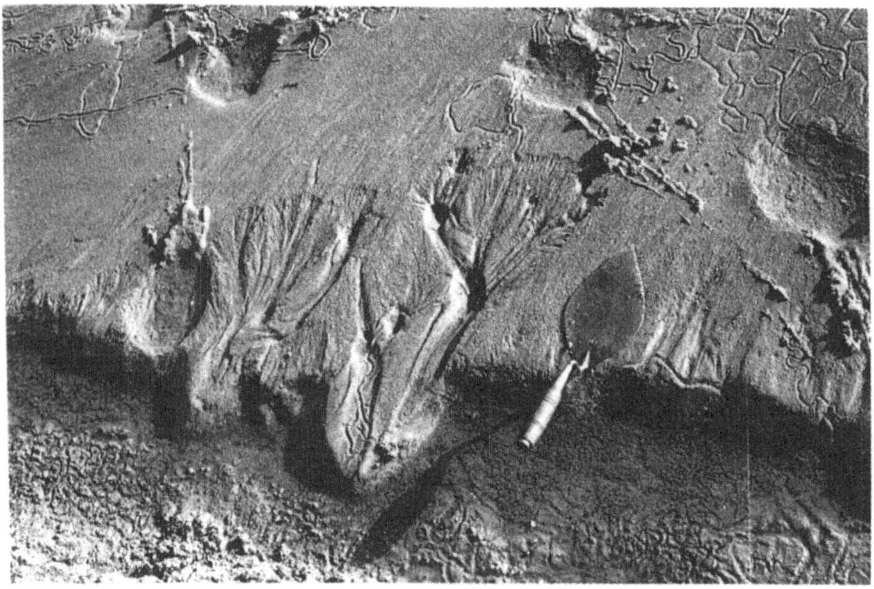

Fig. 5.28: Rill marks and worm trails in the Usri River point bar, Bihar, India.

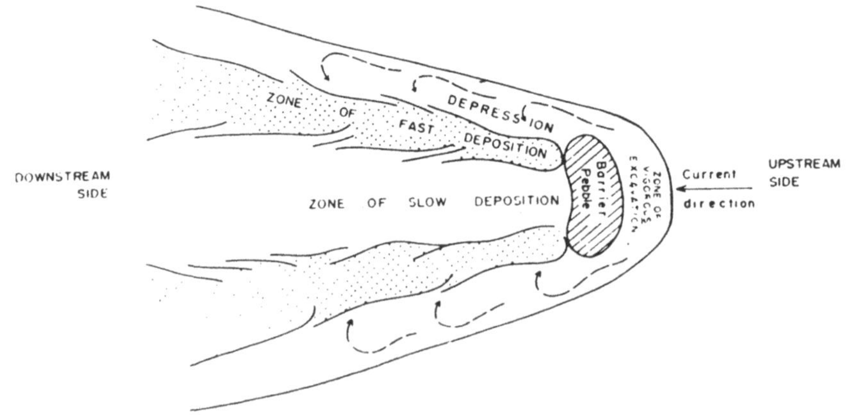

Fig. 5.29: Flow lines associated with a current crescent, plan view (after Sengupta 1966, with permission of SEPM, Society for Sedimentary Geology).

(1973, 1975) demonstrated that flute marks may develop by this process, starting from defects in beds.

Tool marks may be produced by either stationary or moving objects. Objects rolled or dragged by current may leave linear, narrow grooves (groove marks) or V-shaped marks arranged in straight ridges (chevron marks). Asymmetrical, elongated depressions with a pointed, upcurrent part (prod marks) are produced by momentary halt of an object on the sediment surface before being lifted up again. Similarly, bounce marks are produced by bouncing of objects which are quickly lifted up into the current (see Reineck and Singh 1980 for details).

DEFORMED BEDDING STRUCTURES

Slumping, sliding, unequal loading and water escape are some of the inorganic agencies responsible for distortion of layers of sediments. Such deformations occur at the time of, or soon after deposition of the sediments but prior to their consolidation into sedimentary rocks. Hence these structures are designated as penecontemporaneous deformation structures ('pene' means nearly). The process of ingestion of unconsolidated sediments by organic agencies leading to their deformation is called bioturbation. Penecontemporaneous structures have been grouped into the following types by Reineck and Singh (1982): load structures, ball-and-pillow structures, convolute bedding, water-escape structures, dish structures and slump structures.

Sinking of sand layers into unconsolidated, hydroplastic mud may produce sand lobes varying in size from a few millimetres to a few decimetres. The underlying mud may be pushed upward in the form of tongues. At the sediment surface they may appear as bulges or knobby bodies. Casts of these load-generated features (load casts) may be preserved on the undersurface of beds as sole marks. When the sand layer is broken up into ellipsoidal, pillow-shaped masses and the underlying clay is pushed upwards into the sand in the form of tongues, the feature is designated as ball-and-pillow. Kuenen (1958) generated such features in the laboratory by subjecting a layer of sand overlying unconsolidated clay to vibrations. The sand layer in this experiment broke into discontinuous, kidney-shaped bodies, with the intermediate space occupied by clay (Fig. 5.30). Vertical movement of this type could have been responsible for most of the ball-and-pillow structures observed in nature. In some cases combined vertical and horizontal displacements due to slumping could have also been the cause.

Convolute beddings are characterised by folds with sharp, irregular crests (often resembling flames) and broad, round troughs (Fig. 5.31). They are restricted to single sedimentation units, indicating that they are synsedimentary in origin. Sedimentary features of this type have been

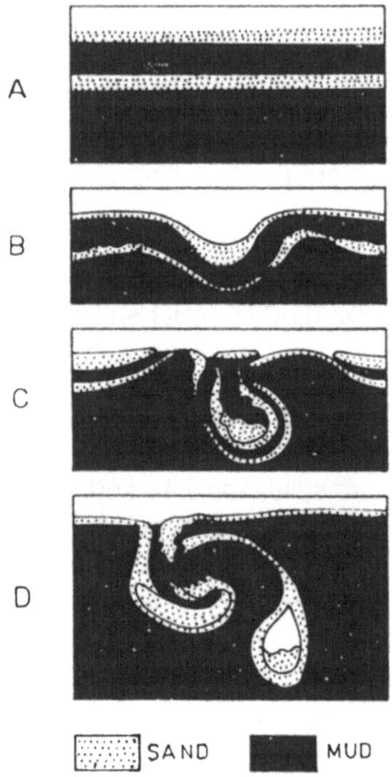

Fig. 5.30: Results of experiments on generation of pseudonodules by vibrating a layer of unconsolidated clay overlain by thin sheets of sand (redrawn and simplified after Kuenen 1958).

reported by many authors, both from ancient and modern sediments (see for example, Ghosh and Lahiri 1983).

Convolute laminations occurring at regular intervals were interpreted by Kuenen to have generated out of sand ripples whose crests were sucked up by water currents. The sucking was inferred to have been caused by differential pressure at the crests and troughs of ripples. Differential loading due to deposition of sand from above (McKee and Goldberg, 1969), vertical expulsion of fluid leading to increase in pore pressure (Allen 1977), upward movement related to vertical expulsion of water or gas and vortex motion of turbulent cells (Chakrabarti 1977), are some of the other interpretations offered for these structures. The contortions running across planes of stratification (named 'interpenetrative convolutions' by Allen 1977) look similar to the box-shaped forms produced at boundaries of liquids of varying densities when heated from below. This also led to the suggestion that some interpenetrative convolutions are generated by transfer of material by processes similar to convection currents. Some researchers have also assigned con-

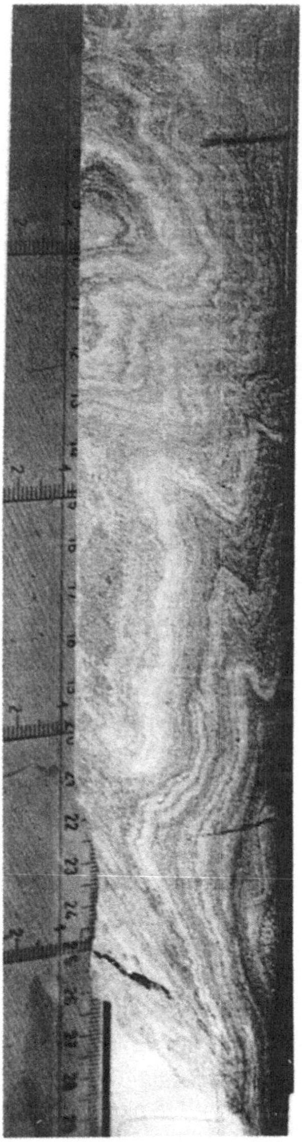

Fig. 5.31: Convolute lamination in a core sample from a Gondwana sandstone, Barakar Formation, Barjori coalfield, W. Bengal, India. (Courtesy of K. Dasgupta, Central Mine Planning and Design Institute, Asansol, India). Scale in cm.

volutions to deposition of layers of sand over water-saturated mud or silt in the initial stage, followed by downward movement of sand and a compensatory upward flow of silt or clay (see Pettijohn 1975). Such movements, it has been suggested, could be caused by a sudden shock, as in an earthquake. In reality, however, evidence of downward movement is rarely found in natural convolutions. The process seems to involve only unidirectional upward movement of silt and clay caused by squeezing out of material from below.

When sand and mud are deposited together at a high rate, the sediment layers may be disturbed by escape of water unless the water can seep out of the system without disturbing the grains. The disturbances produced by water escape are of two kinds: (i) liquefaction caused by temporary suspension of sediment followed by settling consequent to collapse of the framework and (ii) fluidisation involving lifting of the grains by the drag exerted by the fluid. Four types of water escape structures may be produced from the process (Lowe 1975): 1) Hydroplastic or liquefaction layers produced by internal rearrangement of sediment layers. Not much physical movement is involved in this process. Some of the convolute beddings, deformed cross-bedding and pillar structures come under this group; 2) Clastic sills and dykes produced by intrusion of hydroplastic, liquefied or fluidised sediments into the adjacent bodies of sediments; 3) Various types of deformed bedding or cross-bedding produced by downslope movement of sediment layers; 4) Dish structures produced by particle rearrangement. These are generated when sediment-water slurries are forced into a nearly horizontal passage until they reach points of vertical escape. Dish structures occur parallel to the sandy bedding planes as concave-upward laminae a few centimetres thick.

Slump structures are caused by gravity flows of unconsolidated sediment layers. These may range from small-scale intraformational folds to large masses of slumps. Overturned or recumbent folds with associated thrust-like structures are common within such slumps. Such features have been reported from oversteepened point bars, gullies of intertidal flats, glacio-lacustrine and fluvio-glacial sediments. In the latter case an overriding mass of ice may cause the sediments to slump. Subaqueous cross-beddings may also be folded with or without overturning. A combination of folding and faulting is common in eolian dune sediments. Avalanching of masses of cohesive sediments on the slip face of eolian sand dunes may also lead to slumping. In coastal dunes deformation is aided by the cohesive nature of sand. Graben-like features, a few centimetres in width, may be produced by slump-related gravity faults (for details see Reineck and Singh 1980).

6

SEDIMENTARY ENVIRONMENTS AND FACIES

INTRODUCTION: FACIES MODELLING

The term facies, derived from a Latin (also French) word, implies a facial expression characteristic of a particular condition. In modern sedimentology the term is used as an expression for 'the total field aspect of the rocks themselves .... The key to the interpretation of facies is to combine observations made on their spatial relations and internal characteristics (lithology and sedimentary structures)...' (Middleton 1978a). The process of construction of a facies model for an environment involves identification of the diagnostic or common characters of the particular environment. This is done by studying a large number of modern and ancient situations and then formulating general principles about that environment after elimination of the local details. The purpose is to obtain 'a norm for comparison, a guide for future observation and a predictor for new geological situations' (Walker 1984a). With this end in view, considerable effort has been made during the last few decades to collect information about the modern environments of sedimentation. The following pages summarise the present-day knowledge of the more important environments—marine, non-marine and mixed. A genetic explanation for an ancient phenomenon should come by way of analogy with these modern environments.

MARINE ENVIRONMENTS

Traditionally the ocean-floor is divided into three zones according to depth: neritic (200 m and above), bathyal (200–2000 m) and abyssal (below 2000 m). The deep-sea trenches are located in the hadal zone at depths exceeding 6000 m. The more modern schemes use ocean-floor gradient as the principal criterion for classification (Fig. 6.1). The subdivisions thus obtained are: continental shelf (average slope 0.1°), continental slope (4°), continental rise (0.05°–0.6°) and abyssal plain (0.05°). The facies associations in these areas are briefly discussed below after Walker (1984b & c).

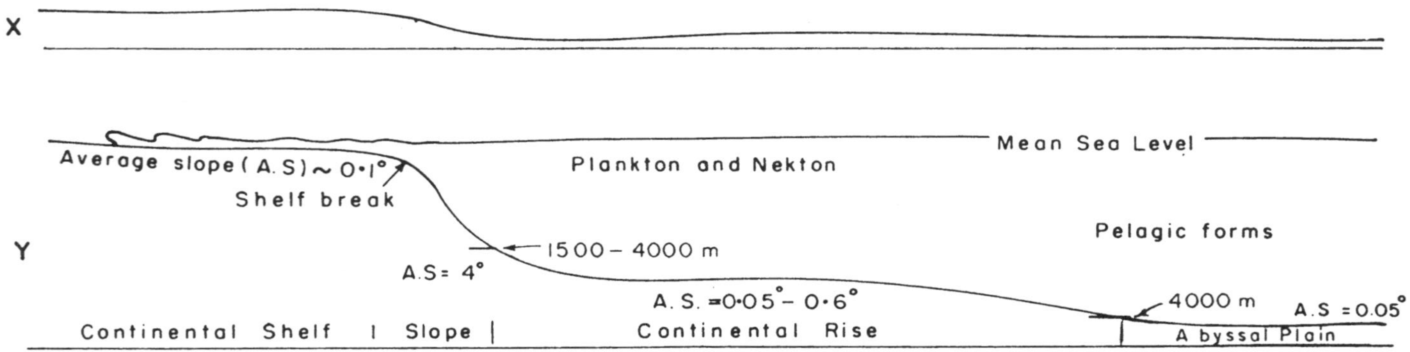

Fig. 6.1: Morphological classification of ocean floor according to gradient (based on Cook et al. 1982).
(X) vertical exaggeration 1X (Y) vertical exaggeration 20X approximately.

Continental Shelf

The continental shelf is that area of the sea-floor lying between the shoreline and the shelf break. The sediment distribution in the present-day continental shelves is controlled mainly by three types of currents: meteorological (storm) currents, tidal currents and intruding ocean currents. The modern shallow marine shelves are classified into three major groups depending on the predominant type of current.

Siliciclastic Sediments on Shelves

Storm-dominated Shelf. About 80% of the continental shelves of the world today are frequented by storm waves. These waves agitate the ocean-floor above the storm wave base, 'liquefying' the sediments and generating suspension currents of high density (turbidity currents) in the process. Deposition from these suspension currents below the wave base takes place in the form of turbidites. Hummocky cross-stratifications are found below the storm wave base in ancient deposits. The features which dominate modern shelves are sharp-based, cross-bedded, linear sand ridges a few kilometres wide, several kilometres long and 3–12 m thick, running subparallel to the shorelines. Ancient counterparts of these sand waves are lacking.

Tide-dominated Shelf. Seventeen per cent of the world's continental shelves today are dominated by tidal waves. In the North Sea, adjacent to the coasts of England and the Netherlands, tidal waves have generated large-scale subaqueous sand waves over extensive areas ranging between 5000 and 15,000 km^2 (Fig. 6.2). The crests of these sand waves, which are about 35 m high, lie below 10 to 15 m of water, with their steeper sides dipping at low angles. Smaller sets of cross-bedding, dipping up or down the master bedding surfaces, constitute the internal structures of these sand ridges. Similar features have been reported from the Atlantic coast of North America also.

Shelves Dominated by Oceanic Currents. The continental shelf south-east of Africa, which is under the influence of a major oceanic current called the Agulthus current, is an example of such a shelf. Extensive elongate strips of sand waves, measuring 20 km × 10 km × 17 m approximately, occur on this shelf at a depth of about 40 m below the Indian Ocean surface. These sand waves have been generated by reworking of the sediments already present on the shelf. No comparable ancient example is known.

Carbonate Sediments on the Shelf

The deposits of the shallow sea are not restricted to siliciclastic sediments alone. Carbonate secreting organisms such as corals, algae and oysters thriving in shallow, turbulent, well-oxygenated, warm oceanic water generate extensive reefal colonies. These colonies may assume many different forms—fringe reefs, barrier reefs, atolls and table reefs. Each such

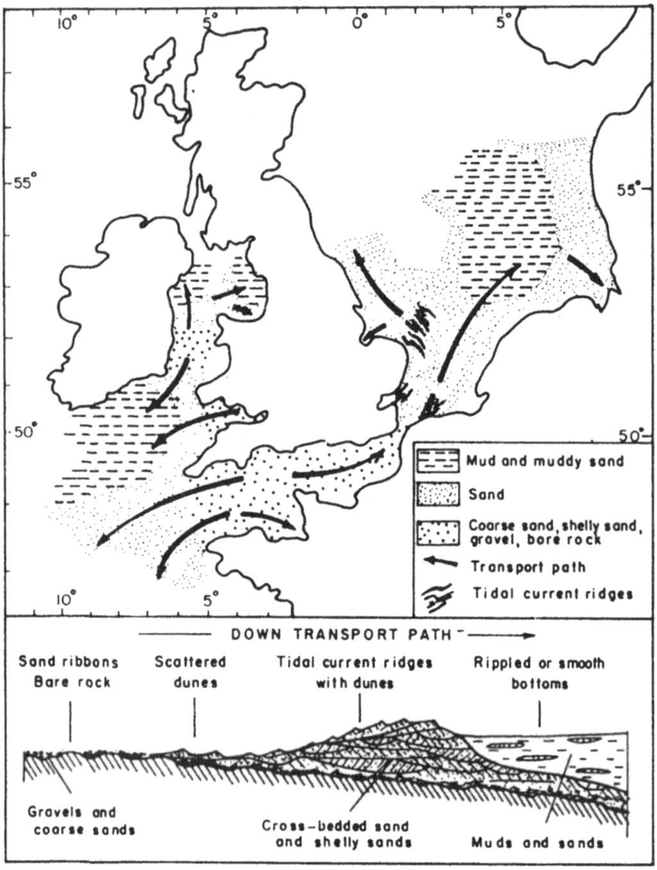

Fig. 6.2: Sedimentary structures in tide dominated shelf off the coast of Britain. The profile shows the structures typically encountered along a transport path (modified after Allen 1970 with permission of the author).

reef is composed of a group of wave-resistant species which thrives on the windward side and several calm-water forms on the opposite side. Each reef can be divided into the following zones: forereef, reef front, reef crest, reef flat and back reef. The best known modern example of reefal growth is the Great Barrier Reef off the north-eastern coast of Australia. This reef runs for a length of about 1200 miles and is, on average, about 60 miles wide.

Continental Slope

The slopes bordering the continental shelves are marked by networks of submarine canyons and gullies. Water-wet terrigenous sediments are carried by gravity *en masse* into the deep sea through these canyons. The phenomenon is called 'mass-transport'. Three major types of mass-transport are recognised—rockfalls, rockslides and sedimentary gravity flows. Rockfalls cause talus accumulations at the toe of slopes. Rockslides can be either translational (glide) or rotational (slump). Several types of sedimentary gravity flows, from debris or mud flows to turbidity current flows, are known. In debris flows the clasts are supported by the cohesive strength of the mud matrix. The water-wet mud behaves like a plastic body. In a turbidity current the clasts are supported by fluid turbulence and the whole mass behaves as a fluid. This also includes liquefied flows in which the sediment is supported by upward displacement of fluid (dilatance), and fluidised flow in which the sediment is supported by upward motion of escaping pore fluid. Intermediate between plastic and fluid flows are grain flows in which the sediment is supported by dispersive pressure (see Fig. 6.3).

Continental Rise

Terrigenous sediments, transported through feeder channels and canyons, splay out on the ocean-floor to produce the deep-sea fans. These fans constitute the major part of the continental rise. The flow processes through the feeder channels include slump creep, slurry flow, grain flow, fluxoturbidity current, turbidity current and fluidised flow (Fig. 6.3). The volume of these mass flow deposits can be of the order of many cubic

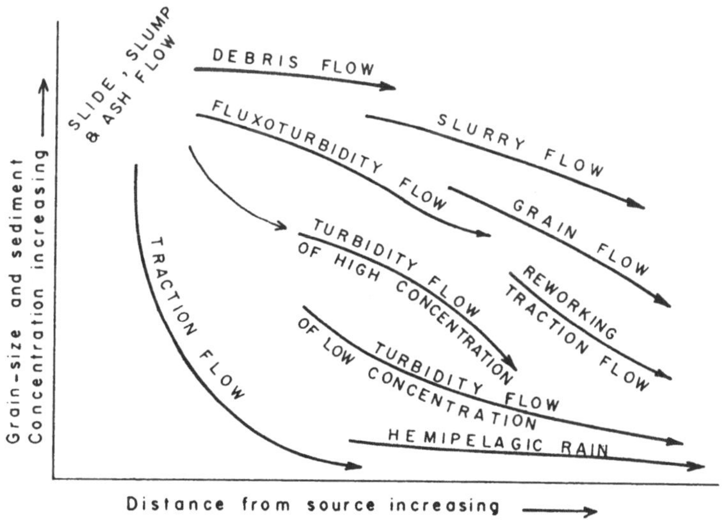

Fig. 6.3: Inferred flow processes in turbidity and related flows (modified after Howell and McLean 1976, Howell and Normark 1982, with permission of American Association of Petroleum Geologists).

kilometres. The enormous amount of material produced by shelf edge slumps due to earthquakes may be transported at an astonishing velocity. A maximum velocity of the order of 20 m/sec was estimated for the turbidity currents caused by the Grand Banks Earthquake of 1929 (Dunbar and Rodgers 1957).

The sequence of sedimentary structures produced within a submarine fan by the deposits of mass flow is as follows (see Fig. 6.4):

Top e. pelitic division
d. upper division of parallel lamination
c. current ripple lamination
b. lower division of parallel lamination
Bottom a. graded or massive beds

This sequence, idealised by Bouma (1962), has been named after him. The complete Bouma sequence (T_{a-e}) is rarely found in a geological record but even a partially preserved Bouma sequence may help in the identification of an ancient fan deposit.

The following generalised facies model for deep-sea fans was formulated by Mutti and Ricci-Lucchi (1975) on the basis of studies by Normark on modern fans and by Mutti and associates on the ancient fans of the Apennines, Italy. An idealised fan, according to this model, consists of the following components (Fig. 6.5).

An inner (or upper) fan with one or more feeder channels. Typical lithofacies: conglomerates, coarse-grained to pebbly sandstones. Some remobilised deposits suggesting mass slumping may be present. Basal scouring and channelling are common.

A middle fan consisting of a number of shallow, branching channels feeding a depositional lobe (the 'suprafan' of Normark 1978). Typical lithofacies: basal gravels and mud chips overlain by coarse to medium sandstones with interbedded fine-grained sandstone and mudstone. The sandstones deposited by turbidity currents are of uniform thickness for a long distance and exhibit a classical Bouma sequence.

An outer (lower) fan, which begins beyond the limits of the distributary channels, contains tabular mudstones with interbedded, graded, thin sandstones. The upper part of the Bouma sequence is preserved in these deposits. Sole marks are common. The depositional lobes, which are topographically smooth, grade laterally into a basin plain.

The basin plain (lower fan of Normark 1978) consists essentially of pelagic and hemipelagic mudstones with a few graded sandstones. The deposits are transitional with the outer fan fringe.

Abyssal Plain Deposits

The abyssal depths of the ocean are dark and cold. Blue-grey to lead-coloured mud, transported from land to the ocean-bottom by suspension

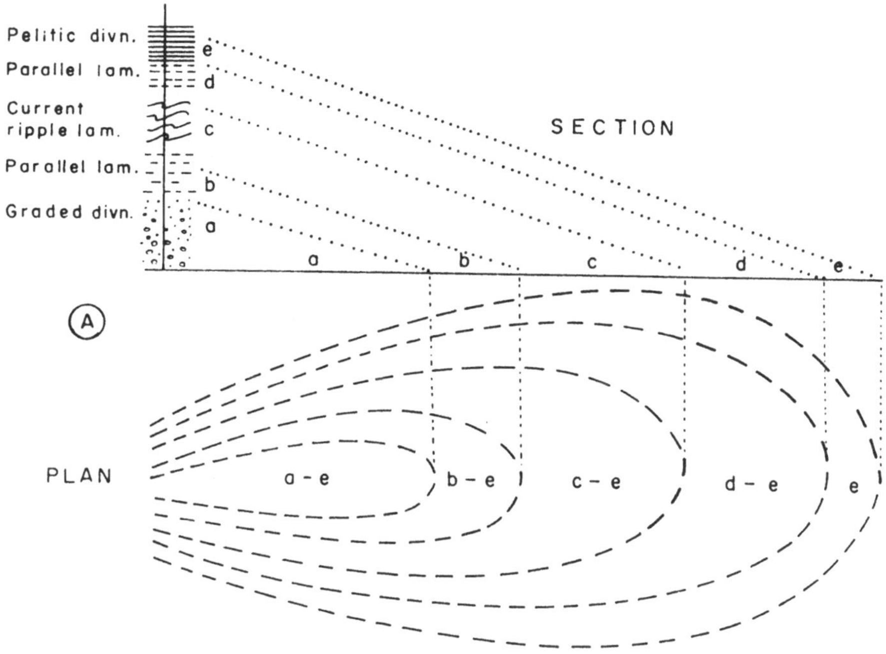

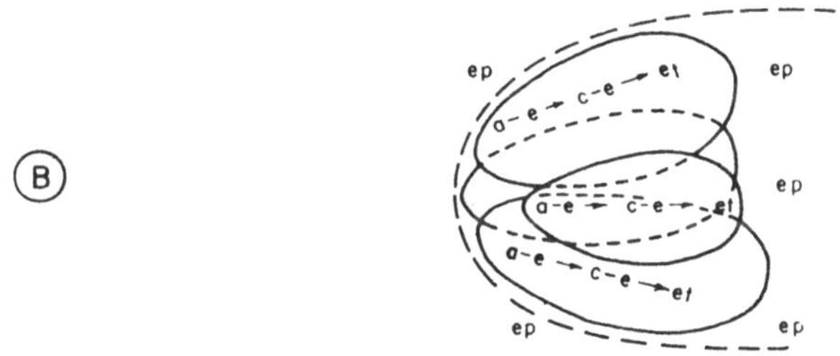

Fig. 6.4: (A) Schematic representation of an ideal Bouma sequence in plan and section. (B) Hypothetical pattern of filling of a basin by turbidites. Note that each cone does not fill the entire basin (modified after Bouma, by Howell and Normark 1982; reprinted with permission of American Association of Petroleum Geologists).

currents, floor thc bathyal zone of the ocean; contrarily, the abyssal zone is covered with red-coloured mud. This red mud, having the chemical composition of an average igneous rock, is believed to have been blown into the deep sea from volcanic vents located in the adjoining land areas. Prolonged oxidation of fine-grained terrestrial particles during slow settling through

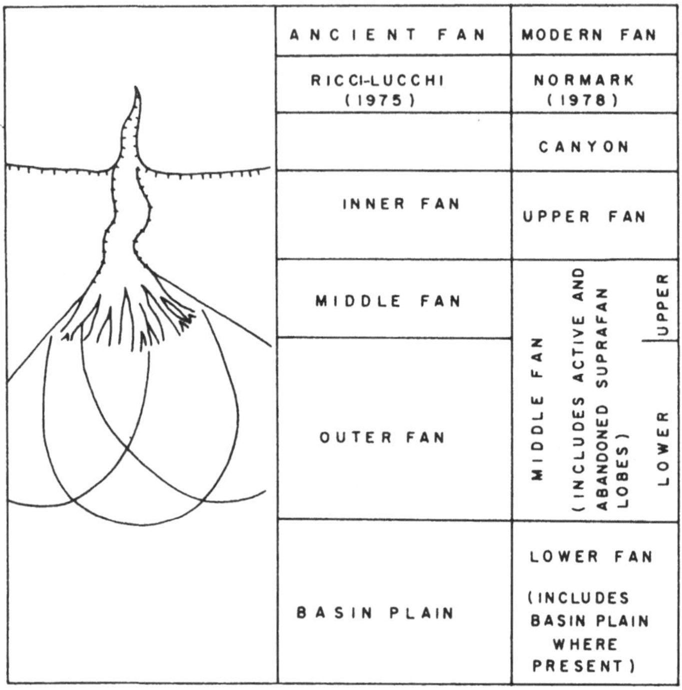

Fig. 6.5: A simple classification of submarine fans proposed by Normark (1978), Ricci-Lucchi (1975) (modified after Howell and Normark 1982, with permission of American Association of Petroleum Geologists).

oxygenated water columns might also have been responsible for the red colouration.

Planktonic organisms such as foraminifera, pteropods and coccolithophorids settle on the ocean-floor to contribute to the pelagic ooze. These are reworked by burrowing organisms on the ocean-floor. Below the Carbonate Compensation Depth (CCD, see Chapter 4) much of the carbonate deposit is dissolved, leaving only 'ghost' structures. This depth is around 2000 m for aragonite and about 5000 m for calcite but the actual depth of compensation varies from place to place, depending on the local climate, geographic conditions and the nature of the carbonate sediment involved. The biochemical precipitates below the carbonate compensation depth are of siliceous nature. Local concentrations of ferromanganese concretions ('manganese nodules') are found in some pockets of the deep-sea floor. These are ascribed to exhalations from suboceanic volcanic or hydrothermal vents but the precise mode of their origin is not known.

Marine Sediments in Stratigraphic Record

The typical facies associations developed in a marine shelf, continental slope and continental rise are briefly described below.

Shelf Sediments

Sedimentation on continental shelves can take place under three conditions: transgressive, prograding and aggradational (Galloway and Hobday 1983). Transgressive sequences may be either tide-dominated, or storm-dominated. Both of these produce fining-upward sequences. The transgressive sequences start with lag gravels at the bottom and end with bioturbated silts and muds on the top, passing successively through coarse and fine-grained, cross-bedded sands (Fig. 6.6A). The storm-dominated transgressive sequences start with lag gravels at the bottom and end in silty and muddy sediments on the top, passing successively through silts and muds (with lenticles of sand), multistorey sets of hummocky cross-stratification and thinly graded, parallel laminated sands (Fig. 6.6C). The hummocky cross-stratifications are storm generated. The thinner sets are deposited as the depth of water increases due to transgression. Laminated mud, rich in organic carbon, glauconitic and phosphatic muds may be laid down on top of the succession of a deep shelf deposit. Many of the outer shelf mudstones are phosphatic with a very high organic carbon content. Some of these have been correlated with the phases of worldwide eustatic transgressions by Vail et al. (1977).

The prograding shelf sequences typically coarsen upward. Two prograding sequences, one dominated by storms and the other, a mixed energy shelf succession, have been identified. The storm-dominated prograding succession is almost the reverse of the storm-dominated transgressive sequence (Fig. 6.6B). A progressive seaward fining of sediments, caused by the storm-induced sorting processes, is represented as a coarsening-upward vertical sequence. The typical sequence produced in a mixed energy (storm and tidal) prograding shelf is represented in Fig. 6.6D.

Aggradational shelf sediments are generally fine grained. Silt and sand are deposited by traction currents. Mudstone is deposited when the rate of sediment supply is slow due to a gradual rise in sea level or alternatively, due to subsidence of the land area. Periodic transgression and progradation may cause sandy sediments to alternate with thick sequences of mud. In ancient geological record mixed energy aggradational sandy shelf successions are represented by thickly cross-bedded quartz arenites.

Slope and Fan Deposits

The deposits of the continental slope and talus include debris sheets, rudite sheets and sedimentary breccias of very large dimensions. The latter are called olistostromes (Greek: *Olistomai*—to slide, *Stroma*—layer). Derived

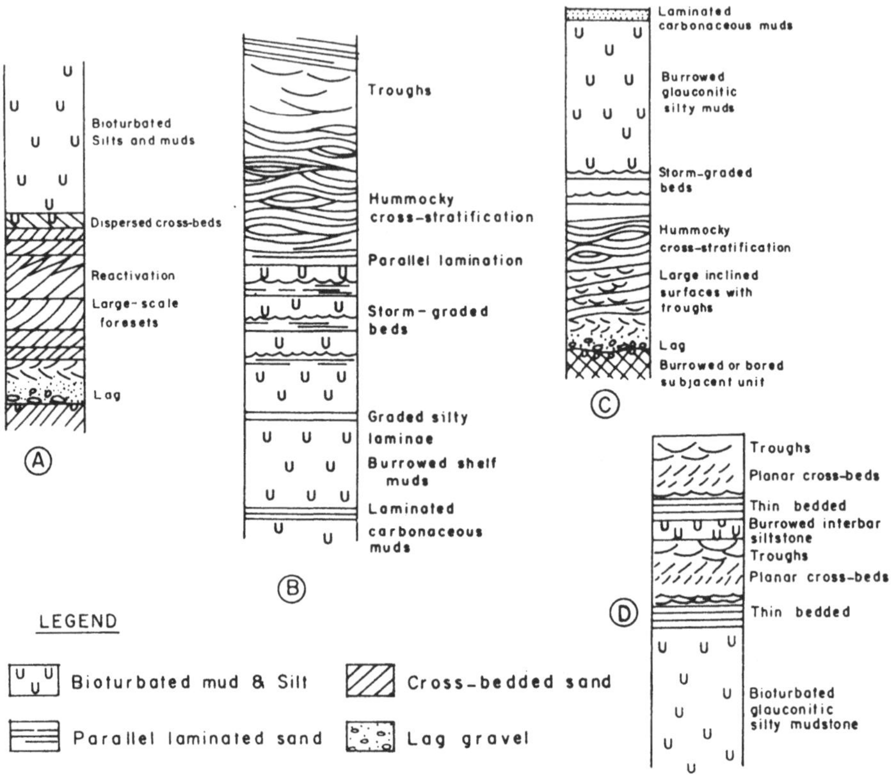

Fig. 6.6: Idealised depositional sequences in (A) transgressive, tide-dominated shelf, (B) prograding, storm-dominated shelf, (C) transgressive, storm-dominated shelf and (D) prograding shelf dominated by storm and tidal actions (modified after Galloway and Hobday 1983, figs. 7-14 to 7-19, with permission of Springer-Verlag, Heidelberg).

from shallow water siliciclastic and carbonate deposits, the slope breccias may be of diverse composition, varying from well-bedded calcareous sands and calcarenites to rounded blocks of reefal limestones resembling bioherms. Downslope they may grade into deep-water pelagic or hemipelagic sediments. The pelagic sediments are open sea deposits containing calcareous ooze, skeletal remains of foraminifera, pteropods and coccolithophorids. The terrigenous component is small (< 20%) in these sediments but very fine-grained pyroclastics (tephra) are common. Siliceous (radiolaria, diatoms) ooze and skeletal remains are common below the Carbonate Compensation Depth. Hemipelagic sediments are admixtures of fine-grained pelagic and coarser grained terrigenous components.

The largest submarine fan known today covers an area of about 3,000,000 km^2 on the Bay of Bengal floor. About 2 billion tons of sediments are washed into this fan annually through the 'Swatch of No Ground' submarine canyon, to be distributed through a channel-levee system up to the northern Indian Ocean. The biggest of these channels is about 27 km wide and 180 m deep (Fig. 6.7).

A submarine fan migrates through space and time to produce a vertical sequence which is preserved in the geological record. A progressive migration from the proximal to the distal areas (from the inner to the outer fan) may be caused by uplift of the source area, or by any other process by which the rate of sedimentation is increased, or also when the sea level is lowered. In the vertical succession produced from this process the coarse-grained inner fan deposits overlie the finer-grained outer fan sediments (Fig. 6.8A). When the rate of sediment influx decreases, or the sea level is raised, a retrogradation of the submarine fan takes place. In the vertical sequence the coarser inner fan deposits are successively overlain by the progressively finer middle fan and fan fringe sediments (see Fig. 6.8B).

Many examples of marine shelf deposits exist in geological literature. The Lower Palaeozic sections of the upper Mississippi valley and the Great Lakes of North America are often cited as typical examples. In both the

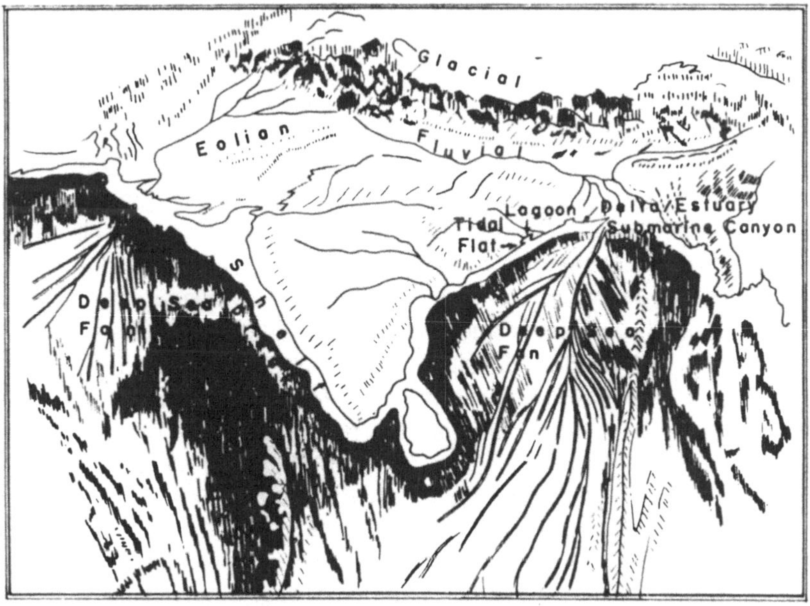

Fig. 6.7: Sketch map of India showing the Bay of Bengal and the Indus submarine fans and several other major depositional environments.

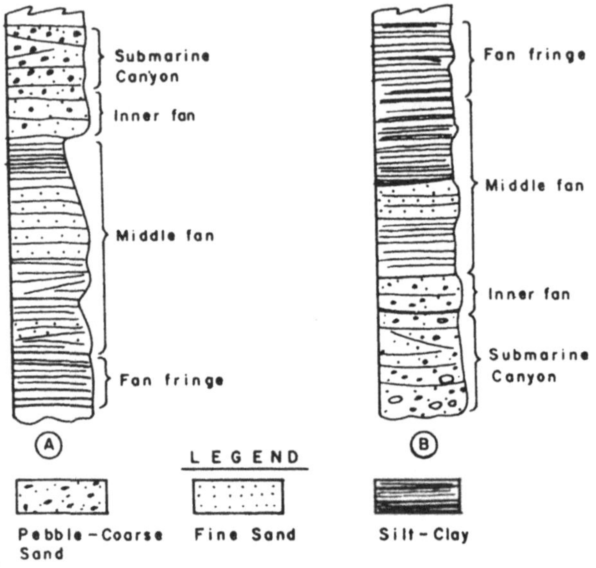

Fig. 6.8: Examples of vertical sequences produced by migration of submarine fans: (A) a progradational sequence, (B) a retrogradational sequence (modified and generalised after Howell and Normark 1982, with permission of the American Association of Petroleum Geologists).

areas shelf sediments were deposited on the southern extension of the Canadian shield when the conditions were very stable (Pettijohn 1975). The Jura Quartzite of Scotland and the Gog Quartzite (Lower Cambrian) of the Southern Rocky Mountains are ancient examples of tide-generated sand bodies (Walker 1984b). Reefoidal complexes are recorded from all geological ages but coral reef builders evolved on a large scale from the Silurian onwards. The Silurian reefs of Gotland, Sweden, the Devonian reefs of Alberta, Canada, Carboniferous reef complexes of North England, the Permian reefs of New Mexico and Western Texas, Triassic reefs of the Dolomite Alps, Italy, and the Tertiary reefs of the Mediterranean and the Persian Gulf are some of the most famous examples of carbonate reefs in the geological record.

In India the shale-quartzite-arenite sequence of the Kaimur Formation in Rajasthan is believed to have been deposited during the Proterozoic by an epiric sea on a very gently dipping, storm-dominated continental slope. Many sedimentary structures, including hummocky cross-stratifications, have been reported from these sediments (Bose, Chaudhuri and Seth 1988). The Cretaceous Nimar Formation of M.P., India, was also interpreted as the product of sedimentation on a storm-cum-fair weather-dominated continental shelf (Bose and Das 1986).

Extensive marine sediments have been reported from the Himalayas by many authors. In one of the more recent studies 'a complex jumble of volcanic and sedimentary rocks' occurring between Lamayuru Formation (Triassic to Cretaceous) in Ladak, and the Triassic flysch in southern Tibet, has been described as an 'ophiolitic melange' (Thakur and Virdi 1979; Virdi 1986). The Lamayuru Formation in Ladak contains exotic masses of limestones which are believed to have accumulated at the toe of an ancient continental slope by slumping. Olistoliths and grain-flow deposits have also been reported here (Virdi 1986).

NON-MARINE ENVIRONMENTS

Glacial Environment

Glaciers are masses of moving ice produced by recrystallisation of precipitated snow. Two types of glaciers are recognised: continental ice sheet and valley glacier (Figs. 6.9A, B). On crossing the snow line a valley glacier starts losing its volume by melting and evaporation and eventually reaches a zone where the amount of wastage exceeds the amount of accumulation in the upper reaches. The annual loss of volume of a glacier is called ablation. As the snout of a glacier vanishes by melting, the glacier is said to recede.

A moving glacier shatters the bedrock by exerting pressure on it. The shattered rock fragments, frozen at the base of the glacier, scratch and abrade the pavement below, producing striations. Gouging and scratching also produce 'chattermarks'. As a glacier slides downhill the pavement is sculptured to asymmetric mounds called *roche moutonnée* (French for 'rock sheep'). A series of irregularly spread mounds, called 'drumlins', is also produced by smoothening of a glaciated terrain. Such features form an ubiquitous part of the landscape of North America and Europe, which were covered by extensive sheets of land ice during the Quaternary.

The deposits ('drifts') produced out of glacial action are of two types: stratified and unstratified. Varves, outwash deposits and eskers are examples of stratified drifts. Unsorted mixtures of morainic gravel, sand and mud (designated as 'till' in older literature) are called 'diamicts'. Lithified equivalents of diamicts are 'diamictites' (meaning a thorough mixture of mud and gravel, from Greek: *diamignymi*—to mix thoroughly). Glacial facies are described here under two headings: continental and marine, following Eyles and Miall (1984).

Continental Glacial Facies

These include grounded ice facies, glacio-fluvial, glacio-lacustrine and periglacial facies. Ground ice grows into ice sheets of continental proportions. Two types of facies associations are possible within ground ice sheets, depending on whether the ice margin is inactive, that is, frozen to the sub-

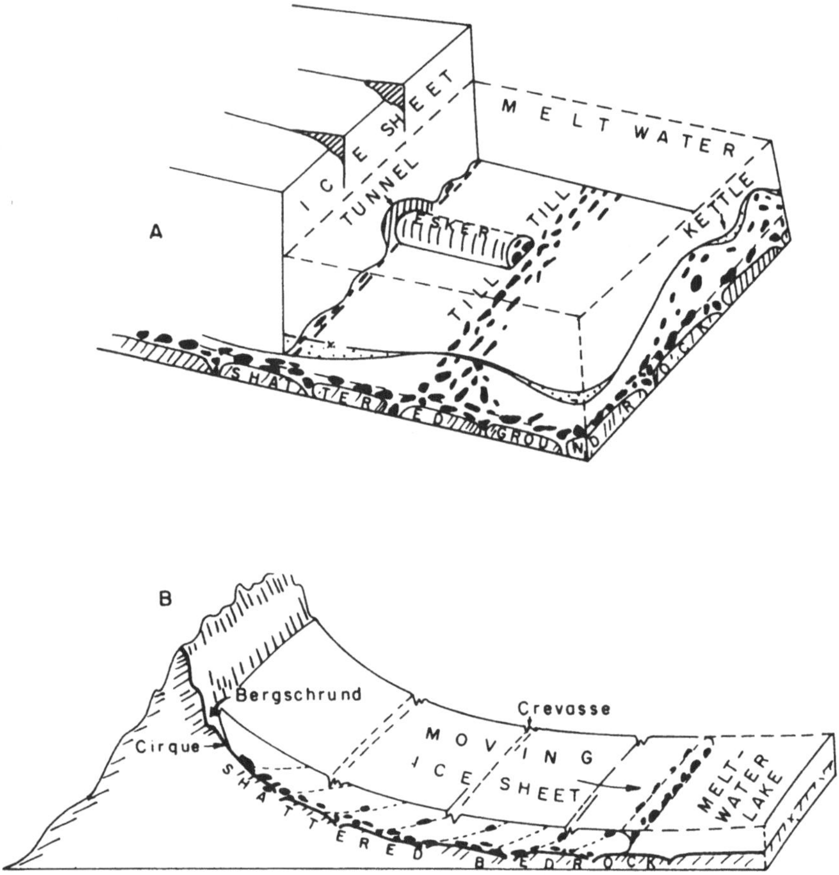

Fig. 6.9: Schematic representation of the deposits associated with (A) land ice sheet and (B) valley glacier.

strate, or actively sliding over the bed. In the case of an inactive margin, the frozen basal debris is draped by the diamicts of melt-out englacial debris. These may be resedimented to fill basins elsewhere. These basin fills are eventually exposed as 'hummocks' after melting of the surrounding ice cover. Typical vertical profiles of these hummocks show unstratified, *in situ* aggregates of meltwater diamicts at the bottom, overlain by graded and stratified diamicts which are the products of re-sedimentation (Fig. 6.10B).

When the glacier continues to slide over the bed, being continuously lubricated by the basal meltwater, faceted clasts result from the intense

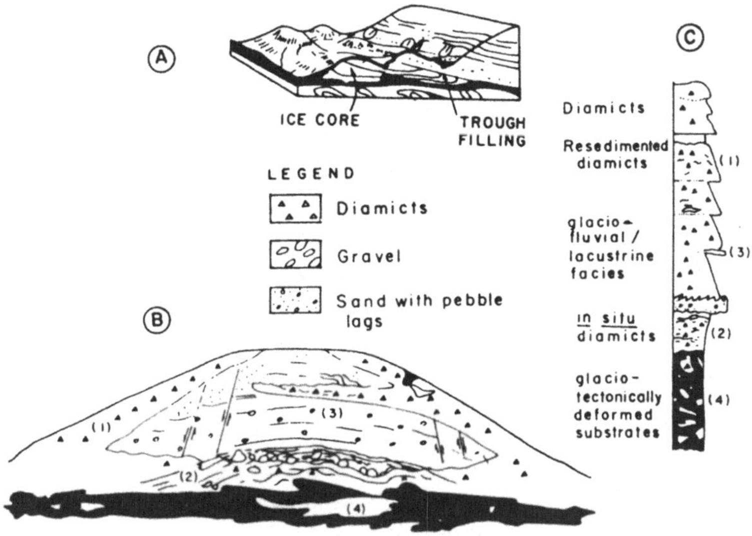

Fig. 6.10: (A) Sediment gravity flows moving into troughs of meltstreams, (B) continental glacial basin-fills exposed on 'hummocks' after melting down of the surrounding ice cover, (C) cross-section through a typical 'hummock' (modified after Eyles and Miall 1984).

abrasion (see Figs. 4.1 and 4.2). These clasts, with a strong preferential alignment to the direction of ice movement, are incorporated into the basal diamicts. They may be overlain by channel fills produced from subglacial stream drainage. Eskers, the deposits of the subglacial streams, may be of various types and forms, depending on the site of deposition and the nature of the conduit through which the glacial stream flows. In channels bordered by high walls, antidune structures may develop due to high-flow velocities, whereas within tunnels, parallel or cross-bedded structures dominate. These structures are preserved in vertically stacked cycles. Deltaic sedimentation at the mouth of the tunnel produces subaqueous fans. The proximal parts of these delta are marked by cobbles and pebbles. The grain-size decreases downstream (Banerjee and McDonald 1975).

A glacio-fluvial facies is produced from the deposits of the meltwater draining the outwash plains. These deposits, which may extend to hundreds of kilometres, are called *sandar* (plural for *sandur*, an Icelandic term). Glacial rivers which drain *sandar* are braided and their deposits are typically coarse grained. These deposits show a progressive decrease in grain-size downstream, ending, ultimately, in coastal fans. In a humid climate the swamps within these outwash plains may produce peat (or even coal) by decomposition of the subarctic vegetation. In a dry climate fine silt may be blown out of the outwash fans and dropped in the downcurrent areas as extremely well-sorted deposits called loess.

Lakes may be produced within the continents in the areas evacuated by ice sheets. Two types of glacio-lacustrine facies are identified: periglacial and proglacial. The former, not in direct contact with the ice margin, is fed by braided stream systems. Varves with a sandy or silty lithofacies deposited during the meltwater (summer) season, followed by 'winter clays', are common in periglacial lakes. Varves, traditionally believed to be seasonal deposits ('annual couplets'), have also been interpreted as graded units of a Bouma sequence deposited by turbidity currents within periglacial lakes (Lambert and Hsü 1979). The silt unit ('summer layer') of a varve may be graded by density underflows with pelagic materials incorporated from suspension (Banerjee 1973). A diamict lithofacies forms a minor constituent of the periglacial sequence.

Proglacial lakes, which are in direct contact with the ice sheets, are fed by meltwater conduits, subaqueous fans and by ice rafting. The accumulation of suspended sediments and stratified rafted debris (diamictites in these lakes) are overlain by cross-bedded, deltaic silty sand. Dropstones are common.

The Talchir (Lower Permian) deposits in eastern India may be cited as examples of continental glacial facies. The sequence at Dhakoi, for example, starts with a conglomeratic facies at the base and is followed upward successively by massive sandstone and a fine sand to shale facies (Fig. 6.11). The conglomerates were deposited as longitudinal bars within proglacial outwash plains during retreat of the Talchir ice lobe. Blocks of diamictites enclosed within these conglomerates represent ice-rafted morainic materials. The overlying sandy facies, which is often cross-bedded, represents channel sands and migrating bars of braided streams. The finer clastics were deposited in protected or abandoned parts of glacial outwash plains. A fining-upward cycle was produced by shifting of the braided channal bars within the glacial outwash plains (Casshyap and Tewari 1982).

Marine Glacial Facies

Two different types of glacio-marine sediments, proximal and distal, are recognised. In the proximal areas dense, sediment-laden meltwater produces bottom currents and submarine fans. Diamicts with ice-rafted clasts are common deposits at the fan apex. Downstream the deposits are much finer cross-bedded sand forming the main fan body with mud deposited from suspension. Pebbly diamicts may be incorporated into this deposit by gravity flows. In the steeper channel banks horizontally stratified and inversely graded gravels are deposited by gravity flows.

A generalised model for marine glacial facies was proposed by Edwards (1986), based on Pleistocene examples. Four depositional areas within ma-

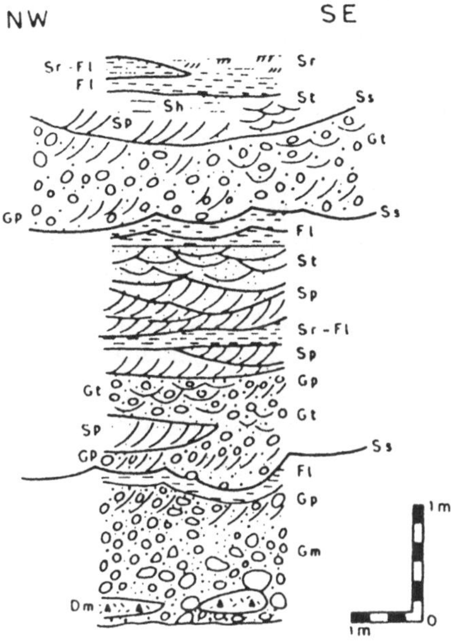

Fig. 6.11: A typical Talchir sequence exposed at Dhakoi, Bihar, India (after Casshyap and Tewari 1982, with permission of SEPM, Society for Sedimentary Geology).

rine glacial facies were recognised: isostatic, submarine retreat, marginal and proglacial. These deposits, Edward assumed, are controlled by the relative effects of rise and fall in the sea level due to eustatic and isostatic causes.

Recognition of Ancient Glacial Environment

Geologically young glaciated terrains may be recognised by the presence of such geomorphic features as *roches moutonnées*, eskers and drumlins. But these surfacial features are rarely preserved in older geological records. An association of striated pavements, chattermarks, varves and diamictites with typically pentagonal, flat-iron-shaped, faceted pebbles indicates glacial activity.

Glacially transported quartz grains are characterised by conchoidal breakage patterns, semi-parallel steps, arc-shaped microchattermarks, parallel striations, imbricated breakage blocks and irregular indentations. These surface textures, identifiable under transmission electron microscope, are diagnostic of glacial transport (Krinsley and Donahue 1968).

Tillite beds of Upper Carboniferous to Lower Permian age have been reported from all continents of the Southern Hemisphere. Parts of the Dwyka of South Africa, Tubarao of Brazil, Kamilario of Australia, and the Talchir of India may be mentioned as examples. The glacial origin of the Talchir Boulder Bed of India, proposed by pioneer workers of the Geological Survey of India, was later confirmed by a host of geologists. At places such as Irai, central India, glaciated striae were incised on the underlying Precambrian limestone pavement by Lower Permian ice movement, allowing faithful preservation. The direction of plucking and chattermarks on the underlying limestone, elongate troughs and striae enabled determination of the direction of ice movement (towards NE and NNE) at Irai (Smith 1963).

The Talchir glacio-lacustrine shales and siltstones in the Raniganj coalfield in eastern India are punctuated by sandy beds with a graded, solemarked base and laminated top. Slump structures, parallel current-ripple and convolute laminations, mark these laminated beds. Varve-like rhythmites containing current-generated structures such as current-ripple lamination are also found within these glacio-lacustrine sediments. All these are interpreted to have been produced by turbidity current activity. These density currents are envisaged to have been triggered within the outwash glacial lakes by slumps caused by oversteepening and overloading of an ancient delta front (Banerjee 1966).

Eolian Environment

Eolian activity is strong in an arid climate—in sand deserts located in rain-shadow zones within continents, and in glacial outwash plains. The climatic conditions in these areas do not permit growth of vegetation and wind blows unabated for long distances. Eolian activity in coastal dunes is independent of climate. Vegetation plays an important role in stabilising these dunes. Disintegration and exfoliation of rocks in a desert environment is caused mainly by exposure to sunrays. Sediment transportation is essentially by suspension and saltation. The impact of the saltating grains may cause the finer particles to creep over the surface without losing contact with the ground (surface creep). The impact of the sand grains in motion (deflation) is another phenomenon responsible for erosion in deserts.

Bedforms of various dimensions are produced by the action of wind: ripples (wavelength λ = 0.01–10 m), dunes (λ = 10–500 m) and draas (λ = 0.5–5.0 km). Each of these may be further subdivided into transverse and longitudinal elements leading to a total of eight subgroups, all apparently independent (Wilson 1972). A more comprehensive scheme of classification of eolian bedforms is given in Table 6.1 (after McKee 1979). Wilson's draas are recognised in this scheme as the result of combination of two or more bedforms of the same type. Sand seas of still larger dimensions

Table 6.1. Morphology and classification of eolian bedforms (after McKee 1979)

Morphology	Name	Association
Sheet-like	Sheet sands	
Thin elongate strips	Streaks	Compound—two or more of the same type combined by overlap or superimposition (Wilson's draa).
Circular to elliptical mound, dome-shaped	Dome	
Crescent in plan	Barchan	
Connected crescents	Barchanoid	
Asymmetrical ridge	Transverse (reversing)	Complex—two different basic types occurring together, either superimposed (Wilson's draa), or adjacent.
Symmetrical ridge	Linear (seif)	
Central peak with arms	Star (pyramidal)	
U-shaped	Parabolic	

($\sim$ 10,000 km^2) are named ergs. These are produced by redistribution of sand initially deposited elsewhere.

The external morphology and internal structures of the more important eolian bedforms are shown in Fig. 6.12 (A and B). With increasing complexity of wind regimes the dune form changes from simple transverse to complex star dunes and draas (see Fig. 6.13). In the sand roses shown in this diagram the arm length of a rose is proportional to the amount of sand that can be moved by the wind from a given direction towards the centre of the rose. (Stereonet plots of wind movement prepared from of the sand roses will give a diametrically opposite picture.) With a decrease in Resultant Drift Potential (RDP) and the RDP/DP ratio, the pattern of sand accumulation changes from lateral migration to vertical accretion. The Drift Potential (DP) is a measure of the total sand-moving capability of the wind expressed in vector units (v.u.) without regard to wind direction at the locality. RDP is a measure of the resultant or net sand-moving capability of wind at a locality in vector units (Fryberger 1979). Thus RDP/DP measures wind directional variability. By plotting RDP/DP against Equivalent Sand Thickness (EST), Wasson and Hyde (1983) could discriminate the four major dune types: linear, barchan, transverse and star (Fig. 6.14).

Eolian Facies Models

Two different models, one applicable to transverse bedform migration and the other to longitudinal bedform migration, have been proposed (Brookfield 1984). In the transverse model, unidirectional wind flows, sand patches and barchan dunes are initiated with the onset of aridity. With increasing aridity, these change over to transverse dunes and compound transverse draas.

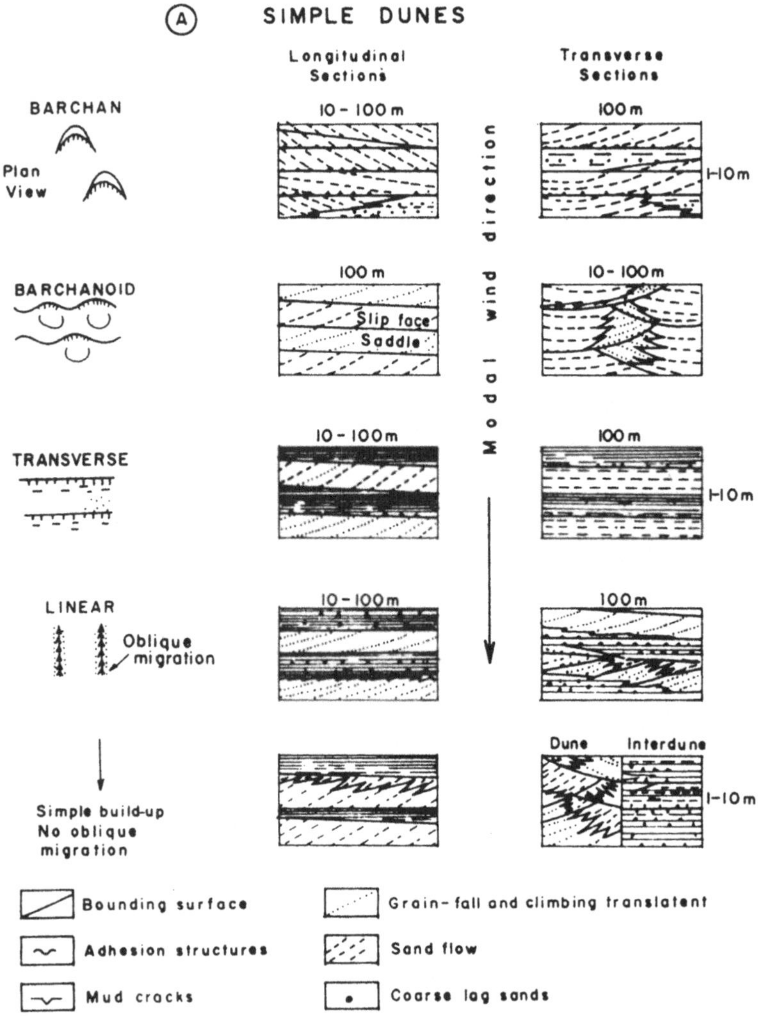

Fig. 6.12: Schematic representation of stratifications found in (A) simple and (B) complex eolian dunes of different types. The longitudinal sections are parallel to the resultant wind direction, the transverse sections are perpendicular to it (modified and redrawn after Brookfield 1984 with permission).

These forms decrease in size with decreasing aridity and sand supply. In the longitudinal model the initial sand patches and barchans are followed subsequently by linear dunes, linear complex draas and star draas. These longitudinal patterns are rarely found in ancient eolian dunes.

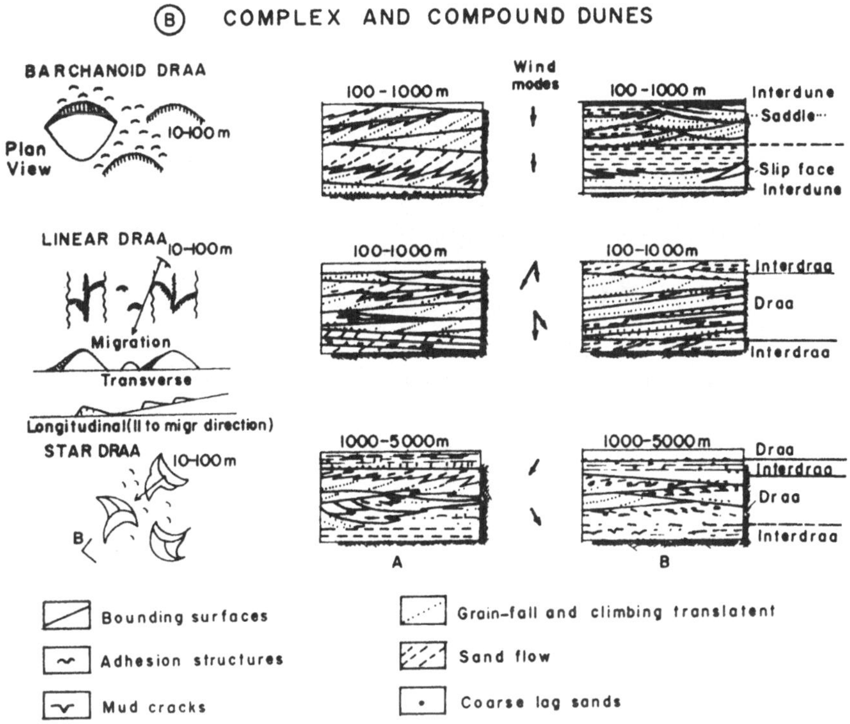

Fig. 6.12 (continued)

Vertical sections of eolian deposits worked out in the Permian Thornhill and Dumfries basins of Scotland show barchan dunes marking the onset of a desert. These are overlain by large barchanoid draas corresponding to the main eolian phase when arid climate predominated. The overlying small dunes mark the dwindling eolian phase, which were superimposed by alluvial fans during semi-arid climate (Brookfield 1984).

Recognition of Ancient Eolian Deposits

Forms and magnitudes of dunes of eolian and subaqueous origin overlap to a large extent. These cannot be used as criteria for recognition of ancient eolian dunes, therefore. The presence of lag gravels on avalanche toes, convex foresets, arching either up or down, deformation and reactivation surfaces, and progressive flattening and thinning of cross-bedding surfaces towards the top mark many of the present-day eolian deposits. Modern eolian dunes are also characterised by the absence of trough cross-beds. The upper parts of slip faces are generally steep (29°–34°). A combination

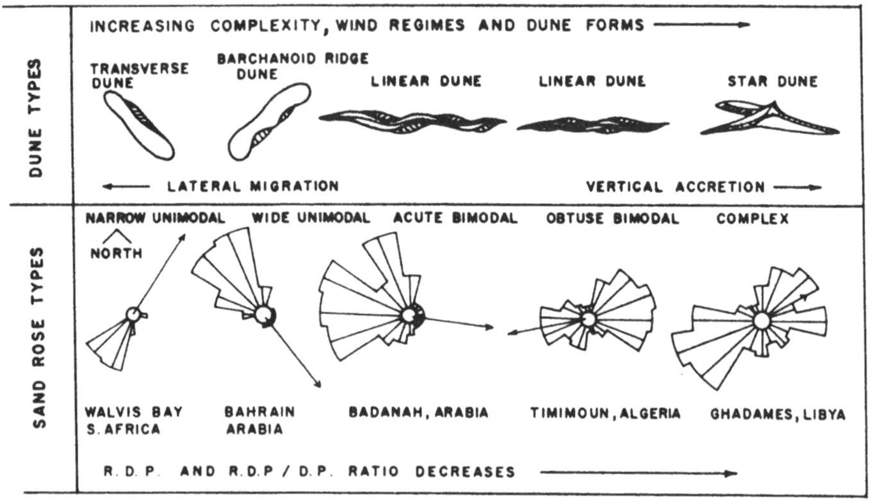

Fig. 6.13: Relationship between eolian dune forms and wind regimes. DP = Drift Potential, RDP = Resultant Drift Potential, RDP/DP is a measure of lateral migration. The proportion of vertical accretion increases with decrease of RDP and RDP/DP (modified after Ahlbrandt and Fryberger 1982, fig. 4, with permission of American Association of Petroleum Geologists).

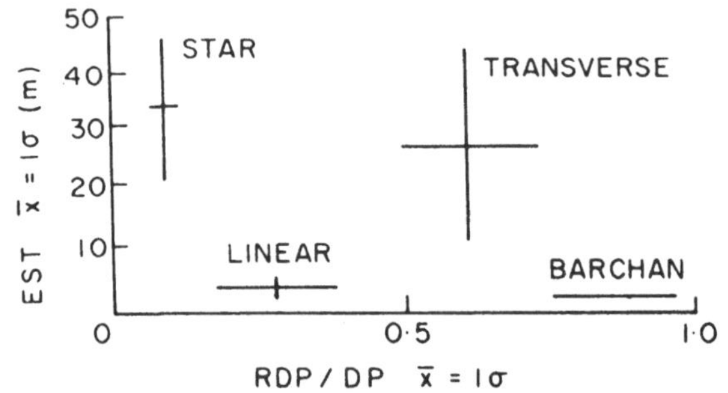

Fig. 6.14: Plots of equivalent sand thickness (EST) vs. wind direction variability (RDP/DP) as a tool for discriminating different types of eolian dunes (after Wasson and Hyde 1983).

of all these features might indicate eolian sedimentation although no single criterion can be taken to be diagnostic.

Eolian sands, because of their selective mode of transportation, are better sorted than sand of any other environment. The presence of fines in these sands skews the grain-size distribution positively. Rounding, pitting and frosting are common on the surface of quartz grains of eolian

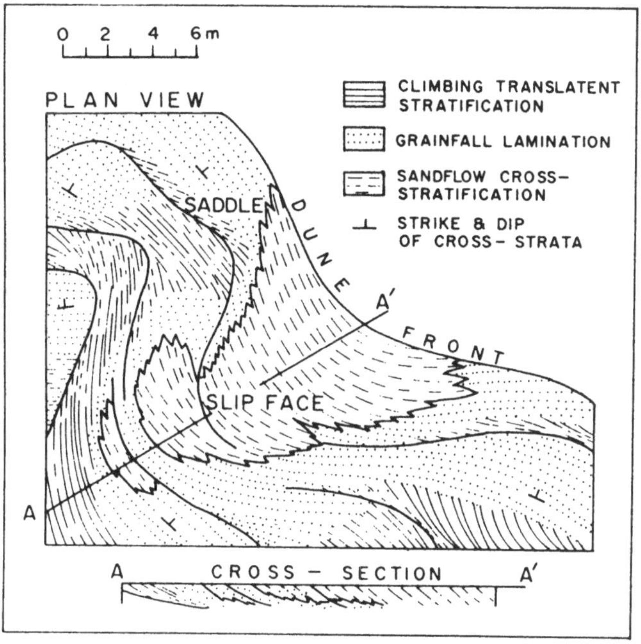

Fig. 6.15: Plan and sectional views of a barchanoid dune showing different types of stratification and cross-stratification (modified after Hunter 1977, fig. 7, with permission of the International Association of Sedimentologists).

origin. These may be detected by scanning electron microscopy (Krinsley and Doornkamp 1973).

A combination of sand-flow cross-lamination, climbing translation stratification and grain-fall lamination is indicative of the eolian process (Hunter 1977; see discussions under sediment structures). Two types of eolian climbing ripple lamination are recognised: translatent laminae and ripple-form laminae. In the former only the bounding surfaces between the ripples can be seen but in the latter the foresets are also visible. Grain-fall lamination is produced by deposition from suspension in the zone of flow separation. Sand-flow lamination is caused by downslope avalanching on the lee face. Most of these laminations can be identified on the eroded surface of barchan dunes (Fig. 6.15). The degree of porosity may also help in the identification of eolian sediments. Sand-flow cross-strata are loosely packed and highly porous. Climbing translatent strata are tightly packed and less porous. The grain-fall laminae are intermediate between the two in porosity and grain packing (Hunter 1981).

The Permian Toroweap Sandstone of Arizona, the Mesozoic Navajo Sandstone of Utah and Arizona, the Aztec Sandstone of Nevada, and the Lower Bunter Sandstone of England are some of the well-studied ancient eolian deposits. The Late Proterozoic Venkatpur Sandstone of southern India is interpreted to be of eolian origin on the basis of (i) grain-flow strata, (ii) inverse graded translatent strata, (iii) adhesion laminae and (iv) grain-fall strata. Nine facies, representing both eolian and related aqueous environments, have been recognised within what is believed to be a well-developed erg. Cosets of large-scale cross-strata are present within this sand sea which terminates at an alternate sequence of eolian and marine sediments. The horizontally stratified units within the Ventatapur sandstone have been variously interpreted to represent interdune, inland sabkha, sand sheet and coastal sand flat deposits (Chakraborty 1991).

Lacustrine Environment

Lakes occupy natural depressions surrounded on all sides by barriers or rims. These rims may be of tectonic or volcanic origin. Damming of river valleys, solution, glacial excavation and meteoric impact may also generate lakes.

Organic activity is an important part of lake sedimentation. The upper, well-oxygenated layer of water in lakes, where sunlight can penetrate, is replete with biological activity. The lower layers are cold and depleted of oxygen. The zone of rapid temperature variation between the two layers is called the thermocline (Fig. 6.16A). A salinity layering may also develop in inland lakes connected to the sea. Ice capping on the lake surface in cold climate does not allow mixing of the upper and lower layers, but in spring, when the upper layer warms up to 4°C (the temperature at which water is heaviest), the lake 'turns over' due to settling of the heavy layer to the bottom. The phenomenon is repeated again in autumn. During periods of turnover, nutrients carried to the upper layer from the lake bottom enhance biological activity.

Organisms settling down from the upper layers produce organic debris within the lake mud. Photosynthesis by phytoplanktons during the summer months removes much of the CO_2 from the upper layer of water. The dissociation of bicarbonate that follows increases alkalinity, thereby aiding in precipitation of calcium carbonate. The presence of carbonate indicates a warm, alkaline condition at the lake-bottom, a cold, acidic condition being non-congenial to carbonate precipitation. Together with fine clastics, the lake-bottom carbonate may produce marl.

In the productive stage of a lake, floating microscopic algae lead to the development of a brown to black, lipid-rich organic sediment called gyttja. During the low-water stage, the zone of aquatic vegetation at the lake margin widens and a floating mat of sedges grows within the lake, covering

A.

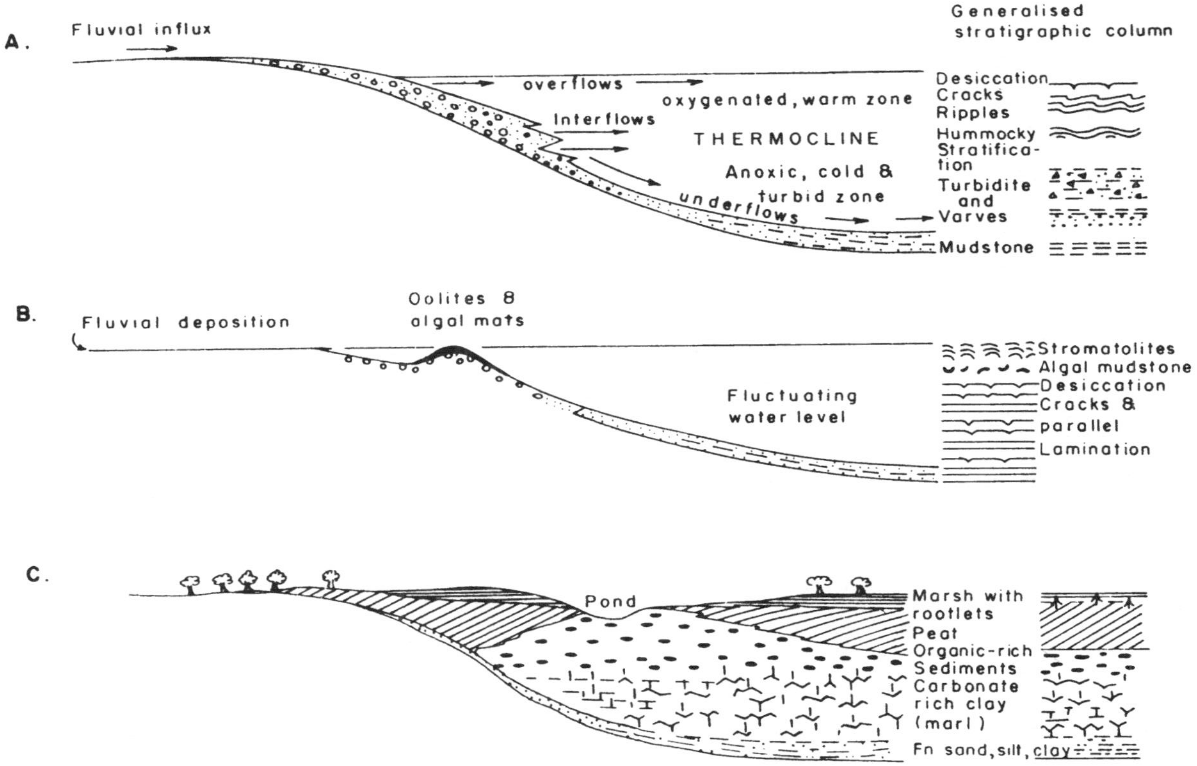

Fig. 6.16: Schematic representation of lacustrine sedimentation: (A) with high sediment input during low water level, (B) during high water level and (C) during low sediment input (modified and redrawn after Galloway and Hobday 1983, and Fouch and Dean 1982).

the gyttja, reducing the lake area and making it smaller. A reducing and acidic condition converts this vegetable matter into peat (Fig. 6.16C). A rise in water level, on the other hand, produces oolitic shoals and algal mats (Fig. 6.16B).

Unlike sea-water, complete evaporation of lake-water is possible, and a variety of salts such as sulphates, chlorides, borates and fluorides, normally not expected from evaporation of sea-water, precipitate within the lake sediments. Authigenic silicates such as feldspars and zeolites may also develop. Reduction of iron at the lake-bottom may cause precipitation of pyrite (FeS_2). Reducing and restricted basin conditions are needed for such precipitation.

Reducing (*Euxinic*) *Basins*

Hydrologically closed lakes, as also all other basins cut off from open sea by barriers or sills, enjoy reducing environment. The term euxinic, derived from the Latin name of the Black Sea (*Pontus Euxinus*), implies a 'cut-off or silled' condition. Being cut off from the Mediterranean by a barrier the Black Sea enjoys an euxinic condition (Fig. 6.17, X and Y). Other well-known areas enjoying euxinic conditions today include the Drams Fjord of Norway, Saanich Inlet of British Columbia and the Baltic Sea.

To be euxinic a lake should be located in a climatic zone wherein seasonal 'turnover' or upwelling is not possible. The Black Sea is euxinic because it is located in a humid area. The Mediterranean and the Red Sea are not euxinic because they are located in arid regions where evaporation exceeds inflow of water and, consequently, the bottom waters are well oxygenated (Demaison and Moore 1980).

The conditions prevailing below an active wave base in restricted basins are acidic as well as reducing. Temperature increases towards the lake-bottom. Organic matters, settling down to anoxic bottom conditions produce a black ooze (sapropel) which is rich in ferrous sulphide. The anaerobic organisms thriving in lake-bottoms absorb trace elements such as molybdenum, copper, vanadium and uranium. The colour of the shale deposited at the lake-bottom is black due to the presence of dispersed manganese granules (Goldhaber 1978). These black shales contain much larger proportions of trace elements than any other shale (see Table 6.2).

Black shales are known from a wide range of geological horizons but they are perhaps most common in the Palaeozoic and the older rocks. The world's well-known euxinic deposits include the Precambrian pyritiferous and carbonaceous shales of the Iron Belt, Michigan, the Kupferschiefer of the West European Permian, the Permian Black Cherts and the Phosphoria Formation of the Rocky Mountains. In the eastern part of the Precambrian Vindhyan basin in India, an isopach map shows a stratigraphic basining caused by a local sagging of the basement. Pyritiferous black shales (the Bijaigarh shales) found within this depression are interpreted to have been

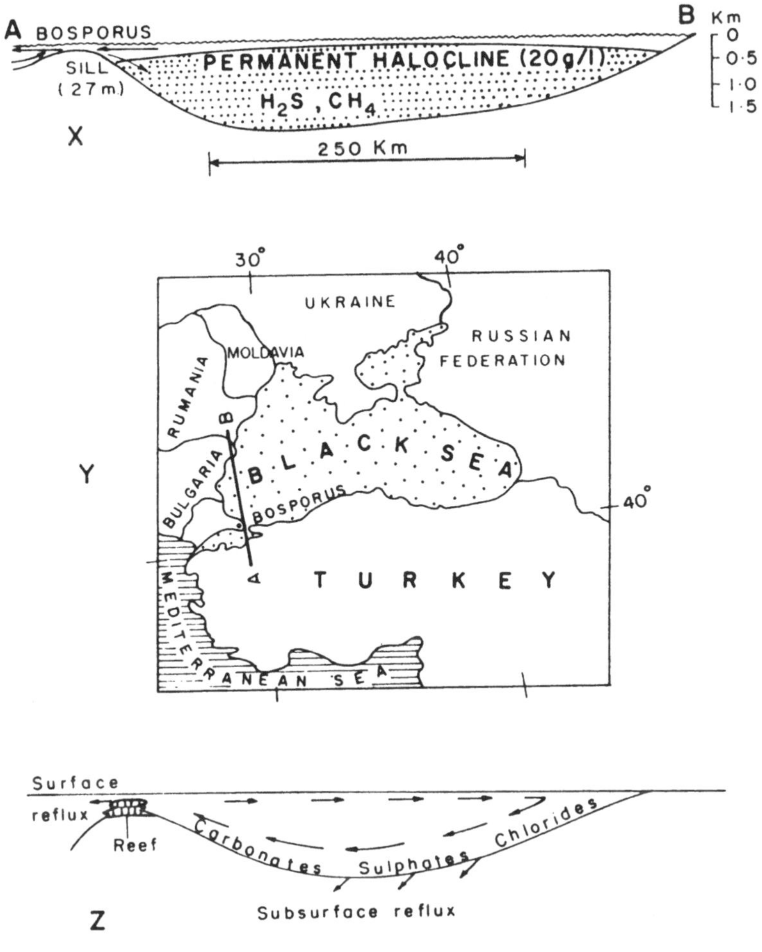

Fig. 6.17: (X) A-B section across the Block Sea (Y) sketch map of the Black Sea showing a modern euxinic environment (modified after Demaison and Moore 1980); (Z) locales of deposition of carbonates, sulphates and chlorides in a typical euxinic basin (modified after Krumbein and Sloss 1963).

deposited in the 'lagoonal pond' which enjoyed euxinic conditions during the Precambrian. The sandy facies overlying and underlying the shale were deposited at the lagoon fringe during transgressive phases of the Vindhyan sea (Banerjee and Sengupta 1963; Singh 1980). The trace element proportion of the Bijaigarh shale is less than that of an average black shale but the concentrations of Ni, Co, Cu and Mo in these shales compare favourably with those of oxygen-free sediments (see Table 6.2).

Table 6.2. Trace elements in shales and black shales (average estimates in ppm)

	Shale*	Black shale**	Bijaigarh***	
			Shale	Sandstone
Ba	800	300	85	79
Sr	299	200	26	39
Sc	10	10	4	15
W	-	-	6	2
Li	-	-	3	-
Cr	160	100	-	-
V	130	150	12	77
Ni	21	50	7	1
Co	12	10	12	9
Cu	38	70	62	29
Y	33	30	31	-
Mo	0.74	10	16	9
Tn	-	-	34	22
Ga	40	20	17	24
Pb	20	20	7	17
Sn	-	-	17	-
Zr	200	70	47	-

* Analysis of average shale (Green 1959, cited by Vine and Tourtelot 1970).
** Analysis of average black shale (Vine and Tourtelot 1970).
*** Analysis of average black shale and sandstone, 'Bijaigarh Sub-Stage', Vindhyan 'System' (Precambrian), Sinha and Singh 1964.

Recognition of Lacustrine Facies

Morphological and sedimentological characters of large lakes and shallow ocean basins overlap in many respects. Rivers debouching into large lakes produce fan-shaped deltas similar to those bordering seas. In cross-section these show simple, steep foresets overlain by flat topsets (typical Gilbert-type delta). The finer sediments deposited at the lake centre below an active wave base are generally thin bedded and parallel laminated. Varves may develop as seasonal deposits within lakes located in a temperate climate. Storm surges produce hummocky cross-stratifications that are smaller in dimensions than those formed under marine conditions.

There is no single criterion which is distinctive of a lacustrine deposit. Identification of an ancient lake is possible only by taking several of the following criteria together (Pickard and High 1972). Lakes have overall circular or elliptical shapes and are surrounded by rims or barriers ('positive structural elements'). They are of limited size compared to shallow seas. In plan lake deposits show concentric patterns with gradual increase of sediment thickness towards the centre. Narrow shore and nearshore deposits surround these sediments. Mean grain-size of the sediments decreases while the proportion of organic content increases towards the lake centre. Non-marine invertebrates (bivalves and gastropods), plants, algae and stromatolites are common.

Other significant characters of lake deposits are the presence of chemical and biochemical precipitates in a terrestrial setting such as evaporites, chert-banded sedimentary iron ores, freshwater limestones and marls. Finely laminated, graded turbidites at the centre; large-scale delta foresets, flaser beddings in the fringe areas and occasional hummocky cross-stratification within the deposits may also indicate a lacustrine environment. As the coarse, gravelly and sandy sediments of the lake fringe approach the finer sediments at the lake-bottom during progradation, a coarsening-upward sequence is produced. Such a sequence in a non-marine environment, together with some of the other features described above, point to an ancient lacustrine environment.

Fluvial Environment

Rivers originate in elevated areas and flow through flat plains before debouching into the sea or large lakes. On leaving the mountain gorge a river deposits most of its coarse sediment load near the areas of high relief. This relief is often a faulted mountain front. Sparse vegetation and copious rainfall cause quick denudation and redeposition of the sediments in the form of alluvial fans. Humid, tropical climate is not congenial to the development of alluvial fans. Such a climate causes chemical rather than mechanical weathering and also produces dense vegetation which protects the slopes. In glaciated valleys paraglacial fans develop where tributaries join the main glacier. Two types of alluvial fans are recognised (Schumm 1977): dry or mudflow fans and wet fans. The former are the result of ephemeral streams and the latter are produced by perennial flows.

On entering the plains a river can take four different forms: straight, braided, anastomosing and meandering. Of these, the braided and the meandering types are more common. Rapid fluctuations in water discharge, high gradient, a large proportion of coarse sediment load and readily erodable banks facilitate braiding. In a braided stream midchannel bars develop due to deposition of coarse-grained sediments which the stream is unable to transport. These deposits further decrease the competence of the stream by increasing the roughness of the channel. The enhanced rate of sediment accumulation which follows leads to the development of alluvial islands. The channel constantly subdivides and rejoins in an attempt to bypass these islands (Fig. 6.18). These multiple, low sinuousity, relatively unstable channels characterise braided stream patterns (Rust 1978).

Figure 6.19 shows the typical longitudinal profile of a meandering stream. As a stream adjusts its course with respect to its water discharge and sediment load, its longitudinal profile changes constantly until it assumes a

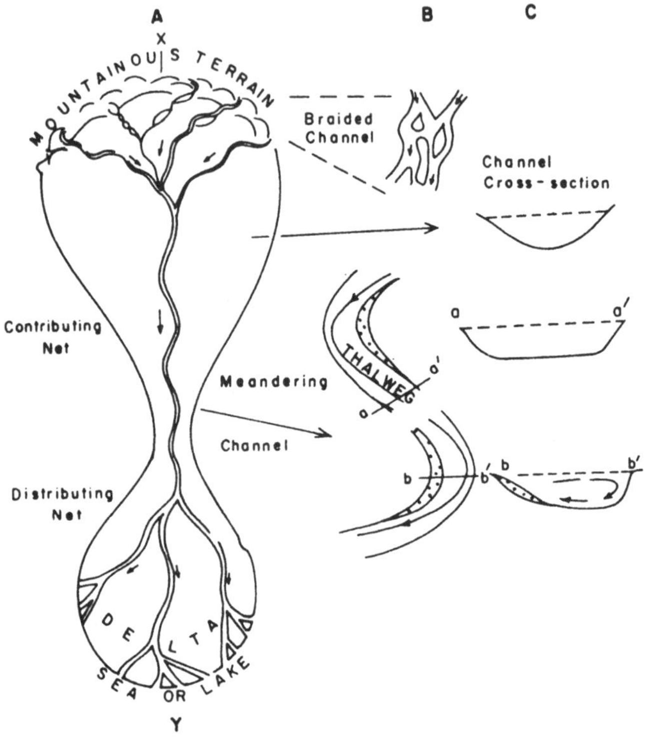

Fig. 6.18: Schematic representation of fluvial patterns and cross-sections of fluvial channels (based on Allen 1965 and Leopold, Wolman and Miller 1964).

smooth (graded), concave-up pattern. During the process of adjustment an overloaded stream deposits a large part of its load and also straightens its course by downcutting. The limit of downcutting, called the base level of erosion, is the sea level. When a stream reaches the base level and cannot cut further, it meanders in an attempt to decrease its gradient by increasing its length. Non-erodable clay banks and bed loads of fine-grained material facilitate meandering.

The transverse cross-section of a stream channel is dependent on its sediment load and water discharge (Fig. 6.18C). The cross-section, which is parabolic near the head of the stream, changes to a rectangular shape downstream. This change is effected by an increase in shear stress downstream due to an increase in water discharge. The banks, unable to adjust themselves to the increasing shear stress, cave in, thereby changing the channel cross-section (Leopold, Wolman and Miller 1964).

Within meandering channels the upper layer of water moves faster than the lower layer, which is in constant friction with the bed. The fast-moving

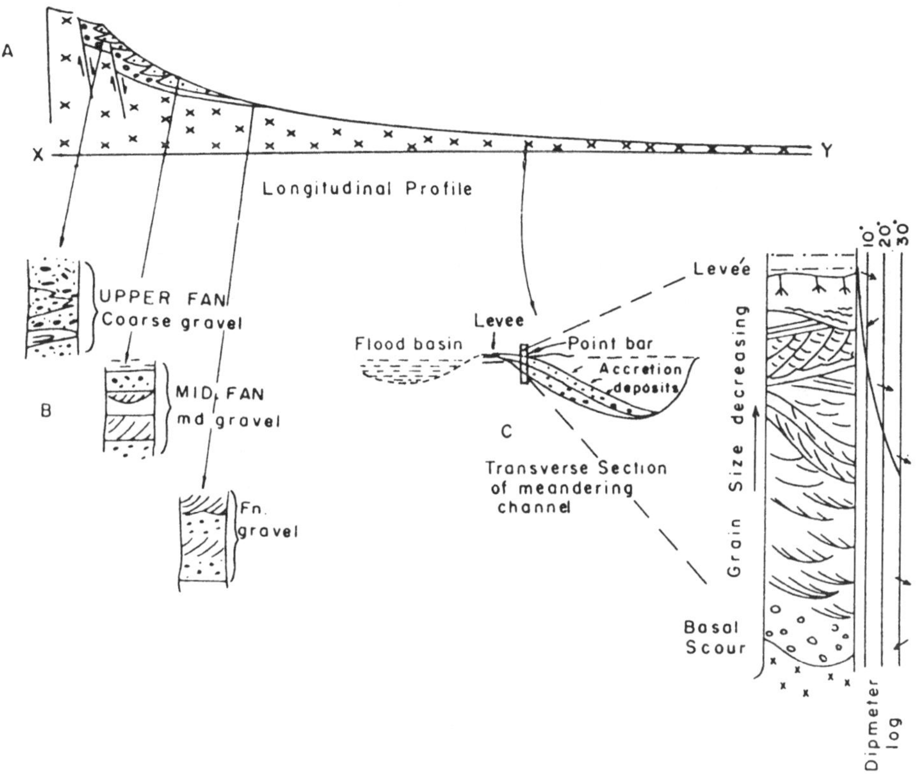

Fig. 6.19: Typical longitudinal profile of a meandering stream showing the vertical sequence of sediments at different locations (redrawn after Galloway and Hobday 1983). X-Y of (A) refers to Fig. 6.18. Note 'red pattern' in dipmeter log in C.

upper layer of water, impinging on the concave bank due to centrifugal action, starts eroding it. As the surface water plunges downward on the concave bank, some water from the lower level emerges at the surface on the opposite bank to maintain continuity. This circulatory motion, superimposed on the downstream motion of water in the stream channel, generates a helical flow. The sediments caught up in this flow are transported downstream to be eventually deposited by lateral accretion at the inner convex bends of the stream. These locations, called point bars, are the main areas of deposition within a meandering stream channel. A fining-upward sequence develops in these bars as the coarser grained materials, which are the first to be deposited, are followed upward by the finer deposits (Fig. 6.19C). Lateral accretion may produce 'ridge-and-swale' topography at the surface.

From time to time a meandering channel may abandon a portion of its loop by locally increasing its slope. Chute cut-offs are produced when a new meandering channel is excavated on the bar enclosed by the loop. When the abandoned loop is left as an ox-bow lake, and the stream continues to flow through the shortest possible channel, it is called a neck cut-off (Fig. 6.20). The deposits formed within the chute and the neck cut-offs are important features of an alluvial flood basin. These prismatic, U-shaped or triangular (in cross-section) deposits are essentially coarse except in the uppermost parts. They reveal a discordant relationship with the older alluvium on the concave side. On the inner, convex margin they grade laterally to the point bar sediments.

Natural levees and crevasse splays are the other two important features associated with a meandering stream channel. Levees are natural embankments bordering the concave banks with slopes towards the flood basins. They are constituted of alternating coarse and fine sediments deposited by sediment-laden water overflowing the channel. During high floods the levees are breached at places by crevasse. Coarse sediments, which find easy routes from the main stream channel to the flood basin through these crevasses, constitute the crevasse-splay deposits. These deposits are lobate or tongue-shaped in plan. The flood basins are occupied by finer grained sediments accumulated by the process of aggradation (Fig. 6.19C).

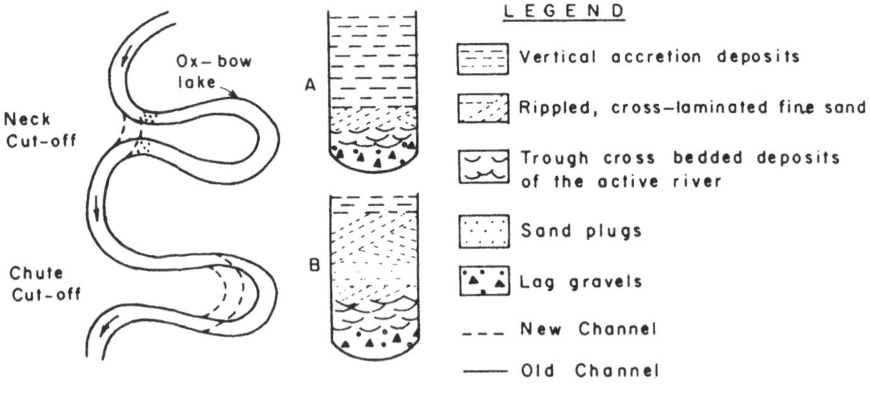

Fig. 6.20: Neck and chute cut-offs in meandering channels. Vertical sections of the deposits produced in neck and chute cut-offs are shown in A and B respectively. In chute cut-off the channel is abandoned gradually, but in neck cut-off the meander loop is suddenly abandoned and sealed off by deposition of (stippled) sand plugs leaving the main channel as an ox-bow lake. Hence in the latter case the cross-laminated sandy deposits are thin but the vertical accretion deposits are thick (based on Walker and Cant 1984, fig. 6).

Anastomosing rivers have attracted the attention of sedimentologists relatively recently. An interconnected network of low-gradient channels characterises these rivers. The channels are deep and narrow with stable banks composed mostly of silt and clay. The channels, which may be either straight or sinuous, are separated by vegetated islands, natural levees and wetlands (Smith and Smith 1980).

Fluvial Facies Models

Two distinct facies models are possible within fluvial systems: a gravelly facies associated with alluvial fans and braided rivers and an essentially sandy facies associated with meandering systems. The characters of these two facies assemblages are summarised below.

GRAVELLY ALLUVIAL FACIES

The facies characterising fans and braid-plain deposits are presented in Table 6.3. The sediments containing at least 50% gravel have been termed coarse grained (prefixed 'G') in this scheme. Only small quantities of sand are present in this facies.

Alluvial Fan Facies: Debris flows are characteristic of 'dry' alluvial fans. Downstream decrease in grain-size and radiating sediment transportation patterns are common in these flows. An accumulation of a threshold amount of loose muddy detritus is essential for such flows. The flows are initiated when an abundant water supply is available. In areas of steady rainfall the threshold amount of debris cannot accumulate because of the continuous rather than abrupt process of erosion prevailing there (Schumm 1973). Expections to this rule, however, are not unknown. Debris flow deposits are common in both semi-arid and paraglacial environments (Rust and Koster 1984).

Downstream decrease in grain-size within a fan is mainly due to a decrease in the competency of the stream. The deposits change gradually from coarse gravel in the proximal part of the fan to sand in the distal part, through clast-supported fine-grained gravel and sand-supported gravel in the intermediate areas (Gm → Gp → Sp). A 'sieve texture' may be produced when the gravel lobes have a scanty supply of sand. The deposits are horizontally stratified or imbricate in the proximal parts and horizontally laminated in the distal sandy or muddy areas. When adequate proportions of mud are available in the distal areas, the clasts, being supported by a muddy matrix, may be oriented subhorizontally or even vertically (Rust and Koster 1984).

Braided River Facies: Two different facies, proximal and distal, are recognised within braided rivers and braid-plains. In the proximal areas gravels (facies Gm) accumulate by vertical accretion on longitudinal bars. These are either horizontally stratified or imbricate. Next in abundance are the cross-stratified gravel deposits (Gp) with minor amounts of sand (facies Sp and

Table 6.3. Facies typical of fans and braid-plain deposits (Miall 1977; as modified by Miall 1978a and Rust 1978a)

MAJOR FACIES	Gm:	Clast-supported, commonly imbricate gravel with poorly defined subhorizontal bedding.
	Gms:	Muddy matrix-supported gravel without imbrication or internal stratification
	Gt:	Trough cross-bedded clast-supported gravel
	Gp:	Planar cross-bedded gravel, transitional from clast-supported gravel through sand matrix-supported gravel to sand (Sp)
MINOR FACIES	Sh:	Horizontally stratified sand
	St:	Trough cross-stratified sand
	Sp:	Planar cross-stratified sand
	Fm:	Massive fine sandy mud or mud
	Fl:	Laminated or cross-laminated very fine sand, silt or mud
	P:	Pedogenic concretionary carbonate

Sh). The distal braid-plains have autocyclic, fining-upward sequences. These sequences have cross-stratified gravels at the base, sandstone in the middle and mudstones at the top. Many such widespread braid-plains formed in the geological past in response to tectonic uplifts.

The deposits of the ephemeral Markanda River in Haryana, India provide a well-studied example of a terminal fan. A network of radiating, abandoned distributaries mark the surface of this fan (Fig. 6.21). The surfacial sediments of the fan consist of fining-upward channel and overbank deposits. The erosional surfaces of the channels are alternately overlain by coarser and finer sand facies. The former is horizontally bedded or cross-bedded. The latter is cross-laminated and flaser bedded. The channel sand occasionally passes onto horizontally laminated silt and massive mud facies upwards. From the proximal to the distal part of the fan there is a distinct decrease in the thickness of the channel sand. The overbank sediments consist of thin, climbing ripple-laminated sand, trough cross-laminated sand, interbedded mud, silt and sand facies passing into thick, massive mud facies with rootlets and desiccation cracks (Parkash, Awasthi and Gohain 1983).

Sandy Alluvial Facies

Sandy Braided Stream sequences start with channel lag deposits at the bottom. The overlying sandy facies is cross-bedded in trough-lenticular and planar-tabular patterns. This sequence is overlain by finer grained cross-bedded and cross-laminated bar-top deposits, mostly consisting of siltstones and mudstones. Orientations of the successive sets of cross-beddings may differ widely, suggesting quick change in sediment transportation direction in the braided channels. This facies model for sandy braided streams is

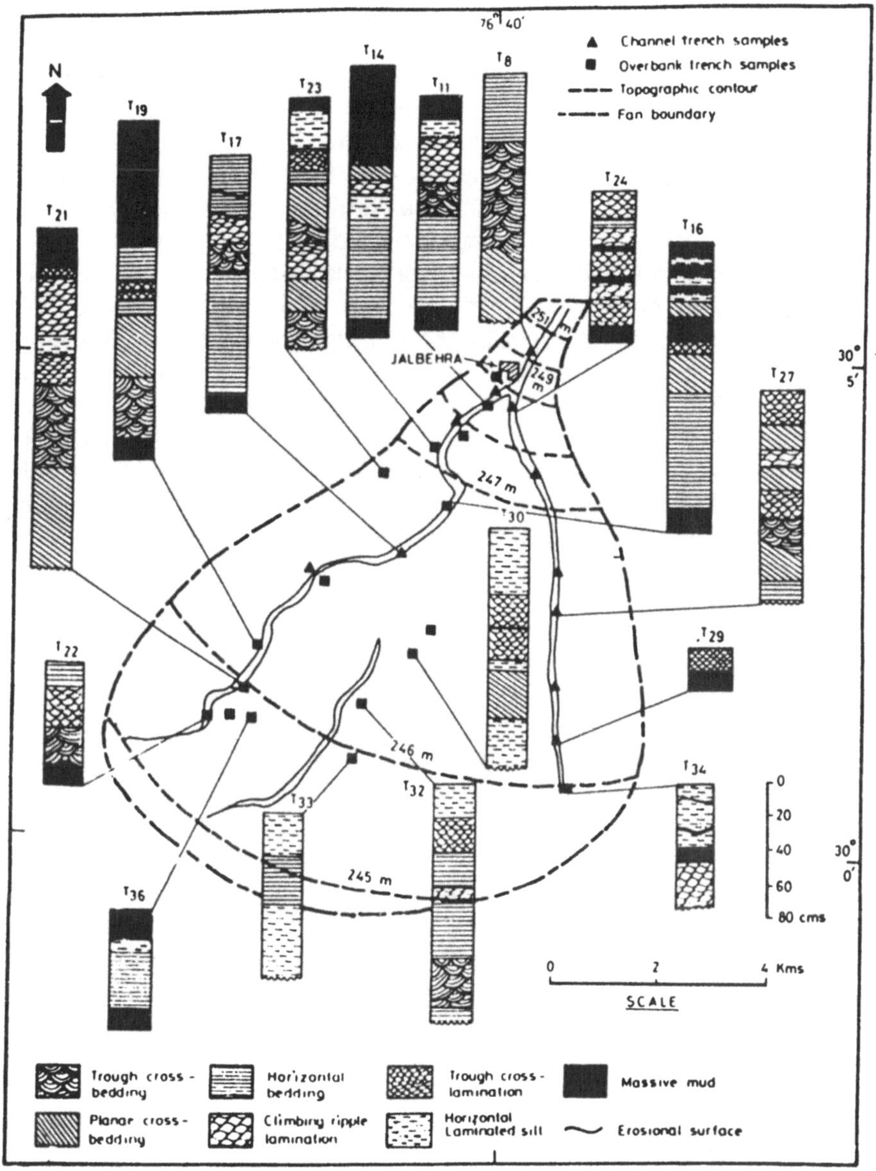

Fig. 6.21: Vertical sections of sedimentary structures at various locations of the Markanda terminal fan, Haryana, India (after Parkash et al. 1983; with permission of the International Association of Sedimentologists).

based on observations on the Devonian Battery Point Sandstone of Quebec (Walker and Cant 1984). Coalescing bars and sand flats of braided streams often result in extensive sand sheets. These, unlike the sand deposits of meandering streams discussed below, are not confined by clays or shales.

Facies of Meandering Streams: Point bars are the main depositional areas within meandering stream channels. Fining-upward sequences are preserved in point bars by the process of lateral accretion. A typical sequence, exemplified by the Devonian Old Red Sandstone, starts with channel lag deposits at the base. This includes gravels too coarse to be moved by the flow as bed load except during high flood, water-logged plant material and chunks of mud eroded from the river bank. Overlying the lag deposits are cross-bedded sands constituting the bulk of the lateral accretion deposits. The cross-beds are typically of the trough-lenticular type. These are overlain by horizontal laminations (plane beds) which are the products of high-flow regime conditions. There is a systematic decrease in the scale of cross-bedding towards the top but occasionally plane beds may occur interbedded with the cross-bedded units due to fluctuations in flow conditions (Fig. 6.19C). Sigmoid surfaces (large-scale epsilon cross-beds), characterised by a steep face, flattened towards the top and bottom, may be preserved when the process of lateral accretion is episodic instead of continuous.

The deposits within chute cut-offs start with a cross-bedded channel facies at the bottom. The latter are overlain by thick sequences of ripple cross-laminated fine sand deposited during the phase of abandonment. The whole sequence is overlain by finer grained, vertical accretion deposits. In neck cut-offs thin sections of ripple-laminated fine sand are deposited during short periods of abandonment. These are overlain by thick sequences of vertical accretion deposits laid down during flood time (Fig. 6.20).

Anastomosing River Facies: Six different facies assemblages have been recognised in the gravelly, anastomosing rivers in western Canada (Smith and Smith 1980). These are: 1) peat bog facies, 2) backswamp facies (silt and clay with organic debris), 3) flood-pond facies (laminated silt and clay with sparse vegetable matter), 4) levee facies (silty sand and sandy silt with rootlets), 5) crevasse splay facies (thin layers of sand with gravels) and 6) channel facies (gravel and coarse sand). Thick deposits of vertically accreted sand bounded by backswamp facies are the essential characters of anastomosing river facies. Erosion-resistant fine sediments deposited within backswamps help in stabilising the channel sands.

Recognition of Ancient Fluvial Deposits

Alluvial fans may be identified in a geological record by their geometry, which is radial in plan and wedge-shaped in section. The deposits are characteristically coarse and unsorted and show a progressive decrease in grain-size downstream (Fig. 6.19B). Organic matter is absent due to the prevailing oxidising conditions. Disarticulated and scattered organic remains may be found. A radial sediment transportation (palaeocurrent) pattern is another characteristic. In the absence of suitable cross-stratification, channel-bar orientation, clast-size variation and pebble imbrication may serve as palaeocurrent indicators in braided stream deposits.

The deposits in meandering channels show typically fining-upward sequences, starting from gravelly deposits at the bottom and ending in silty or muddy sediments through coarse- to medium-grained sand in between (Fig. 6.19C). The channel deposits, on the whole, form shoestring sand bodies in plan, but the chute and crevasse splay deposits are lobate. Levees occur as disconnected lenses tapering towards the flood basin. Large-scale cross-beds of the trough-lenticular type are common near the bottom of the sequence. These are overlain by smaller scale trough lenticles. Current lineation and ripple cross-bedding occur at the top of the sequence.

Of all the ancient sedimentary deposits, the fluvial ones are possibly the most widely studied. These include the classical studies on the Moenkopi Formation (McKee 1954), Old Red Sandstone (Allen 1964, 1965b), Buntsandstein deposits (Mader 1985). Facies associations and palaeocurrent systems of the coal-bearing Indian Gondwana deposits have been the objects of study of many Indian sedimentologists (Banerjee 1960; Casshyap 1970, 1973; Sengupta 1970, to name only a few). In each such study two major facies associations were recognised: a coarse, sandy facies and a finer grained facies. In the Kamthi (Permian) Formation of the Godavari valley, India, the former consists of closely interwoven bands of quartz wackes of both high and low matrix variety. Grain-size frequency distributions of these rocks are comparable to the sediments found in the fluvial bed loads. These coarse sandstones occur as thin, elongate bodies, sinuous in plan and prismatic or lenticular in cross-section. The geometry of these sandstones, which have gradational contacts between them and an intertonguing relationship with the flat sheets of clay and siltstone which overlie and underlie them, is comparable to that of the point bars formed by the lateral accretion of meandering deposits. The majority of the cross-stratifications within these sandstones are of the trough-lenticular type. Consistency of palaeocurrent directions within these channel sands is fairly large (66–75%, Sengupta 1970).

The deposits constituting the finer facies are structureless, thinly laminated bands of clay, silt and silty sandstone. These run parallel to the coarse-grained channel sands. These have a very large length/width ratio. The geometry and grain-size distribution of these sun-cracked, graded-bedded, fine-grained rocks are comparable to the suspension sediments of flood basins. Palaeocurrent directions within these deposits have little internal consistency (Sengupta 1970).

A rhythmic pattern of sedimentation consisting of alternately dominating channel (lateral accretion deposits) and interchannel facies (vertical aggradation) has been reported from the Barakar Formation of the South Karanpura coalfield, India (Banerjee 1960). An idealised, standard sedimentary cycle within the Barakar (Lower Permian) Formation of India consists of the following in a vertically upward sequence—pebbly sandstone, coarse to medium sandstone, fine sandstone, interbedded assemblage of fine to medium sandstone, siltstone and shale, carbonaceous siltstone and shale, coal and shaly coal. Several such cycles are observed. Most of these cycles display a fining-upward (sometimes partially truncated) character. The lower units are characterised by large-scale cross-bedding, while the upper units contain only smaller scale cross-beddings and wavy or parallel lamination. Apparently the flow intensity progressively decreased with time and was considerably reduced during deposition of the carbonaceous unit. The presence of coal seams suggests stagnation of the depositional area during the late phase of sedimentation, allowing vegetal accumulation. This pattern of cyclic sedimentation has been interpreted as the result of channel wandering and intermittent tectonism (Casshyap 1970). The sequence of sedimentation outlined above is a fairly good representation of the general pattern of fluviatile sedimentation.

MIXED ENVIRONMENTS

Barrier-Island and Beach Facies

Bordering many a coastline of the world today are rows of long narrow islands. Between these barrier-islands and the mainland lie lagoons rimmed by tidal flats. Connecting the lagoons with the sea are tidal channels (inlets). Until about 1972 these barrier-island complexes were believed to have developed wholly out of prograding barrier-beach systems. Models for these systems, based mainly on studies in the Galveston and Padre Islands, Texas, were also formulated (Dickinson et al. 1972).

More recent studies show that the barrier-island complex is a composite depositional system involving several discrete environments—beach, tidal inlet, tidal delta and lagoon (Fig. 6.22). A single generalised model cannot do justice to such a complex system. Four separate facies models, applicable to the barrier-beaches, tidal inlets, tidal deltas and lagoons, are discussed below, following Reinson (1984).

Barrier-Beach Facies

The barrier-islands bordering coastlines have shorefaces (beaches) on the seaward side. Four subenvironments of sedimentation are possible within the barrier-beach system. These are: shoreface, foreshore, backshore-dune and washover flats (Figs. 6.22 and 6.23).

Shoreface: Extends from the low-tide mark to a depth of about 10–20 m. Three zones of shoreface—lower, middle and upper—are recognised. The

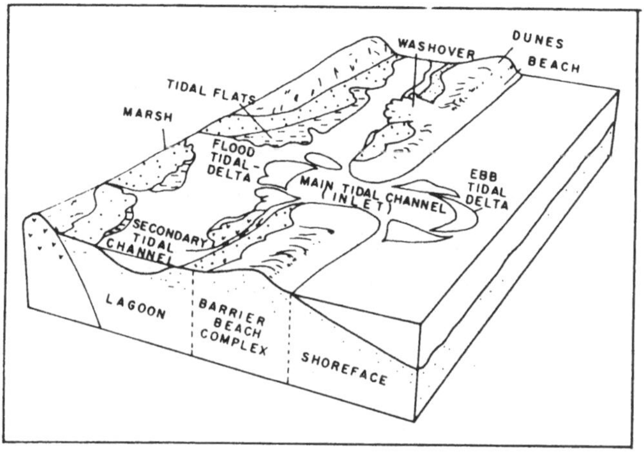

Fig. 6.22: Block diagram showing various subenvironments in a barrier-beach complex (after Reinson 1984, fig. 3, with permission of the author).

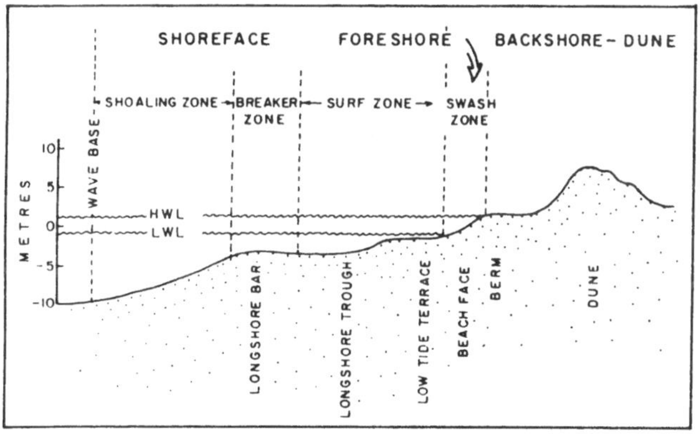

Fig. 6.23: Generalised cross-section of an ideal beach and shoreface (after Reinson 1984, fig. 5, with permission of the author).

middle shoreface, which extends up to the breaker zone, is dominated by wave energy. The typical deposits in this zone are highly bioturbated, fine- to medium-grained, clean sand, with silt and shell layers. Seaward-dipping, wedge-shaped, planar laminae, and landward-dipping ripple laminae are the common sedimentary structures in this zone. Trough cross-laminae may also be present. Occasional storm surges may drastically lower the wave base in the middle shoreface, putting much of the sediment in suspension and subsequently depositing it seaward. This phenomenon is believed to be responsible for the generation of the hummocky cross-stratifications found in geological record.

In the lower shoreface shelf and basinal processes are predominant but some wave action is also present. Plane beds of bioturbated, fine to very fine sand, with layers of silty mud are deposited in this zone. The degree of organic activity decreases towards the upper shoreface where the deposits are coarser. The complex hydraulic condition of the surf zone produces multi-directional trough cross-bedding. Ridges (bars) and runnels (troughs) develop on the upper shoreface when the sediments transported from the beach to the shoreface by 'short-period' wave action are returned and redeposited at the edge of the foreshore under quieter conditions by 'long-period' waves.

Foreshore: The area exposed between tides. It has a steeper gradient than the surf zone. The deposits here are well sorted by swash-backwash action. Thinly laminated, wedge-shaped beach deposits dip gently towards the sea. In some beaches berms separate foreshores from backshores (Fig. 6.23).

Backshore: Occurs on the landward side of the foreshore. The sediments of the backshore include lag deposits and heavy minerals. These dip very gently towards the land. *Eolian dunes* with planar and trough cross-stratifications occur closer towards the land. Rootlets and burrows are common within these dunes.

Thin lobate sheets of sand which flow over foredune barriers during high strom surges and extend into lagoons are called *washovers* (Fig. 6.22). These consist of medium to coarse sand and, rarely, gravels. Parallel lamination is the usual structure of the washover deposits but small-scale delta foresets may form when washovers plunge into lagoons.

Tidal Inlet Facies

Tidal inlets are narrow deep channels connecting the lagoons with the open sea (Fig. 6.22). Water flows vigorously through these channels during flood and ebb. Lateral migration of the tidal inlets causes the tidal deposits to accumulate as in a migrating stream channel. The sand body produced out of this process of migration is elongate parallel to the trend of the barrier islands. Its thickness is of the same order as the depth of the inlet.

The deposits within the inlets show fining-upward sequences, starting with lag gravels at the bottom and ending in finer sediments at the top. The cross-beds within these deposits carry imprints of flood and ebb flows and are naturally bidirectional. Plane beds and ripple laminae occur within the upper deposits. Five major depositional units, starting from lag gravels at the bottom and ending in finer land- and seaward-dipping laminae at the top, were recognised within the Fire Island Inlet, New York state (Kumar and

Sanders 1974). These are the channel floor, deep- and shallow-channel, spit-platform and spit deposits. Each of these deposits is associated with a particular type of sedimentary structure and texture. This model should work broadly for all tidal inlet deposits.

Tidal Delta Facies

Tidal deltas generate at either end of tidal inlets by tidal flows (Fig. 6.22). The ebb-tide deltas, produced at the seaward end on the tidal inlet, are modified by wave action and longshore drifts. The flood-tide deltas occur at the landward outlet of tidal inlets. The deposits of tidal deltas are essentially of sand size. The structures at the base are bidirectional megaripples. These are overlain, in the case of flood deltas, by landward dipping, large-scale ripples with planar-tabular and trough-lenticular internal structures. The cross-bedding set thickness decreases upwards. In ebb deltas, complex bedform patterns may be produced as a result of interaction with ocean waves.

Lagoonal Facies

The lagoonal deposits consist of alternating sandy and silty facies. The former is found in channel fills and washover sheets. The latter is produced within the lagoon under subaqueous conditions or derived from the adjacent tidal flats. The sandy deposits are rich in organic content. These are overlain by coal and peat produced within marshes and swamps at the margins of the lagoon.

Recent continental and marginal marine facies of the Godavari delta on the eastern coast of India provide an example of a mixed environment of sedimentation. The Gautami-Godavari River channel with its secondary distributaries forms a part of the subaerial topset beds of the Godavari delta. The two major distributaries of the Godavari River are interconnected by a reticulate system of tidal creeks. These creeks pass laterally into mangrove swamps. The backwater is a vast shallow lagoon, separated from the open sea on the south by a narrow barrier beach and on the east by a composite system of swamps, tidal creeks and beach. At least five barrier islands occur off the Vrudha-Gautami River confluence. These narrow, elongate islands run parallel to the mainland beach.

The sands of the offshore barrier islands of the Godavari delta are very well sorted and have an almost symmetrical grain-size distribution. Those of the mainland beach are moderately sorted and negatively skewed. The dune sands are well sorted and positively skewed. The silt and clay proportion is lesser in the beach and higher in the dunes (Naidu 1966). These observations might help in the identification of the subenvironments in an ancient barrier-beach complex.

The Proterozoic Vindhyan exposures of central India are arranged in several linear facies belts. Each of the facies in lithology, sedimentary structure and palaeocurrent pattern corresponds to a particular unit of sedimentation within a barrier-coastline model (see Banerjee 1974 for details). The Vindhyan sea was recognised as microtidal (tide range < 2 m), occasionally rising to mesotidal level (2–4 m). On the open shelf to the north-east a shore-normal, bipolar tidal pattern was identified, while within the embayment the pattern was much more complex. Strom surges, wind and wave-generated currents added to the complications (Banerjee 1980).

The parallel and cross-laminated, grey-black, pyrite-rich shales and siltstones of the Bijaigarh Formation (Kaimur Group) within the Vindhyans were interpreted as Proterozoic lagoonal deposits, while the micritic and dolomicritic lime-mud-bearing Bhander Limestone was thought to represent an ancient carbonate tidal flat (Banerjee 1974; Singh 1980, 1980a). Ripple-drift cross-laminations with rhythmically size-differentiated foreset laminae, believed to be characteristic of tidal environment, were also reported from the Bhander Formation (Chanda and Bhattacharyya 1974). Although all these observations point to a barrier-coastline, tidal or lagoonal condition of deposition, alternative interpretations, recognising at least a part of the Kaimur Group of rocks within the Vindhyan Supergroup as fluviatile deposits, are also known (Auden 1933; Bhattacharyya and Morad 1993).

Stratigraphic Sequences

Stratigraphic sequences which have developed out of regression (progradation), transgression and migration of barrier-island systems are as follows.

Regressive Model: A typical sequence developed out of regression of a barrier-island system has finer shoreface deposits at the bottom, grading gradually to coarser dune sands at the top through foreshore and backshore sediments (Fig. 6.24C). Abundant sediment supply is an essential condition for preservation of these deposits. This condition is satisfied in an interdeltaic coastal setting. In fact, many a barrier-island sequence in stratigraphic record are associated with deltas.

Transgressive Model: The facies sequence is variable, depending on the local condition. It is generally characterised by subtidal and intertidal back-barrier facies and does not show a definite fining- or coarsening-upward facies. In the regressive model the lagoonal facies overlie the sandy facies but in the transgressive model they underlie or are incorporated into the lower part of the sequence (Fig. 6.24C).

Barrier Inlet Model: This is a fining-upward sequence with a thinning-upward trend in cross-bedding set thickness. It has an erosional base and is dominated by sand facies of tidal-channel and marginal spit-beach environments (Fig. 6.24C).

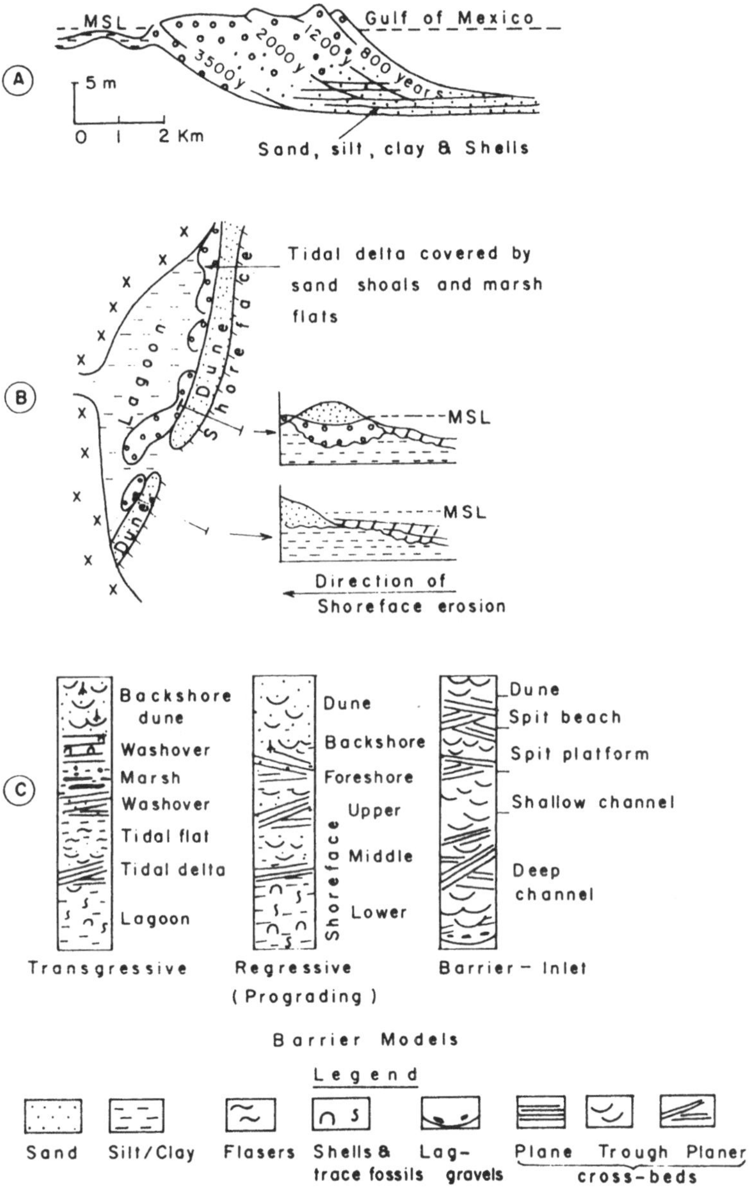

Fig. 6.24: (A) Cross-section of prograding barrier-island at Galveston, Texas (modified after Bernard et al. 1962). (B) Diagrams illustrating disconformity bounded delta sand body produced by inlet migration and concomitant shoreface erosion (based on Reinson 1984, fig. 25). (C) 'End-member' facies models of barrier island stratigraphic sequences (based on Reinson 1984, fig. 26).

Tidal Flats

Tidal flats develop in coastal areas where wave action is not strong. Within the barrier-island system such an environment prevails near the lagoons. The tidal flats of northern Europe, which are sheltered from the wave action of the North Sea by a row of barrier islands, may be cited as examples. Tidal flats may also occur on open sea coasts where the large width and low gradient of the beach dampens wave action. The Mailiano Flat of Taiwan, the Inchon Bay of Korea and the Chandipur Flat on the eastern coast of India are examples of open-to-the sea tidal flats. Tidal flats can be divided into two groups, depending on whether the constituting materials are essentially siliciclastic or carbonates.

Siliciclastic Tidal Flats

Fine sand, silt and clay constitute the bulk of the siliciclastic tidal flats. The nature of sediment distribution in these flats largely depends on local geographic and climatic conditions. In the sheltered intertidal areas bordering the North Sea a mud flat composed of mud, fecal pellets and thin bands of sand occurs on the landward side of the flat. The proportion of sand increases towards the sea until the flat becomes wholly sandy near the sea coast. In-between the mud and the sand flats occur mixed flats composed of muddy sand. The biogenic content of the flat increases as one moves from land to the sea. A generalised cross-section across an open-to-the-sea tidal flat based on observations at St. Michel, France is shown in Fig. 6.25.

The surface structures in sheltered intertidal zones, like rain and hail imprints, desiccation cracks and bird tracks, indicate periodic subaerial exposure. Horizontal lamination in sandy flats and parallel lamination in silty flats are common. Combinations of straight-crested, bifurcating (in plan), tuning fork-shaped ripples, interfering ripple systems produced by current and wave movements, and lenticular flaser beddings are some of the important characters.

The subtidal zones, which remain submerged during low-water, spring-tide level, are the sites of deposition of coarse sediments. The gullies traversing these flats contain coarse, channel-lag deposits. Megaripples and flaser beddings are found here. Escape traces of organisms are common but shells in orientations which they occupied, when alive, occur rarely.

The supratidal areas located on the elevated landward side of the tidal flats normally remain exposed to the atmosphere. They are inundated only during exceptionally high spring tides. Salt marshes consisting of clayey sediments and occasionally peat are common in the supratidal areas. Coarser sediments are restricted to the channels traversing the marshes. The clayey sediments are often stabilised by growth of salt-resistant plants within the marshes.

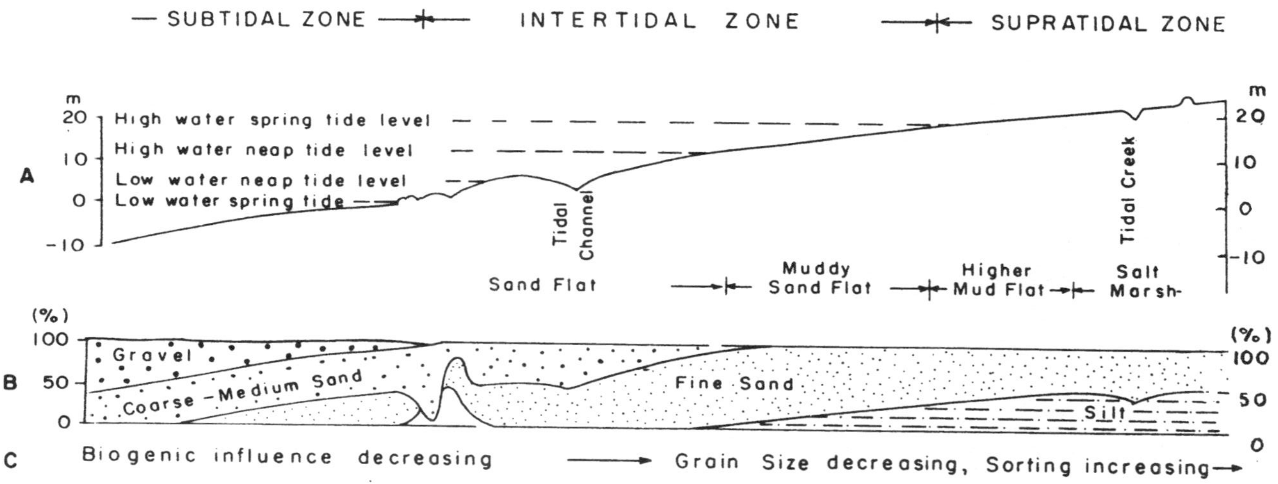

Fig. 6.25: Schematic section across an 'open-to-the sea' tidal flat (modified after Larsonneur 1975).

The Chandipur tidal flat situated on the eastern coast of India provides an example of a beach-related siliciclastic tidal flat bordering the open sea (Fig. 6.26). This tidal flat enjoys tropical climate. The mean tidal range lies between 2 and 4 m; hence the setting is designated as macrotidal. The western part of the intertidal zone enjoys a non-barred setting, while several sand bars occur on the eastern part.

Nearer the land the non-barred planar flat is composed of laminated sand frequented by burrows of ocypodid crabs. As one moves towards the sea the laminated sand grades to silty sand which is occasionally bioturbated by polychaetes. Nearer to the sea occurs a more silty facies. The silty sand has ripple-drift lamination, scour-and-fill, flaser bedding and climbing ripple lamination. The silty facies has ripple-drift lamination, scour-and-fill structures and hummocky cross-stratifications (HCS). The HCS develop only during monsoon weather frequented by cyclonic storms. In the barred flat laminated sand, megaripple lamination and flaser beddings predominate over other structures (Mukherjee, Das and Chakrabarti 1987).

The Rann of Kutch, located on the western coast of India, enjoys intertidal-supratidal environments. The extensive Rann ($\sim$ 30,000 km^2 in area) is flooded for about three months every year by sea-water 2 m deep as a result of monsoon storm surges. Fine silt, clay, gypsum and a small amount of carbonate constitute the deposits in the Rann (Glennie and Evans 1976; Singh 1980).

Carbonate Tidal Flat

The carbonate tidal flats do not differ much from the siliciclastic flats except in composition. Based on studies in such classical areas as the Bahama Banks of the USA, Shark Bay of Western Australia and the Persian Gulf, James (1984) recognised two major carbonate flat facies—those associated with high-energy conditions and those related to low-energy environment. A summary of the characters of tidal flat carbonate facies based on James (1984) follows.

The intertidal carbonate deposits are characterised by the presence of algal mats, irregular or even laminations (cryptalgal laminations) with fenestral porosity and desiccation cracks. The algal mats are leathery sheets of blue-green algae growing on top of sediment surface. The growth of these mats is controlled mainly by rainfall, moisture, salinity and the presence of other organisms. Algal growth is luxuriant in areas of high rainfall and hypersalinity. In areas of normal salinity the algal growth is deterred by gastropods which feed on them. In areas of high rainfall the algal growth may extend even up to the supratidal areas.

Elongate, sheet-like cavities (designated as 'laminoid fenestrae', 'loferites' or 'bird's eyes'; see Fig. 4.16) left by growing algal colonies are preserved in the geologic record even after the algae have decayed. These

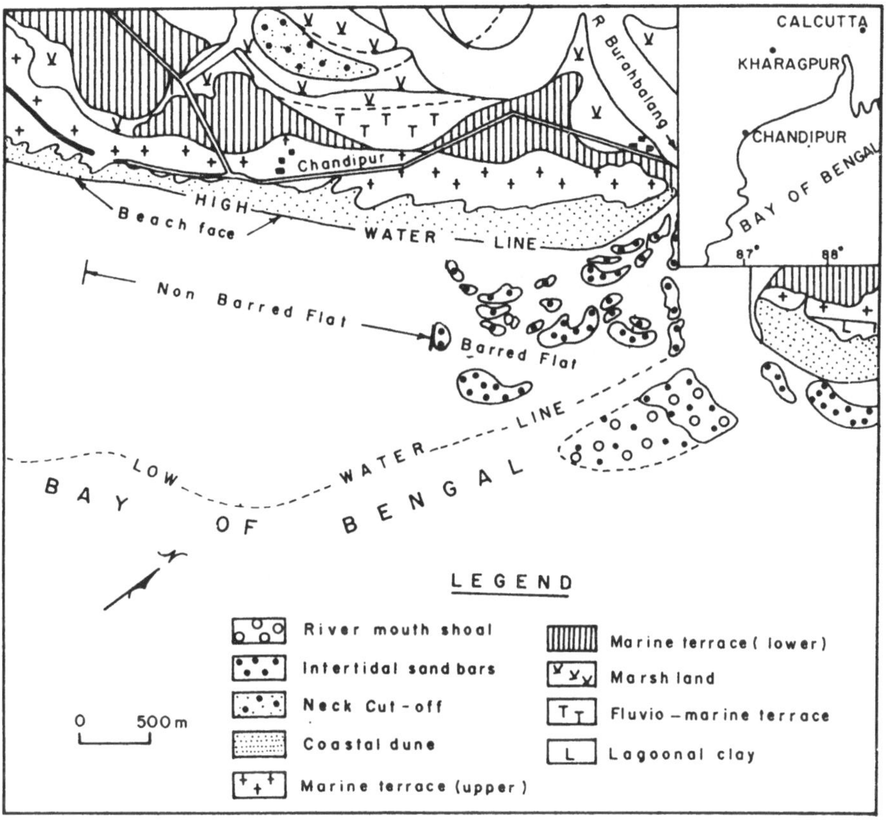

Fig. 6.26: Geomorphic units within the tidal flat around Chandipur, Orissa, India (modified after Mukherjee et al. 1987, and Chakrabarti 1991, unpublished DST report, with permission of A. Chakrabarti).

help in the identification of ancient algal colonies. The presence of algae can also be inferred from the occurrence of stromatolites—stratiform, crenulated, and arched organo-sedimentary structures produced by sediment-binding algae. Desiccation cracks are created by subaerial exposure of the algal mats. The cracks between the desiccation polygons are filled by lime mud. Sediment characters of the different parts of an ideal carbonate tidal flat are as follows.

Lower Intertidal Zone: Characteristics: bioturbated sediments, desiccation polygons and luxuriant growth of algal mat, particularly when the condition is hypersaline. Oolites (spherical objects with concentric internal layerings) formed from aragonite needles and organic matter in the shallowest sea water, may be moved out of place by strong tidal currents and deposited in the lowest intertidal zone.

Middle to Upper Intertidal Zone: Characteristics: sediments are grey-brown due to prevailing oxidising condition. Fenestral pores and desiccation polygons are indicative of prolonged exposure. Burrows of worms, crabs and crustaceans, and rootlets of salt-tolerant plants are common.

Supratidal Zone: Characteristics: Surface crusts ('hard grounds') produced by lithification of sediments due to prolonged subaerial exposure. 'Teepees' are produced by pushing up of the broken polygons of surface crusts due to the force of crystallisation or penetration of plant roots. The clasts are often cemented by carbonates. Caliche (calcrete crusts), algal mats (in areas of high rainfall) and evaporites (in arid areas) are common features in some supratidal areas. The nature of carbonate and evaporite precipitation is largely dependent on the chlorinity of the ground water and the amount of freshwater flushing in the supratidal zone.

STRATIGRAPHIC SEQUENCES IN CARBONATE TIDAL FLATS

Low Energy Flats: Vertical sequences are developed either by progradation of the carbonate tidal flats or by shoaling of carbonate sands of barrier islands. The basal units of such sequences are composed of coarse grained deposits of the surf-zone. These are overlain by bioturbated wackestones and packstones with marine fauna and stromatolites of the subtidal-intertidal areas. The subtidal deposits are well sorted, oolitic, pelletoidal and skeletal lime sand occasionally with oncolites. Planar bedding and herringbone cross-lamination, with set thickness decreasing upward and ending in small-scale ripples on top, are the major sedimentary structures. The structures are often destroyed by worm burrows. In shoaling-upward hypersaline sequences, as in the Shark Bay of Western Australia, stromatolites may abound on top of a skeletal metazoan bioherm or biostrome (Fig. 6.27A and B).

The deposits of the supratidal zone which overlie the subtidal-intertidal facies are evenly laminated. Flat-pebble breccia and desiccation polygons are common features in these deposits. Authigenic evaporites develop in arid regions flushed by hypersaline ground water. This is followed by precipitation of dolomite due to a rise of Mg^{++}/Ca^{++} ratio of ground water and subsequent dedolomitisation. Collapse breccias, produced by partial solution of evaporites, may cap the sequence (James 1984).

High-Energy Carbonate Flats: The characters of high-energy carbonate flat deposits are essentially the same as those of the siliciclastic beaches—coarse-grained, poorly sorted carbonate sand with lime mud matrix in the lower foreshore, and well-sorted lime sands in the upper foreshore. The lower foreshore has small- to large-scale trough cross-bedding, oriented parallel to the shoreline. The upper foreshore has

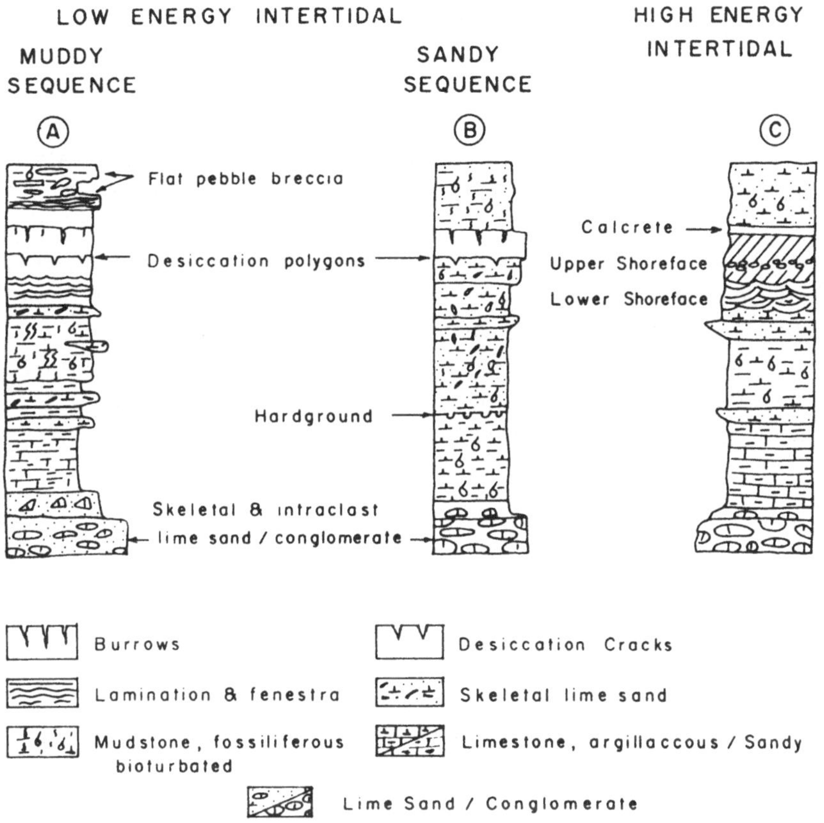

Fig. 6.27: Hypothetical sequences developed on (A) low-energy (muddy) subtidal unit, (B) high-energy (grainy) carbonate sand unit and (C) high-energy intertidal carbonate unit (modified and redrawn after James 1984, figs. 9 and 17 with permission).

seaward-dipping, planar cross-beds. These sediments may also have 'keystone vugs' or 'microcaverns' (similar to fenestral structures) produced by gas escape. These are almost invariably cemented at a later stage. A typical vertical sequence formed in a high energy intertidal situation is shown in Fig. 6.27C. The sequence, which starts with a lime-sand conglomerate at the bottom, is overlain successively by argillaceous limestone and bioturbated mudstone, to be capped, eventually by calcrete.

Deltaic Environment

A delta is defined as 'a deposit, partly subaerial, built by a river into or against a permanent body of water' (Barrell 1912). The morphology of a delta depends on the ratio of sediment supply and the energy of water transporting and depositing it. When the sediment supply is large, deltas appear as protrusions of shorelines into the sea or large lakes. Arcuate, cuspate, lobate, and bird-foot deltas develop progressively as the sediment supply is increased with respect to the wave energy. When the proportion of sediment supply is very low, the erosive power of wave energy predominates and indentations are produced on the coast in the form of estuaries (Fig. 6.28).

As early as 1885, G.K. Gilbert recognised that topset, bottomset and foreset beds are formed when streams debouch into still bodies of water in lakes. The slope of the foreset, under such a condition, is guided by the angle of repose (~ 30°), whereas the top- and bottomsets remain nearly horizontal. Since 1885, deltas showing well-demarcated top-, bottom-, and foreset units have come to be known as Gilbert-type deltas.

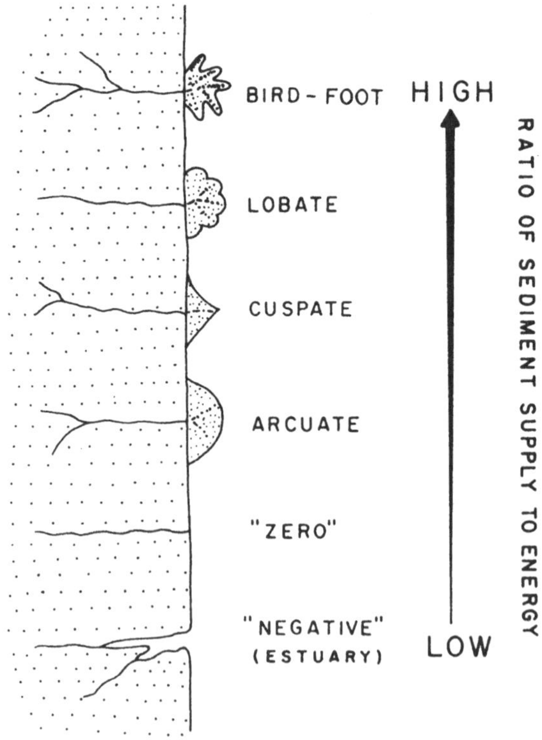

Fig. 6.28: Diagrams illustrating changes in shapes of subaerial deltas depending on wave energy and sediment input (after Curray 1975, in A.G. Fischer and S. Judson, eds.).

In deltas of the classical Gilbert-type it is the foreset that characterises the delta, the top- and the bottomsets being extensions of the fluvial and lake-bottom sediments respectively. The foreset in a Gilbert-type delta consists of coarse sediments deposited suddenly as the flow of the stream was deterred on reaching the quieter body of the lake. The foreset remains undisturbed only when the water level in the lake is stable for a fairly long period of time. When the water level fluctuates irregularly, complex patterns may develop due to interfingering of the top-, bottom- and foresets. When the level of water changes in a series of steps, the delta may take a step-like form. In cases of extremely irregular fluctuation of water level, the foresets, which characterise the Gilbert-type delta, may be completely obliterated. The whole delta deposit in such a case is represented only by a thick wedge of sediment interfingering with the coarser fluvial sediments on one side, and the finer lake bottom deposits on the other side (Dunbar and Rodgers 1957).

Deltas of the Gilbert-type are produced only when the densities of the effluent water in the stream channel and that in the sea or the lake into which this water is released are of the same order, allowing uniform mixing in three dimensions (Busch 1974). When the water in the stream is less dense than the water in the embayment, a sheet flow results because the mixing now is only 'two-dimensional'. When the magnitude of sediment discharge is small, an arcuate sand bar is deposited at the mouth of the stream. When the discharge is high, a cuspate or bird-foot type of delta may result. If the effluent water in the stream, because of its high suspended load, has a much higher density than the water in the ocean, a channel is eroded on the continental shelf in the form of a submarine canyon (Bates 1953, see also Fig. 6.28)

Classification of Deltas

Deltas have been classified by Galloway (1975) on the basis of relative influence of the fluvial, oceanic and tidal processes which operate during delta building (Fig. 6.29). The facies characters of these three types of deltas are described below, following Miall (1984a). Many natural deltas, it must be realised, display characters which are intermediate between these three ideal end-members.

River-dominated Deltas

As a river debouches into a static body of water its competence decreases and the coarser sediment load is dropped at the fringe of the sea or lake in the form of spits, bars and islands. The distributary channels cutting through the alluvial plain are bordered by levees. When the proportion of suspended sediment carried by the stream is high, the river mouth is extended towards the sea by sedimentation. The deposits are oriented at a high angle to the coastline to form bar-finger or shoestring sand bodies,

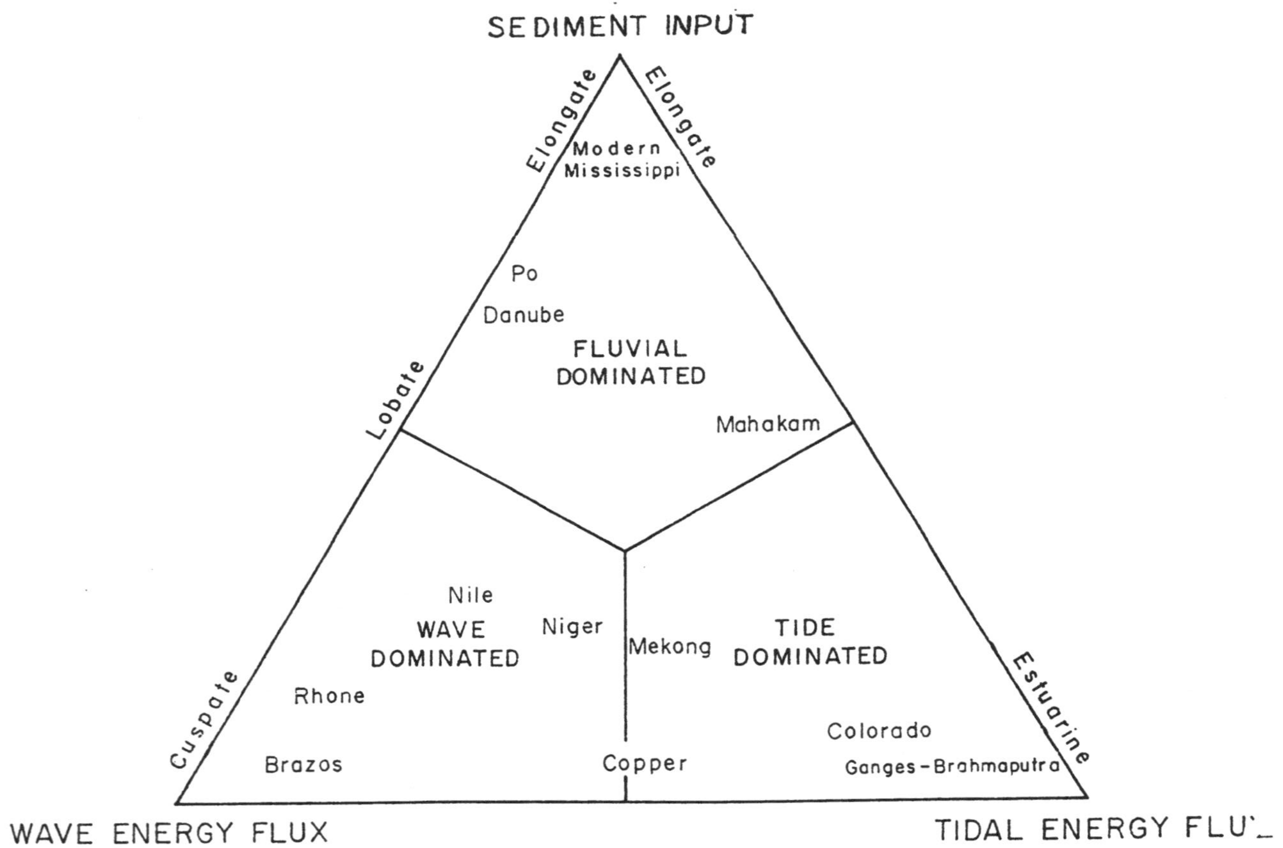

Fig. 6.29: Variation of shapes of deltas as a function of sediment input, wave energy flux and tidal energy flux (modified and redrawn after Galloway, 1975).

as in the bird-foot delta of the Mississippi River (Fig. 6.30A). The space between the distributary channels of such a delta is filled by marshes. At a high stage of flooding the distributary channels may be breached by crevasses. Crevasse splay deposits are formed when coarse sediments, pushed out of the crevasse, splay out on the adjoining marshes. Unstable conditions may arise when the rate of sediment deposition is exceptionally high in a delta. Frequent failures on the seaward side of the delta in such a case might result in faults contemporaneous with sedimentation. Such faults are called 'growth faults'.

A lobate, fan-shaped delta is produced when a heavy load of sediment is discharged by a stream on a plain of low gradient. In the Mississippi delta for example, a major delta lobe was created every time the Mississippi River discharged its sediment load towards the sea. The lobes thus produced are preserved even after the distributary channels responsible for their sedimentation have abandoned their earlier courses to take a shorter one to the sea. Seven such deltaic lobes have been mapped in the Louisiana coastal area (see Fig. 6.31).

While deltaic sedimentation was going on along the Louisiana coast, the sea encroached on the land on several occasions to flood the delta surface. Complex patterns of fluvial, fluvio-marine, paludal and marine sedimentation developed within the Mississippi delta out of the interaction between the delta and the sea. The fluvial sediments were deposited in stream channels, natural levees, point bars and abandoned stream courses. Brackish fluvio-marine deposits were laid down at the mouths of distributary channels. Paludal sedimentation took place in marshes, swamps, tidal channels and lakes. The marine sediments were confined to beach, foreshore and deeper water reefs (Kolb and van Lopik 1966).

WAVE-DOMINATED DELTAS

In areas of strong wave action the delta sediments are continuously reworked and the sand bodies are reoriented into a series of beach ridges paralleling the coastline. These are called 'strike-oriented' ridges. In areas dominated by longshore drifts the entire delta assumes an asymmetric ('skewed') shape. The delta of the Nile is a typical example of wave-dominated delta (Fig. 6.30B).

The Rhone delta, perhaps the best-studied wave-dominated delta in the world, has a thick sequence of prodelta shelf mud at the bottom. This is overlain by delta front sand, often called 'coastal barrier sand' (Galloway and Hobday 1983). Stacking and coalescence of individual sand bodies here resulted to delta front sand sheets of enormous thickness. The thickness has further increased by growth faulting. The delta front sand dips very gently towards the sea and shows a larger degree of marine influence than the sands of river-dominated deltas. The wave-dominated delta front sequence

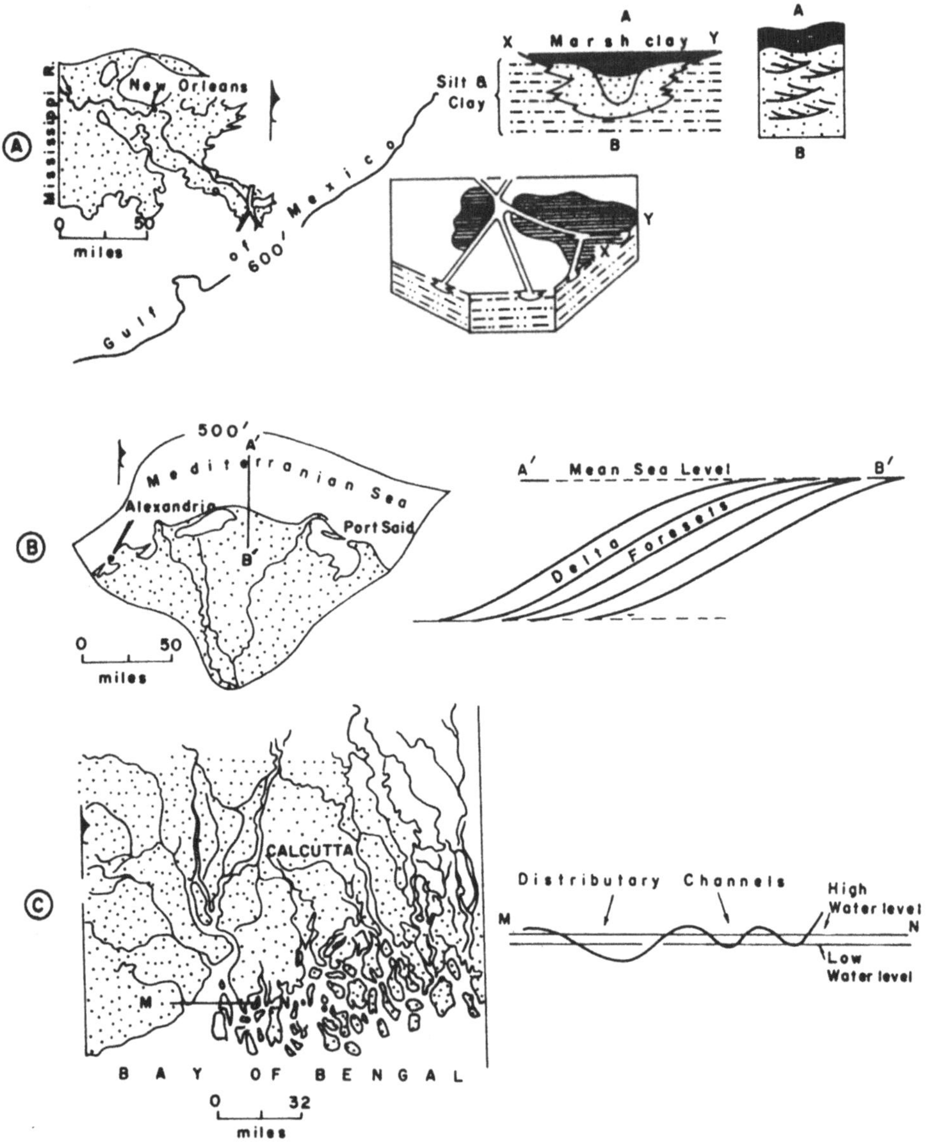

Fig. 6.30: Deltas in plan and profile. (A) the Bird-foot delta, Mississippi, dominated by fluvial action, (B) the Nile delta, dominated by wave energy and (C) the Bengal delta-estuary system, dominated by tidal energy.

coarsens upward like the sequence produced by progradation of a marine shoreface. In areas influenced by strong winds the coarsening-upward facies is covered by finer grained, well-sorted, coastal dune sands. The uppermost part of the Rhone delta sequence is criss-crossed by fluvial distributaries.

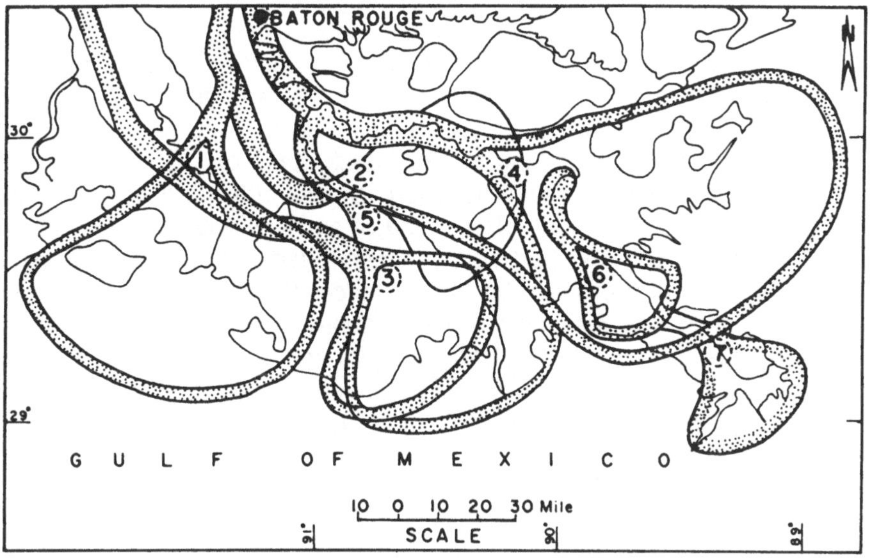

Fig. 6.31: Deltaic lobes of the Mississippi River (simplified and redrawn after Kolb and van Lopik 1966). The flow lasted for 500 to 1000 years in each case. Approximate time of flow before 1950: (1) around 5000 years, (2) around 4000 years, (3) around 3000 years, (4) around 2000 years, (5) and (6) between 2000 and 1500 years, (7) about 500 years to present day.

TIDE-DOMINATED DELTAS

Tidal currents, like waves, rework the deltaic sediments and rearrange them into a series of elongate bars extending ('dip-wise') from the channel mouth to the subaqueous delta platform. This platform is produced by deposition of mud, silt and muddy sand, which are carried downstream by suspension currents from the prodelta deposits. Estuaries develop when the distributaries find outlets into the sea through funnel-shaped channels. On land these tidal estuaries may take sinuous forms, cutting across tidal flats and supratidal marshes (Fig. 6.30C).

The vertical sequence in the Colorado estuarine channel, one of the well-studied tidal deltas of the world, starts with a basal scour filled with lag deposits. This is overlain by well-sorted channel-fill sand containing mud, silt drapes, clasts, macerated plant debris and shells. The sedimentary structures in this sequence include ripple laminations and cross-beddings with bidirectional dips. Laterally, the coarse-grained estuarine sediments grade into finer delta plain deposits. The sediments in the tidal current sand ridges coarsen upward. The bottom of the sequence is occupied by mud of prodelta shelf facies. These are overlain by sand with interbedded mud lenses. Ripples, cross-bedding and parallel laminations are the common sedimentary structures. The sequence is capped by finer grained tidal flat deposits.

The history of sedimentation in the Bengal delta, the largest tide-dominated delta in the world, with an area exceeding 59,000 sq km, is comparable to that of the Colorado delta. The Bengal delta had its origin in the Oligo-Miocene when deposition of a thick sequence of clay and silt (the Bhagirathi Group; Biswas 1963) took place on the Eocene shelf limestone of the Bengal basin. The total thickness of these deltaic deposits in the south-eastern part of the Bengal basin exceeds 5 km (see Fig. 7.22). The thickness of the Bengal deltaic sediments decreases north-west, near the fringe areas of the basin. Marine influence decreases and the section becomes sandier as the basin fringe is approached. The present delta, covering large portions of West Bengal (India) and Bangladesh, is occupied by tidal flats and marshes which are traversed by the distributaries of the Ganga and the Brahmaputra rivers (Fig. 6.30C).

Recognition of Ancient Deltas

Being the largest repository of clastic sediments, deltas are well represented in stratigraphic record. Ancient deltas may be identified by the presence of a thick wedge or lobe of shallow-marine to lacustrine clastic sediments showing typically coarsening-upward sequences. The wedge grades on one side to sediments of deeper water facies and on the other side to non-marine sediments.

The character common to delta deposits is an upward-coarsening progradational sequence leading to a marine, delta margin sand unit. This sequence is overlain by deposits of an aggradational facies—distributary channel sand and crevasse-splay sand bodies. The typical vertical succession in such a delta, as recorded by Galloway (1975), is:

Top 4. Open marine shelf mud or carbonates
3. Thick sand body separated from the coarsening-upward sand body by thin layers of mud
2. Typically coarsening-upward sand body which thins and pinches out laterally
Bottom 1. Prodelta mud blanket

In some cases the sandy zones (2 and 3 above) are replaced by a single, thick sand body which is narrow and elongate in plan and shows a fining-upward profile. In all cases the delta platform and delta plain deposits are criss-crossed by distributary channel sands. These lens-shaped sand bodies show fining-upward sequences.

7

TECTONICS AND SEDIMENTATION

SEDIMENTARY BASINS

Sediments exposed to atmosphere are likely to be eroded soon after their deposition by various natural agencies. For preservation in the geological record it is essential that the sediments be pushed below the base level of erosion. The preservation potential of a deposit is determined by the rate of subsidence of the area of deposition vis-à-vis the rate of sedimentation. The areas of the earth's crust which sink rapidly to allow preservation of a thick sediment load are called basins. Basins occupy large areas measuring thousands of square kilometres. Thinning of sediment layers towards the basin fringe is a sure indication of subsidence concomitant with sedimentation. In modern tectonic parlance a basin is a 'prism of rock forming a thick sedimentary succession' (Dickinson 1974). A model for basin evolution must explain, not only the mechanism of initiation of a depression in the earth's crust but also the process of uplifting of the adjacent areas, in order to provide a source for the sediment fill.

Basin subsidence was long imagined to be the effect of gravitational loading during sedimentation. Precise computations have shown, however, that the crust of the earth would not buckle to the extent needed for generation of large sedimentary basins due to the weight of the sediment load alone, but this could happen in the case of smaller basins. For more than a century the origin of large sedimentary basins was explained in terms of the classical geosynclinal model. Since the advent of the plate tectonics concept in the mid-sixties of the present century, there has been an equally enthusiastic attempt to explain the origin and development of large sedimentary basins in terms of plate tectonics. These ideas are discussed below in the chronological order in which they evolved.

GEOSYNCLINAL CONCEPT

The geosynclinal concept had its origin from the discovery by James Hall in the mid-19th century that the thickest sediment accumulation in the Appalachians is along the line of maximum depression (Fig. 7.1A). It was

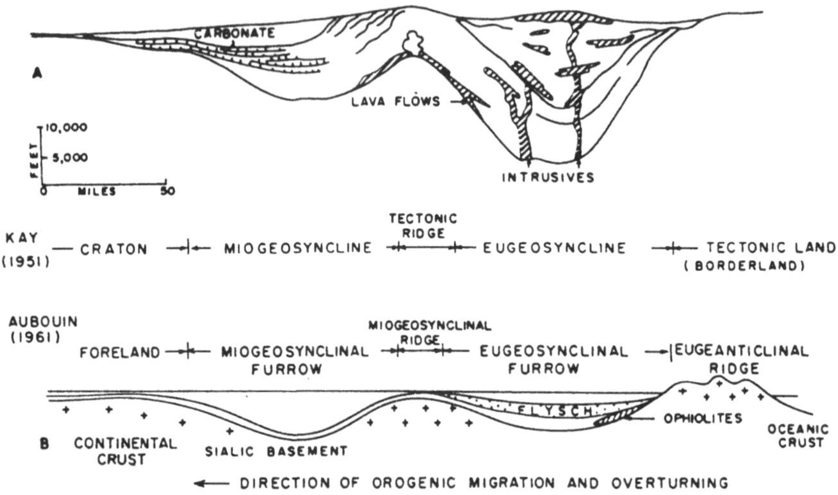

Fig. 7.1: Geosynclinal terminology (A) after Kay 1951, (B) after Aubouin 1965. (A) represents restored cross-section across the Appalachians from Eastern New York to Maine. The deposits are mostly Cambro-Ordovician in age.

recognised that large-scale sedimentation, which eventually led to the development of mountain belts, occurred along linear troughs along the continental borders. These troughs, separated from the ocean basins by elevated areas ('borderlands'), were designated 'geosynclines' by Dana in the late 19th century. A geosyncline need not always be restricted to the continental-oceanic boundary. The Alpine-Himalayan belt, an area of thick sedimentation and intense orogenic activity, is located between two continental blocks. Hence modern workers define a geosyncline as a basin of sedimentation located marginal to, or between, two cratons (Aubouin 1965). Cratons are the tectonically stable areas constituting the nuclei of the continental crust, while geosynclines are the locales of intense tectonic and igneous activity, both syn- and post-orogenic. The essential implication in the concept of geosyncline, therefore, is an idea of mobility. Mobility of the earth's crust in certain long, linear belts was imagined to be responsible for the accumulation of an unusually thick sequence of sediments.

A typical geosyncline of the above-mentioned model consists of an amagmatic zone of relatively shallow-water sedimentation bordering the craton. This zone, which receives sediments mainly from the cratonic source, has been termed the miogeosyncline. Away from the craton lies the deeper part of the geosyncline, the eugeosyncline. Separating the miogeosyncline from the eugeosyncline is a ridge, originally termed the geanticline but renamed the miogeosynclinal ridge by Aubouin in 1965 (Fig. 7.1B). The eugeosyncline receives sediments from the geanticline on one side and from the tectonically active borderland located on the other side. In marked contrast to the miogeosynclinal deposits, the eugeosynclinal sediments contain abundant tuffaceous material, the products of volcanic activity of the island arcs bordering the eugeosynclines. The eugeosynclinal sediments are also permeated by basic flows erupting on the ocean floor in the form of pillow lavas. The eugeosynclinal sediments of extraordinary thickness, traditionally related to orogenic events, are broadly classified into two groups, pre- and post-diastrophic. Since the nineteenth century the former has been designated as flysch and the latter as molasse.

Flysch has traditionally been described as a thick succession of argillaceous sediments of deep-water origin interrupted by dark gritty sandstones. Pillow lavas (altered to 'greenstones') are common within this sequence. Many flysch deposits start with massive graded beds at the bottom and end in pelitic layers on top through a sequence of parallel lamination and rippled beds (the Bouma sequence). The presence of such sequences in beds of deep-water origin, as evidenced by deep-water biota, led to the conclusion that the dark sandstones sandwiched between the argillaceous sequences were transported and deposited by turbidity current activity.

The shale-siltstone-fine-grained sandstone-graywacke sequence of the Simla 'Series' (Precambrian) of the Himachal Pradesh and Garhwal Himalayas were interpreted as flysch (Sen, Bhattacharya and Ray 1969; Valdiya 1970). The Precambrian metasediments of Ghatsila in eastern India, consisting of a rhythmically bedded sequence of quartzites and mica schists, originally deposited as a thick sequence of sandstone and shale, bear all the imprints of a typical flysch. Penecontemporaneous deformation structures such as load casts, pull-aparts, convolute laminations and slump balls have been reported from these metasediments (Naha 1961; Gaal 1964).

Molasse is the product of post-orogenic sedimentation. The deposits, generally several thousand metres thick, consist of many sequences of fining-upward, fluvial and deltaic cycles. The conglomerates contain well-rounded clasts. The fluvial cycles include immature, lithic and feldspathic sandstones, mudstones, coal, caliche, freshwater carbonates and evaporites. In areas of marine transgression the environment may range from paralic to lagoonal. Persistent, well-sorted conglomeratic sandstones dominate the areas of marine transgression.

The extensive Cenozoic molasse of the northern Alps consists of clastic wedges, grading upward from autochthonous flysch, through thin-bedded mudstone, to transitional paralic sequences. The succeeding rocks merge to marine deposits laterally because sedimentation kept pace with subsidence to produce near-sea-level surfaces where exchange of marine and non-marine conditions were possible (van Houten 1974). The Miocene-Pleistocene molasse of the Himalayas (the Siwalik Group of India) consists of more than 6000 m of fluviatile conglomerates, sandstones and shales (Figs. 7.2 and 7.3). The palaeodrainage of the Siwaliks was mostly from north to south but northerly drainage systems, controlled by intrabasinal tectonic highs, have also been identified (Tandon 1976, Tandon 1990; Tandon, Kumar and Singh 1985).

Confusion has sometimes been created in sedimentological literature by varied application of the terms flysch and molasse. The terms have been variously used to denote lithofacies, depositional environments and/or tectonic (syn- or post-tectonic) conditions. The confusion created has prompted some sedimentologists to recommend discontinuation of both terms. It should be remembered, however, that these terms, flysch as well as molasse, have thoroughly pervaded the sedimentological literature and to discard them at this stage would only add to the confusion already created. These terms may thus be retained but the specific sense in which they are used should be fully understood.

Modern Geosynclines

Following the classical studies on geosynclines in the late nineteenth and early twentieth century, many a sedimentary basin was identified as a geosyncline. In fact, Marshall Kay, who proposed a comprehensive scheme of classification of geosynclines in 1951, preferred to designate every sedimentary basin located within, outside or adjacent to cratons as a geosyncline. Such a scheme of classification (Table 7.1) apparently accepts a geosyncline as a universal phenomenon, both in time and space. Following the Principle of Uniformitarianism, one would, in such a case, expect the geosynclines to be present today as extensively as in the geological past. Our difficulty in recognising modern geosynclines, however, has raised doubts about the very basis of the geosynclinal concept.

During the early years of this century the deep oceanic trenches of Indonesia and the bordering volcanic arcs were intensively studied by Dutch geologists and geophysicists following a suggestion from Haug in 1900 that these deep oceanic trenches are the modern counterparts of geosynclines. The concentrically distributed Indonesian arcs (the Banda and Sunda arcs) although accepted by many as the basis for explaining the evolution of the East Indies, failed to provide a model comparable to the large-scale geosynclines found elsewhere in the world. The sediment layers identified within the Java and Japan trenches, for example, are comparable neither in thickness nor in character to those characterising ancient geosynclines.

In the Atlantic Ocean, off the North American coast, thick prisms of submarine sediments (named 'continental terrace') have been interpreted as the modern analogues of orthogeosynclines (Drake, Ewing and Sutton 1959). Similar thick sequences of suboceanic sediments have been termed 'continental rise' elsewhere (Dietz 1963). Some of these submarine deposits are

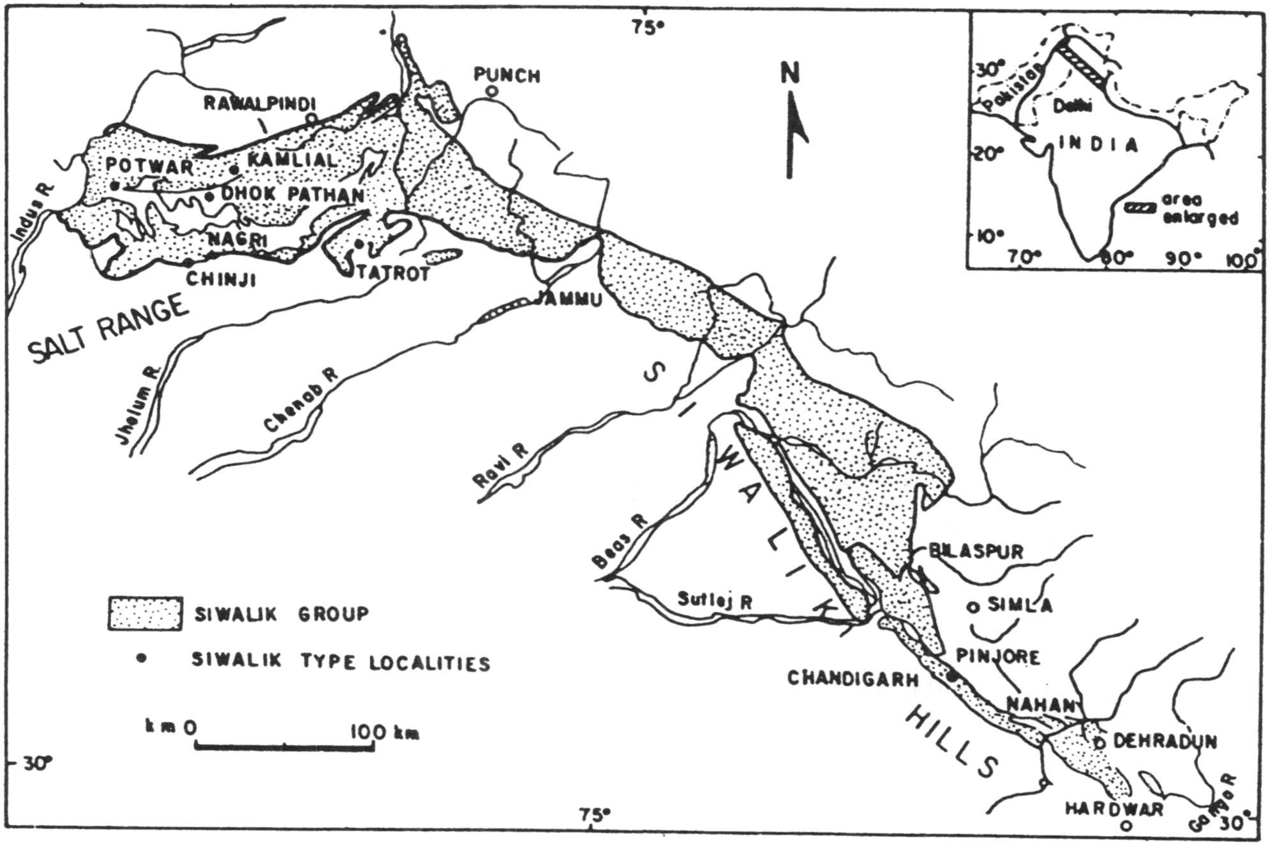

Fig. 7.2: The Siwalik Basin between the rivers Indus and Ganga (after Tandon 1990, with permission of the author).

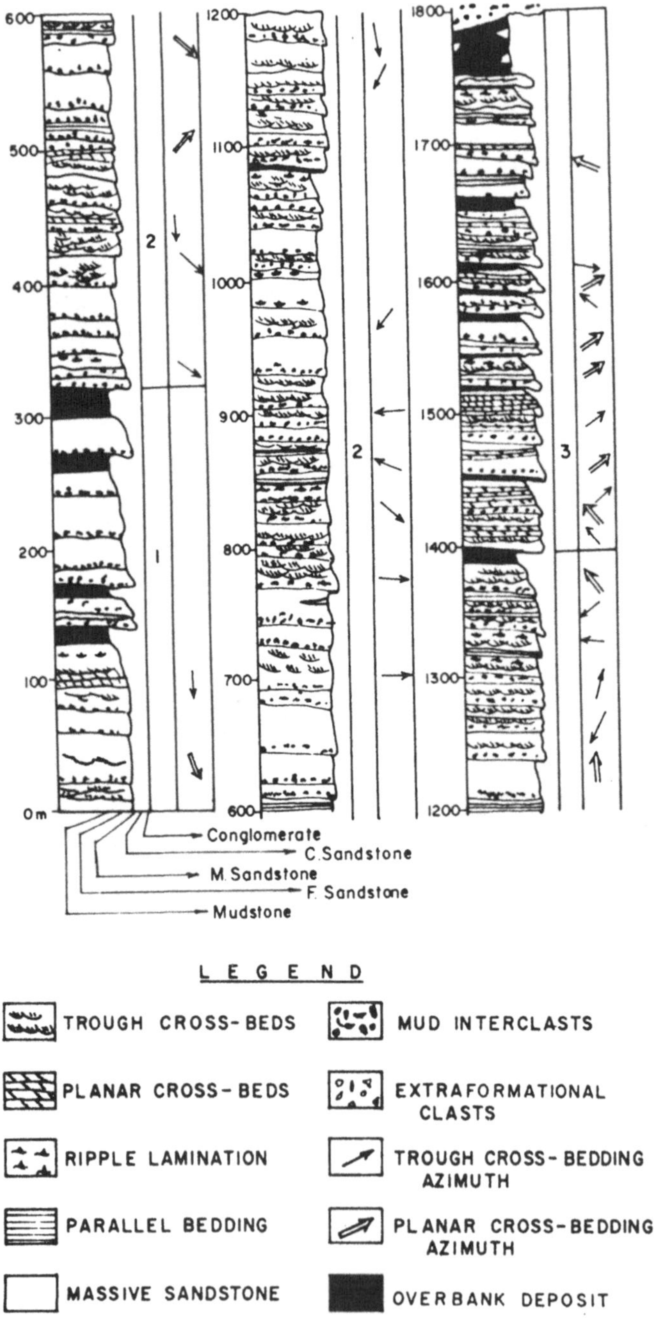

Fig. 7.3: The Middle Siwalik lithological columns, Dehra Dun region (after Kumar and Nanda 1989, fig. 7, with permission of the Geological Society of India).

Table 7.1. A scheme of classification of geosynclines (simplified after Kay 1951)

Geosynclines

1. Orthogeosynclines (linear troughs located between cratons)
   a. Eugeosynclines (actively subsiding areas, volcanics present with sediments)
   b. Miogeosynclines (less active tectonically, volcanics absent)

2. Parageosynclines (ovate troughs located within or adjacent to cratons)
   a. Exogeosynclines (extensions of orthogeosynclines)
   b. Autogeosynclines (isolated basins within cratonic uplifts)

enormously thick—the estimated thickness of the Bay of Bengal deep-sea fan, for example, exceeds 12 km (Curray and Moore 1971). Many authors would like to associate these deep-sea sediments with the phenomenon of sea-floor spreading and would imagine them to be the modern counterparts of classical geosynclines (Dietz 1963; Dietz and Holden 1974). In contradiction to this hypothesis, it has been pointed out that the sediments of the continental rise were derived from the continental side only, while those of the ancient geosynclines were drawn from both the foreland and borderland (Pettijohn 1975).

The idea of whole ocean basins being considered modern representatives of geosynclines being dragged towards mobile belts has been questioned by Pettijohn (1975, p. 577) on the ground that in that case the pelagic carbonates would have ended up in deep oceanic eugeosynclinal sequences. There is no evidence, according to Pettijohn, that this has happened. Of late, several workers have reported the presence of CO_2-rich inclusions in granulites (see for example, Hansen, Newton and Janardhan 1984; Newton 1988). It has been suggested that one of the possible sources of CO_2 is the deeply buried carbonate sediments. This leads one to wonder whether the source behind the carbonate fluid was the pelagic carbonate which was actually carried to the mantle by the process of subduction.

PLATE TECTONICS CONCEPT

A model of global tectonics, called the plate tectonics model, emerged during the two decades of intensive exploration in the sea following the Second World War. During this period palaeomagnetic studies provided evidence of secular changes in palaeolatitudes, thereby confirming large-scale movements of the continents. Detection of newer oceanic crusts with 'magnetic stripes' being added to the ocean-floor on either side of the midoceanic ridges since the Mesozoic, confirmed the idea of sea-floor spreading. The major dynamic features of the earth such as orogenesis and seismicity could be connected to the phenomenon of gliding of the rigid lithospheric plates over the asthenosphere. It was also recognised that the lithosphere is consumed at the Benioff subduction zones, on the landward side of the deep-sea trenches, so that the addition of newer crust at the midoceanic ridges does not change the net global surface area. Following these findings six major and a few minor lithospheric plates were identified and the plate boundaries were divided into three groups, depending on the nature of their movements (McKenzie and Parker 1967; La Pichon et al. 1973): 1) Diverging plate boundaries at the midoceanic ridges. These are the accreting margins where new oceanic crusts are added to the ocean-floor; 2) Converging plate boundaries (called subduction or Benioff zones) at the submarine trenches defined by the plane where the deep foci earthquakes are located. The lithosphere disappears below the mantle at these boundaries; 3) Transform fault or shear boundaries where the plates glide past one another without addition or destruction of the crust.

The continental margins, on the basis of their coincidence (or not) with the plate boundaries, are divided into two groups, 'active' and 'passive'. The western margin of the continent of South America, which is actively involved in plate movement, is an example of an 'active' margin, while the eastern margins of North and South America, not coinciding with any plate boundary, are 'passive' margins (Fig. 7.4).

The setting of a sedimentary basin can now be described in terms of three factors: 1) the substratum of the basin, 2) proximity of the basin to the plate margin and 3) the type of plate margin involved in basin development (Dickinson 1974). In this model the continental and the oceanic crusts are the two end-members of a series of transitional types of substrata. In terms of proximity to plate margin, Dickinson (1974) divided sedimentary basins into two groups: 1) those located within the plates ('intraplate basins') and 2) those lying at the zones of plate interaction. Three different types of plate junctures described earlier, namely, divergent, convergent and transform, are also recognised. For each of the four broad classes of plate settings (intraplate, divergent, convergent and transform), Dickinson recognises four subclasses based on the nature of substratum.

Dickinson's model provided a basis for classification of sedimentary basins from the newer perspective of plate tectonics, but the real situation is even more complex than that stated above. Because of the movements involved over long spans of time, anomalous crusts, inherited from the previous plate junctures, may be incorporated into the substratum of intraplate basins. Furthermore, substrata of both continental and oceanic types may be present in the basins located on the zones of divergent, convergent or transform plate boundaries. For these reasons a direct correlation between the sedimentary basins classified according to the geosynclinal and plate tectonic models is not possible. As Reading (1982) very aptly commented: 'The old geosynclinal terminology can now usually be abandoned provided it is realised that many sedimentary basins fall into more than one cate-

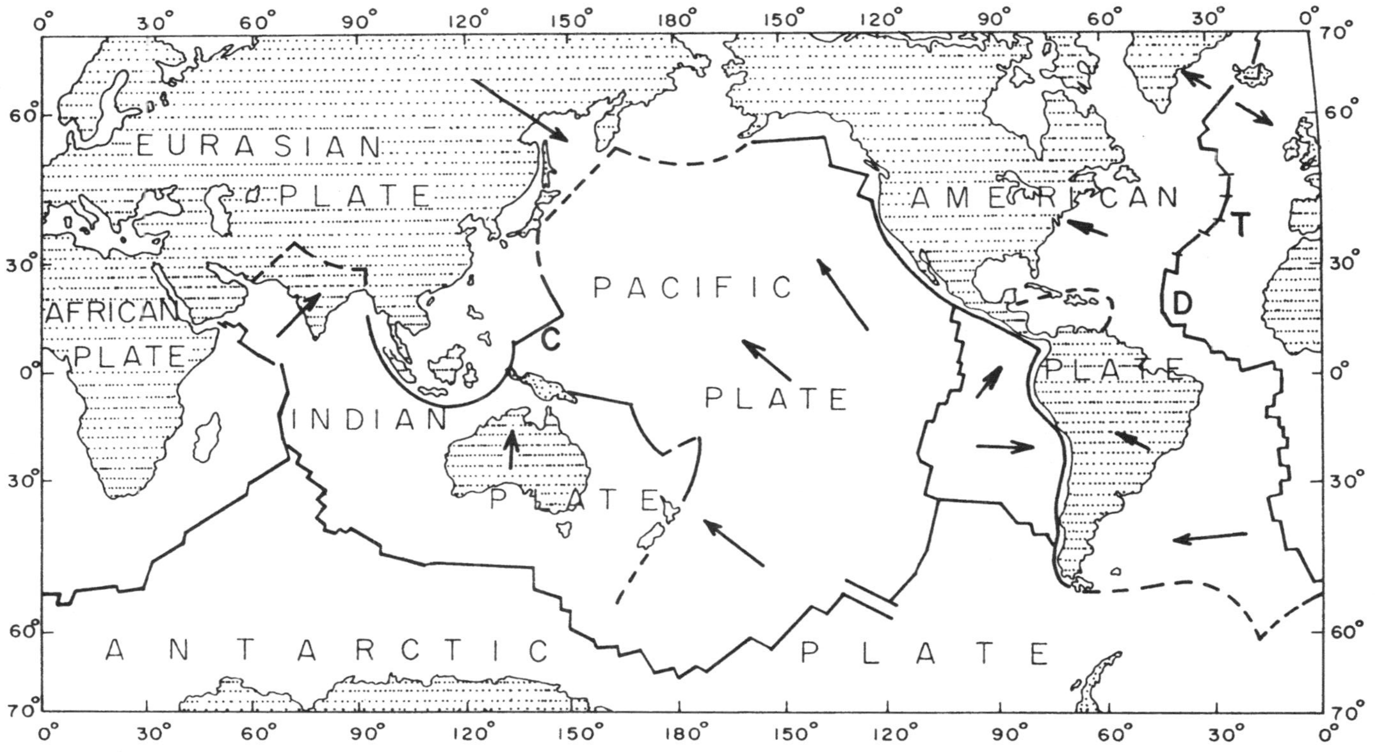

Fig. 7.4: Map showing the major plate boundaries of the world. D—diverging boundary, C—converging boundary, T—transcurrent faults.

gory and that interpretations of the ancient are never more than working hypotheses'.

PLATE MOVEMENTS AND BASIN FORMATION

The mechanism of origin and evolution of large sedimentary basins can now be discussed in terms of diverging (extensional), converging (compressional) and strike-slip plate movements. Some of the current models of basin development under these situations are briefly presented below. The readers may refer to Reading (1986) and Shannon and Naylor (1989) for more detailed discussions. The models discussed here are based largely on theoretical studies and partly on actual geological-geophysical observations on basin architecture, lithic fill and lithospheric conditions. They should be taken only as tentative suggestions. Not one is fully proven.

Basins in Extensional Setting

The simplest type of extensional model envisages upwelling of the asthenosphere during passage of the continents over hot spots leading to doming of the lithosphere. Subsequent cooling of the crust and erosion of the dome leads to generation of sedimentary basins (Fig. 7.5). In another model thinning of the continental crust by rapid stretching of the lithosphere is contemplated to initiate subsidence and thermal anomaly. Slow cooling, following withdrawal of stretching, leads to thermal contraction and subsidence (McKenzie 1978; Jarvis and McKenzie 1980). Seismic evidence of thinning of the crustal layer below the North Sea provides support for this model. The phases of crustal subsidence and marine onlap worked out in the North Sea on the basis of this model match actual observations. At a later stage, isostatic adjustments may introduce complications to the basins developed in this way.

More complex, fault-controlled models for basin development have also been postulated. These models are based on detection in the Rocky Mountains and the Colorado Plateau of major zones of shear traversing the whole lithosphere (Fig. 7.6). These include the uniform simple shear, non-uniform pure shear and mixed shear models. A shear traversing the lithosphere offsets the regions of maximum thinning of the crust and the mantle. The basins developed on such a shear are naturally asymmetric in shape. Moreover, complications may arise due to differential response of the upper brittle crust and the lower ductile mantle to shear movements (Wernicke 1981).

The initial site of rifting is controlled to some extent by the composition of the crust and the mantle. The vertical strength profile of the lithosphere is influenced by these compositions. For a dry, granulitic lithosphere, the strength profile increases downwards, through the crust, only to decrease

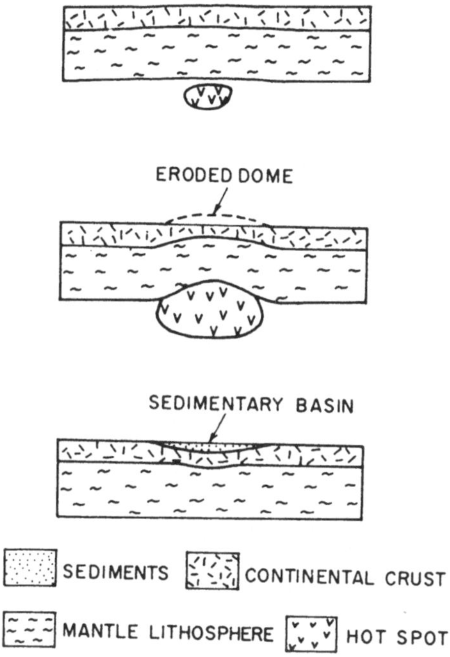

Fig. 7.5: Mechanism of development of sedimentary basins in response to thermal doming (after Shannon and Naylor 1989, with permission of Graham & Trotman).

sharply at the ductile mantle. This makes it difficult for the upper brittle section to stretch. The presence of water pushes up the ductile level, thereby making stretching of the lithosphere easy.

The processes responsible for formation of grabens within the continental crust and also for continental rupture creating new sea-floor are similar in many ways. Both are associated with tensile faulting, crustal thinning, warming of the mantle below the rift zone and updoming of the outer flank of rifts. Thermally controlled subsidence occurs as the anomalously warm mantle below the rift cools. Lateral heat flow may cause transient uplift of the graben flanks (Jarvis 1984). The uplifted basin flanks, in turn, provide sediments to the basin.

In another model of basin development by extensional process, three-armed, radial rift systems ('triple junctions') are contemplated to be produced by convective currents originating in the mantle. Large ocean basins are believed to be produced by spreading of the two active rift arms, while re-entrance of the third arm into continental platforms initiates rift basins. These abandoned rift basins have been termed 'aulacogens' (meaning 'born as furrows') by Russian geologists. The tectonic movements in aulacogens

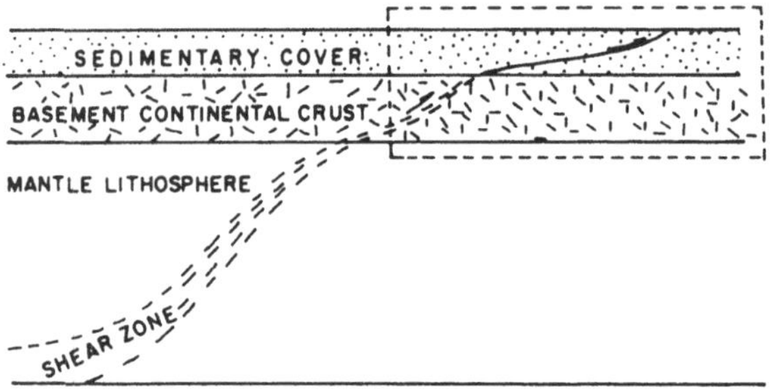

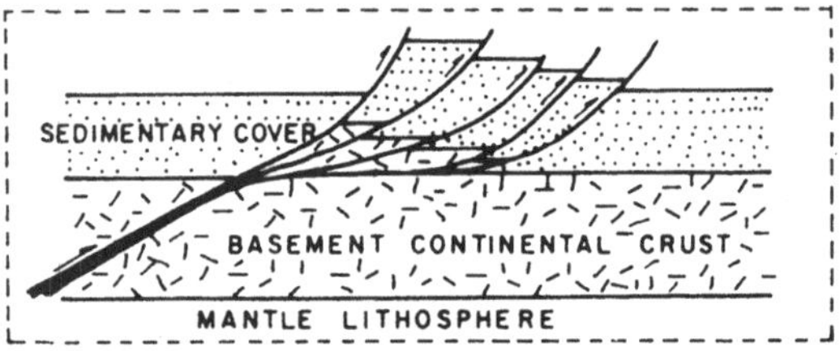

Fig. 7.6: Diagrams illustrating the mechanism of development of 'thin-skinned thrusting' in areas of compression (after Shannon and Naylor 1989, with permission of Graham & Trotman).

are vertical and the sediment transport direction is mainly longitudinal, along the basin axis. Alkalic (basalt-rhyolite) rather than tholeiitic volcanic activity is another characteristic of the aulacogens (Hoffman et al. 1974). The Benue Trough is generally cited as a classical example of a rift arm abandoned during the separation of Africa and South America during the Cretaceous (Fig. 7.7).

Basins in Compressional Setting

Seismically active subduction zones mark the converging plate margins. On the landward side of these subduction zones, which dip towards the continents, lie volcanic island arcs. On the seaward side occur deep-sea trenches. Orogenic belts are produced out of collision of the lithospheric plates. Intrusion of the subducted material into the lower crust causes doming of the igneous and metamorphic terranes at the zone of convergence, into what are called 'arc massifs'. These elevated areas at the basin margins serve as the source of sediments while the sediments themselves are deposited in one of the following five areas: within the ocean, on the downgoing oceanic plate, within the deep-sea trench, in small basins on the slope

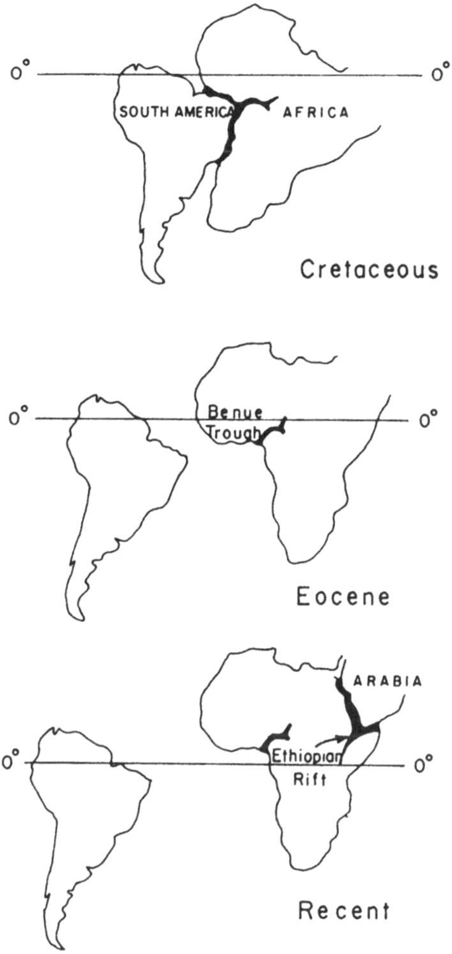

Fig. 7.7: Maps showing stages of origin of Benue trough as a rift arm abandoned during separation of Africa from South America (after Hoffman et al. 1974, fig. 9, with permission of SEPM, Society for Sedimentary Geology).

of the inner wall of the trench, seaward of the volcanic arc (outer-arc or fore-arc basin), and behind the volcanic arc (back-arc basins).

Thick sedimentary piles are stacked in the form of accretionary prisms in the forearcs. As the subduction zone migrates towards the sea the forearc sediments are successively thrust upwards (Fig. 7.8). Small sedimentary basins may develop over these accretionary prisms. Those lying between the volcanic arc and the trench slope break are generally very wide (50–100 km). These are the forearc or outer-arc basins. Behind the volcanic arc occur the back-arc basins.

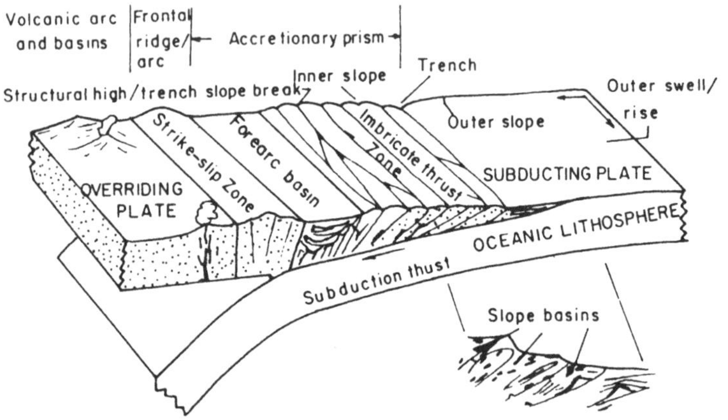

Fig. 7.8: Model of an obliquely subducting margin (compiled by Reading after Walcott 1978, and Lewis 1980, reprinted from Reading 1982, with permission).

The sediments deposited within the accretionary prisms undergo intense tectonic deformation during movements of the subduction zones. Seismic profiles run across many of these deformed areas, such as the Rocky Mountain Province, showed that most of the structural complexities are restricted to the uppermost part of the crust (approximately the upper 10 km) where it is most brittle. These complex structures are replaced underneath by only a few simpler, poorly defined thrusts going down almost to the base of the lithosphere. This phenomenon has been termed 'thin-skinned compressional tectonics'.

Basins in Strike-slip and Transform Fault Zones

Strike-slip faults may occur either at the major plate boundaries or at the border of microplates. Transform faults occur as fracture zones running across the midoceanic ridges. The overall movement in a strike-slip fault is essentially horizontal and parallel to the fault trace but the dip-slip component may also be important locally. This oblique movement causes the strike-slip motion to be either divergent (transtensile) or convergent (transpressive). The former is likely to produce normal faulting, leading to basin development and volcanism. The latter produces reverse faults, thrusts, uplift and folding. With the passage of time, a divergent strike-slip fault may be converted to a convergent one.

Curving, splitting and offsetting of strike-slip fault trends may lead to local zones of extension and compression. The latter produces sedimentary basins while the former leads to uplifted areas which eventually serve as the source of sediments (Reading 1982; also see Fig. 7.9).

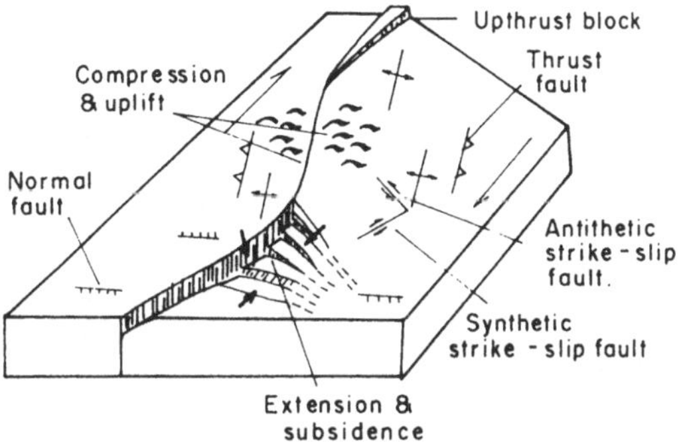

Fig. 7.9: Diagram illustrating the mechanism of basin development from dextral strike-slip fault. Both extensional basins and compressional uplifted areas serving as source of sediments may be produced as a result of such faulting. Direction of sediment transport is indicated by thick arrows (after Wilcox et al. 1973, modified by Mitchell and Reading 1978, reprinted from Reading 1982, with permission).

TECTONIC CONTROL OF SANDSTONE COMPOSITION

The technique of interpretation of tectonics from sandstone mineralogy and chemistry is based on the belief that the detrital framework mineralogy of a sandstone is dependent largely on the tectonic setting of its provenance. This technique of interpretation should be used with caution, however, because the mineralogy of a sandstone may be considerably modified by sorting processes during transportation. Moreover, diagenetic processes may also obliterate the original detrital mineralogy and chemistry of a sandstone.

Quartzose sands from continental cratons are widespread within interior basins, platform successions and opening ocean basins. Arkosic sands from uplifted basement blocks are present in rift troughs and wrench basins related to transform faults. Volcaniclastic lithic sands and volcano-plutonic sands derived from magmatic arcs are found in trenches, forearc basins and marginal seas. Recycled orogenic sands, rich in quartz or chert and other lithic fragments, occur in closing ocean basins, diverse successor basins and foreland basins. These sands are derived from subduction complexes, collision orogens and foreland uplifts. Using triangular QFL and QmFLt plots of quartz (mono- and polycrystalline varieties), feldspar (both plagioclase and K-feldspar) and lithic fragments (volcanic/metavolcanic and sedimentary/metasedimentary), Dickinson and Suczek (1979) could discriminate these tectonic settings (Fig. 7.10).

Bhatia (1983) could discriminate the Palaeozoic sandstones from various tectonic settings of eastern Australia with the help of their major element geochemistry. Plots of Fe_2O_3 + MgO versus TiO_2, Al_2O_3/SiO_2, K_2O/Na_2O,

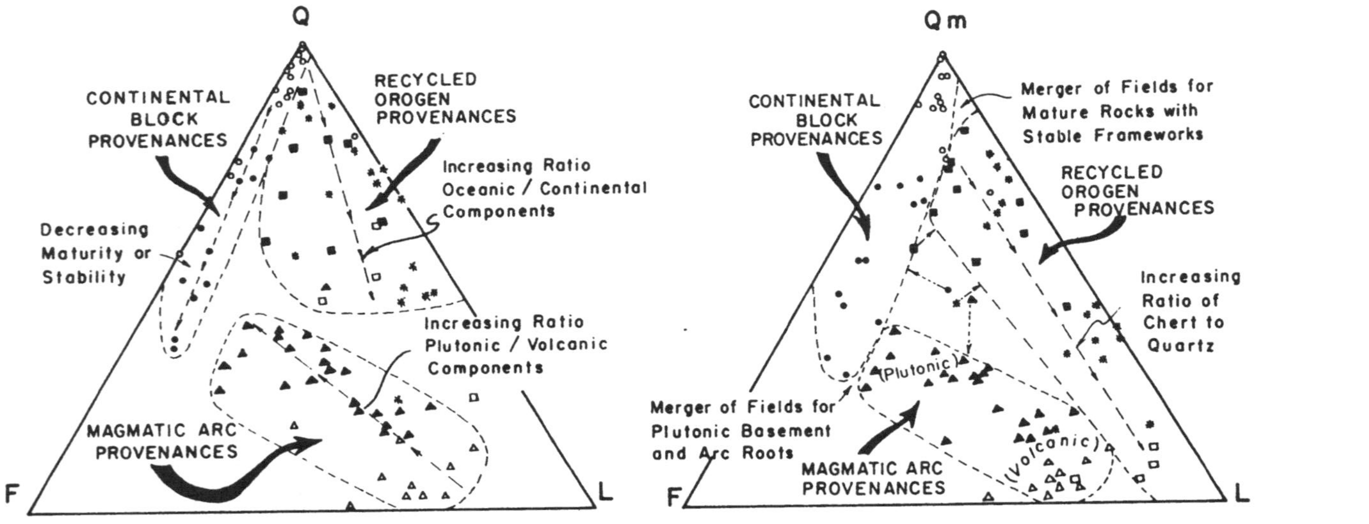

Fig. 7.10: QFL and QmFLt plots showing derivation of sandstone suits from different tectonic provenances (after Dickinson and Suczek 1979, reprinted with permission of American Association of Petroleum Geologists). Q = total quartz grains, Qm = monocrystalline quartz grains, F = total feldspar grains, L = total unstable lithic fragments, all polycrystalline.

and $Al_2O_3/(CaO + Na_2O)$ were utilised for this purpose. The oceanic island arc sandstones, derived from calc alkaline andesites, are characterised by higher proportions of TiO_2, Al_2O_3, Na_2O and Fe_2O_3, and a relatively lower proportion of SiO_2 and K_2O than all other sandstones. The continental island arc sandstones, derived from felsic, volcanic rocks, are characterised by higher proportions of SiO_2, K_2O, and the K_2O/Na_2O ratio ($\simeq 0.6$) and lower $Fe_2O_3 + MgO$ than the oceanic island arc sandstones. The sandstones from the active continental margins (Andean-type and strike-slip basins) are derived mainly from the uplifted basement. These sandstones, reflecting the composition of the upper continental crust, have high SiO_2 and K_2O content with $K_2O/Na_2O \approx 1$. The passive margin sandstones are generally enriched in SiO_2 and depleted in Al_2O_3, TiO_2, Na_2O and CaO. The K_2O/Na_2O ratio of these sandstones is more than 1, but these sandstones may show large variations in composition, sometimes even overlapping the compositions of active continental margin sandstones (Table 7.2).

In another study the framework mineralogy and volatile-free chemistry of graywackes were related by Crook (1974) to specific tectonic settings. Using data from deep-sea sands, taken to be the modern counterparts of arenites in ancient flysch, and sands from rivers debouching into the sea, Crook demonstrated that the quartz-poor graywackes (quartz $< 15\%$, average $SiO_2 = 58\%$, $K_2O/Na_2O \ll 1$) are indicative of magmatic island arcs. The average composition of these rocks approximate the average composition of these arcs and that of tholeiitic andesites. Quartz intermediate graywackes

Table 7.2. Abundances and proportions of various oxides in sands of various tectonic settings (after Bhatia 1983)

	Oceanic island arc (OIA)	Continental island arc (CIA)	Active continental margin (ACM)	Passive margin (PM)
$Fe_2O_3^{(t)} + MgO$	8–14 (11.73)	5–8 (6.79)	2-5 (4.63)	(2.89)*
TiO_2	0.8–1.4 (1.06)	0.5–0.7 (0.64)	0.25–0.45 (0.46)	Depleted
Al_2O_3/SiO_2	0.24–0.33 (0.29)	0.15–0.22 (0.20)	(0.18)	(0.10)
K_2O/Na_2O	0.2–0.4 (0.39)	0.4–0.8 (0.61)	(0.99)	(1.6)
$Al_2O_3/CaO + Na_2O$	(1.72)	(2.42)	(2.56)	(4.15)
SiO_2	(58.83)	(70.69)	(73.86)	(81.95) Enriched

*figures in parentheses indicate the average values.

(quartz = 15–65%, average SiO_2 = 68–74%, $K_2O/Na_2O < 1$) are indicative of Andean-type continental margins and have approximately the same composition as the upper level of the continental crust. Quartz-rich graywackes (quartz > 65%, average SiO_2 = 89%, $K_2O/Na_2O > 1$) are characteristic of Atlantic-type continental margins and are similar in composition to the sand fraction of the continental platform cover.

Quartz content (19% average) and K_2O/Na_2O ratio (< 1) of representative graywacke samples from the Chitradurga Schist Belt (> 2600 m.y.) of India (Tables 7.3 and 7.4) indicate that these belong to the quartz-intermediate type of graywacke of Crook (1974) and were, in all probability, derived from the upper level of the continental crust (Srinivasan et al. 1989).

The graywackes derived from various tectonic settings can also be distinguished on the basis of their immobile trace element content. From oceanic island arc through continental island arc, active continental margin to passive margin settings in eastern Australia, a systematic increase in the light rare earth elements (La, Ce, Nd), Th, Nb and decrease in V and Sc were noted. The ratios of Ba/Sr, Rb/Sr, La/Y, and Ni/Co increase while the ratios of Ba/Rb, K/Th, and K/U decrease in the same direction (Bhatia and Crook 1986). Using a similar approach Srinivasan et al. (1989) suggested an island arc setting and a mixed felsic and mafic source (as indicated by the high proportion of Zr as well as Cr and Ni; see Table 7.5) for the

Table 7.3. Modal composition of the Keshavapura graywackes from Chitradurga Schist Belt (KD1 samples after Srinivasan et al. 1989, table I)

	3	3A	4	5	6	7	M (Avg)
Quartz 'Q' as clusters and polycrystalline augens	18.2	12.9	20.7	25.9	16.7	19.3	18.9
Polycrystalline detrital grains	5.9	10.9	15.7	5.5	7.3	9.9	9.2
Monocrystalline grains	5.2	1.6	-	2.1	1.4	2.0	4.5
Plagioclase (P)	6.1	14.3	4.7	4.3	10.9	8.2	8.0
K-Feldspar (F)	4.7	1.9	1.6	1.2	1.0	1.9	2.0
Rock fragments, volcanics + glass (R)	2.1	2.7	3.4	1.0	-	-	1.5
Biotite	30.9	6.5	9.0	47.8	36.9	44.1	29.2
Chlorite	25.0	8.9	5.5	5.0	9.9	6.3	6.1
Sericite/Muscovite	-	36.6	34.8	-	-	-	11.9
Calcite	-	-	1.6	3.3	13.6	6.0	4.1
Opaque minerals	1.5	2.9	1.9	3.0	1.1	1.0	1.9

Table 7.4. Chemical composition of the Keshavapura graywackes from Chitradurga Schist Belt (KD1 samples after Srinivasan et al. 1989, table IIA)

Wt %	3	3A	4	5	6	7	8
SiO_2	61.05	58.05	61.16	67.66	66.70	64.20	58.80
TiO_2	0.64	0.68	0.64	0.46	0.46	0.57	0.80
Al_2O_3	14.10	14.33	14.13	13.93	14.05	14.62	14.63
FeO	6.44	6.48	5.20	3.36	2.92	4.20	7.12
Fe_2O_3	3.01	4.02	3.93	3.94	3.32	3.83	4.04
CaO	4.70	4.81	4.71	3.72	4.24	2.57	1.94
MgO	6.05	6.25	4.55	1.72	1.42	2.42	4.38
Na_2O	3.05	3.30	3.79	2.70	3.66	3.43	3.21
K_2O	1.50	1.15	1.09	1.75	2.00	2.41	3.04
MnO	0.08	0.10	0.06	0.03	0.03	0.04	0.08
P_2O_5	0.13	0.13	0.12	0.08	0.08	0.09	0.10
LOI	0.52	1.19	0.66	0.53	0.71	0.71	0.70

Keshavapura graywackes of the Chitradurga Schist Belt of Karnataka, India. Discrimination between oceanic and continental island arcs was, however, not possible.

BASIN CLASSIFICATION AND DESCRIPTION

The concept of plate tectonics prompted classification of sedimentary basins in terms of plate movements (see, for example, Miall 1984; de V. Klein 1987). The comprehensive scheme of classification of Shannon and Naylor (1989) reproduced in Table 7.6 relates sedimentary basins to crust type, nature of plate movement, dominant structural movement and location of the basin with respect to the plates. The characters of some of the major basin types are discussed below with reference to this scheme of classification, drawing examples from India (Fig. 7.11) and elsewhere. Many of these groupings, based on scant information, are necessarily tentative. Many of the older sedimentary basins are polygenetic because they passed through several stages of development at different times of their existence.

Downwarp Basins

These large, elongate basins develop by downwarping (sagging) of the cratonic shelves bordering oceans. In these basins sheets of stable shelf quartz arenites and carbonates overlie deeper water argillaceous sediments. The West Canada Basin of North America, the Saudi Arabian Platform of the Middle East, and the Vindhyan Basin of central India are some of the typical examples of downwarp basins.

The Canadian Shield slopes smoothly towards the west until it reaches the eastern boundary of the Rocky Mountain thrust belt. This downwarping,

Table 7.5. Trace and rare earth element composition of Keshavapura graywackes from Chitradurga Schist Belt (KD1 samples after Srinivasan et al. 1989, table IIB)

(ppm)	3	3A	4	5	6	7	8
Zr	141	153	172	221	228	217	188
Y	32	24	27	44	42	45	44
Sr	395	382	467	687	466	378	288
Rb	53	40	41	65	70	76	82
Ba	406	293	327	389	461	448	599
V	140	147	121	72	76	106	155
Cr	423	414	280	114	122	161	193
Ni	69	80	50	34	24	24	56
Cu	47	68	48	55	67	51	51
Zn	105	106	102	91	84	103	108
La	-	-	23.40	-	52.14	-	33.65
Ce	-	-	51.18	-	81.77	-	67.34
Pr	-	-	5.8	-	8.17	-	6.95
Nd	-	-	21.62	-	30.83	-	26.42
Sm	-	-	4.16	-	4.79	-	4.84
Eu	-	-	0.76	-	0.78	-	0.76
Gd	-	-	4.72	-	5.11	-	4.56
Tb	-	-	0.66	-	0.65	-	0.64
Dy	-	-	2.94	-	3.12	-	3.31
Er	-	-	1.70	-	1.54	-	1.27
Tm	-	-	<0.03	-	0.30	-	0.24
Yb	-	-	1.12	-	1.40	-	1.03
Lu	-	-	0.22	-	0.30	-	0.24
K/Rb	232	235	217	220	234	259	304
Rb/Sr	0.13	0.10	0.08	0.09	0.15	0.20	0.28
Ba/Sr	7.66	7.32	7.97	5.98	6.58	5.89	7.30
Zr/Y	4.4	6.37	6.37	5.02	5.42	4.82	4.27
V/Ni	2.02	1.83	2.42	2.10	3.16	4.40	2.76
Cr/Ni	6.13	5.17	5.60	3.30	5.08	6.70	3.44

which took place in the Late Proterozoic, accommodated sediments ranging in age from the Cambrian to the Lower Jurassic. The deposits, which wedge out on the Canadian Shield, attain a maximum thickness of about 6 km on the western side bordering the Rockies. Marine and nearshore sediments, ranging in age from the Cambrian to the Lower Jurassic, were initially deposited on this downwarped cratonic shelf. These were followed by the deposition of thick marine and non-marine sediments during the Middle Jurassic to the Lower Tertiary on a foreland basin which developed east of the orogenic belt. The whole sequence, from the Cambrian to the Lower Tertiary, is punctuated by four major unconformities caused by uplifts related to westward movement of the North American plate. The sedimentary facies and structural elements of the West Canada Basin were controlled by these tectonic pulses (Shannon and Naylor 1989).

Table 7.6. Basin classification based upon genesis and tectonic location (after Shannon and Naylor 1989, with examples of basins inserted)

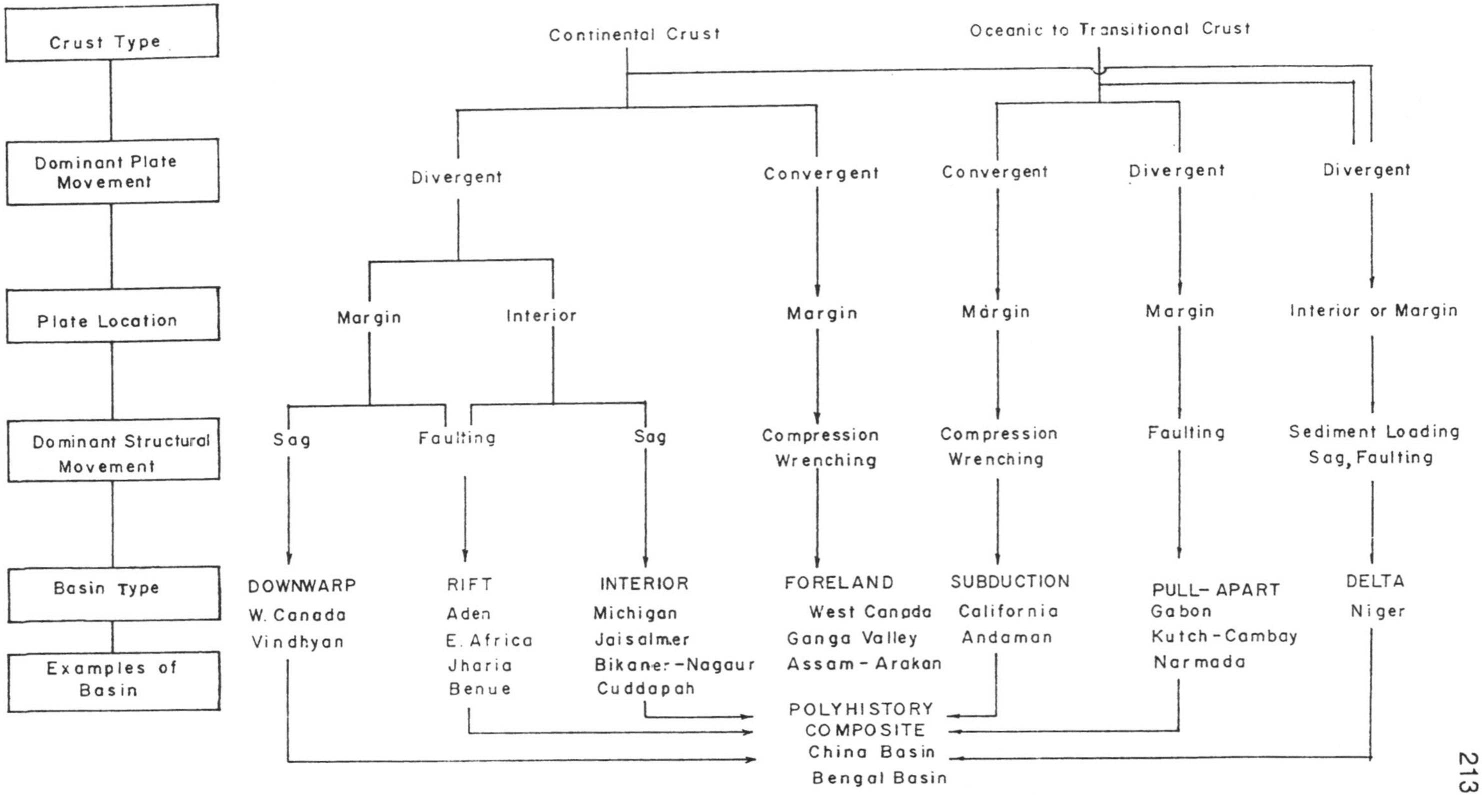

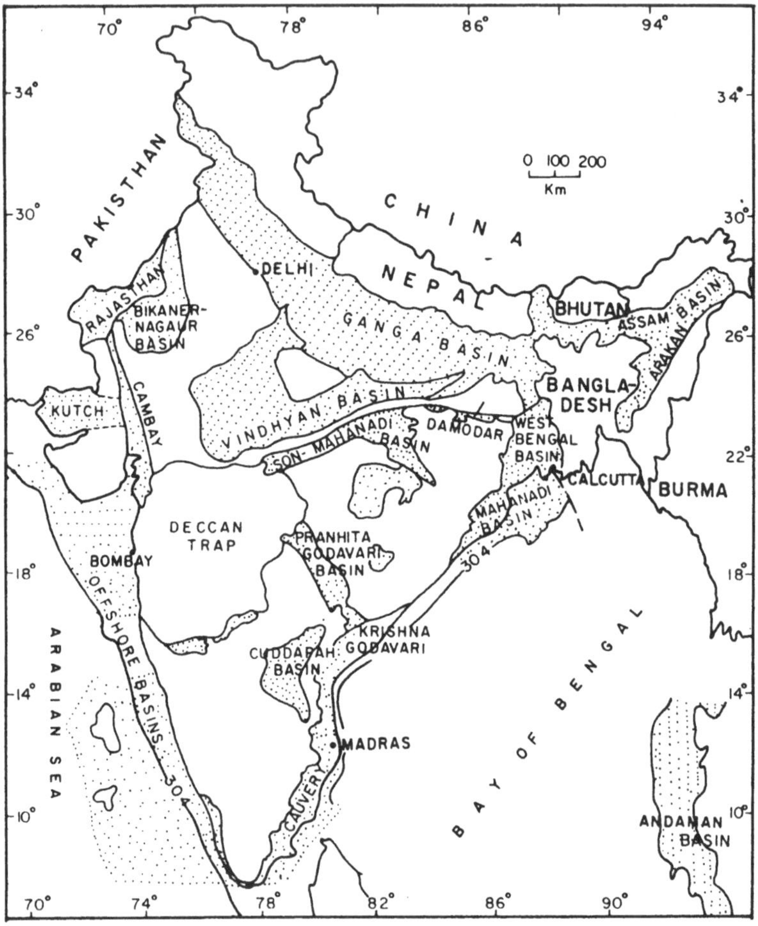

Fig. 7.11: Sedimentary basins of India (redrawn from Berger et al. 1983, with permission of Schlumberger Technical Services, Inc.).

The Vindhyan basin of central India, which is morphologically similar to the West Canada Basin, apparently started as a downwarp on the northern fringe of the Indian Shield. Like the Canadian basin the floor of the Vindhyan basin slopes smoothly to the north until it abuts against the thrust belt of the Outer Himalayas (Fig. 7.12). The exposed part of the Vindhyan basin, which occupies an area of more than 120,000 sq km is composed of 5250 m of Precambrian (ranging to early Cambrian) sediments in the thickest part. Another 1000 m of sediments are believed to be concealed below an area of about 40,000 sq km in the Ganga valley (Fig. 7.12). The crustal warping leading to the formation of this basin followed the Proterozoic Delhi orogeny.

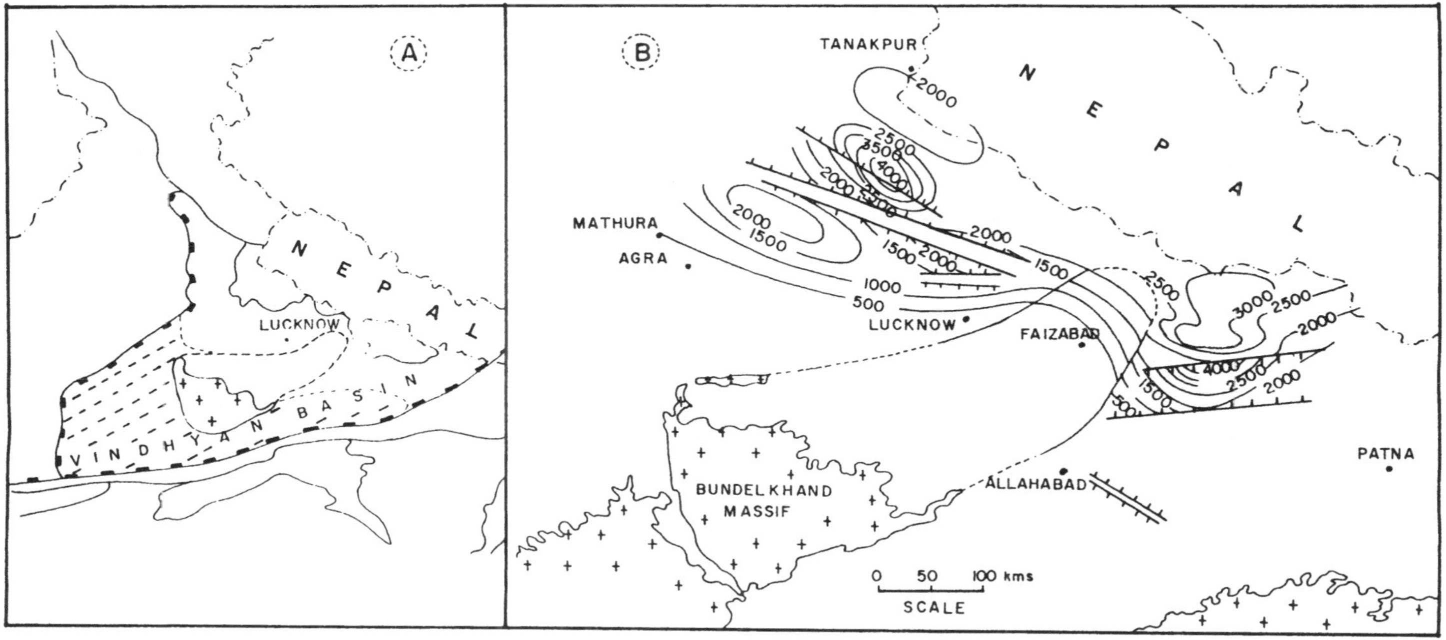

Fig. 7.12: (A) Exposed areas of the Vindhyan Basin. (B) Isopach map of the Vindhyan deposits in the subsurface of the Ganga plains based on seismic-cum-geological information. Contours indicate thickness of the Vindhyan sediments in metres (generalised and redrawn after Narain and Kaila 1982, fig. 3, with permission).

Following encroachment of the sea from the north, the Vindhyan basin enjoyed beach, shallow marine, barrier bar, tidal flat and, locally, lagoonal environments at various times. An essentially quartz arenite-carbonate type of sediment (with minor amounts of shale, chert and pyroclastics) is found in the lower part of the sequence. In the upper part a much larger proportion of shale occurs, together with some limestone and quartz wackes. The lower sequence, which is separated from the upper one by an unconformity, is structurally more deformed than the upper one.

An Archaean granitic massif (the Bundelkhand granite) separates the exposed part of the Vindhyan basin into the eastern and the western sub-basins (Fig. 7.12). The Bundelkhand massif extends to the north, below the Gangetic alluvium, almost up to the Himalayan foothills. Geophysical surveys have indicated the presence of a number of basement faults subdividing the Ganga basin into several horsts and grabens (Narain and Kaila 1982). The major palaeocurrent pattern was northerly or northwesterly during most of the time. The sediments in the eastern part of the basin were supplied by the older Bijawar rocks bordering the southern fringe of the basin (Banerjee and Sengupta 1963; Banerjee 1974; Srivastava et al. 1983).

Following collision of the Indian and the Eurasian plates, the present-day Ganga valley was converted into a foreland depression and the northern part of the Vindhyan sediments was covered by predominantly non-marine, Upper Miocene to Pleistocene Siwalik sediments derived from the newly developed Himalayan orogenic belt.

Rift Basins

The Red Sea and the Gulf of Aden are examples of rift basins formed during early stages of sea-floor spreading. The cause of rifting in these areas was lithospheric extension accompanying thermal doming. The rifted blocks were faulted and tilted landwards as the Red Sea basin began to form during the Tertiary. Basalt erupting through the fissures as a result of mantle upwelling floored the Red Sea and the Danakil depressions. About 3000 m of volcanics and volcaniclastic rocks have accumulated in these depressions. These pass laterally into marine pyroclastics and evaporites (Reading 1982; see Fig. 7.13).

The East African rift system, the most important of all the continental rift systems, extends for a length of about 3000 m and is about 40–50 km wide (Fig. 7.14). The rift, flanked by steep escarpments, is often broken and off-set by cross-faults, volcanoes and horst blocks rising well above the plateau. The drainage flows mostly away from the rift valley, keeping the valley starved of sediments. Only small quantities of clastic sediments, derived from the fault scarps and uplifted blocks, are transported by streams along the rift. Alluvial fans and lakes constitute the main areas of sedimentation. A number of lakes (e.g., Lake Albert, Lake Tanganyika) occur along the rift.

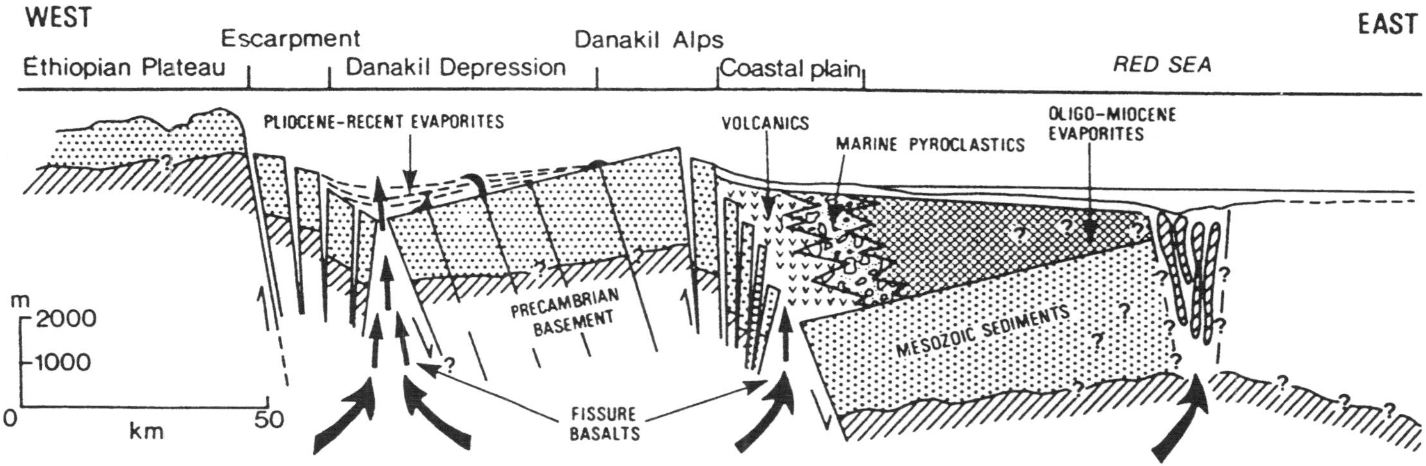

Fig. 7.13: Section across the Red Sea and Danakil depression (modified by Reading after Hutchinson and Engels 1970, reprinted from Reading 1982 with permission).

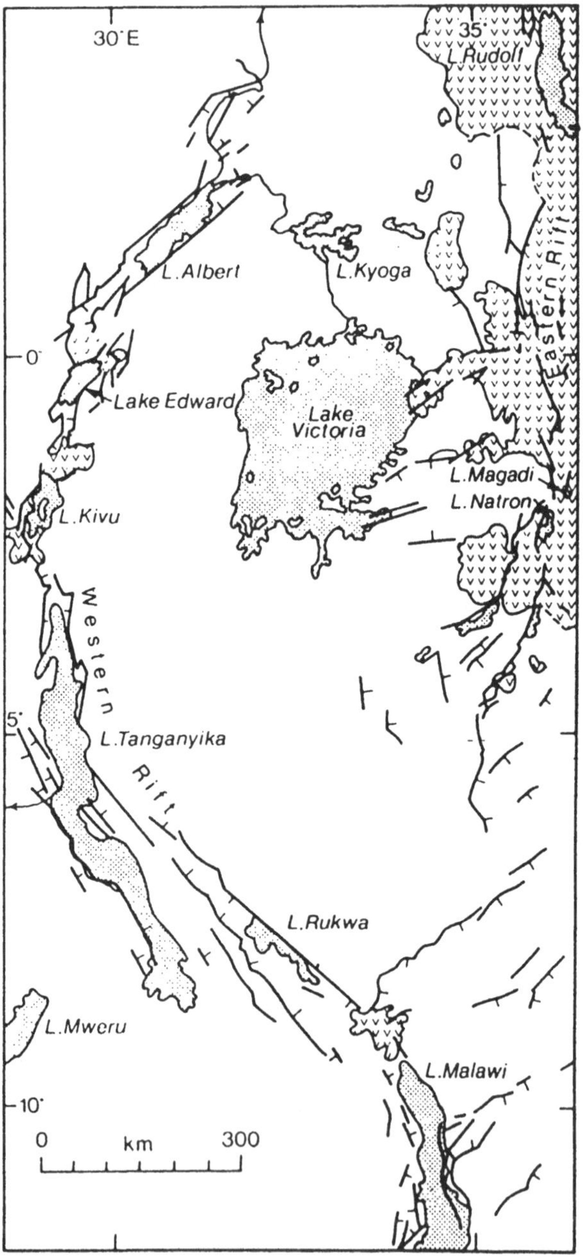

Fig. 7.14: Map of the East African Rift system (modified by Reading after King 1970, reprinted from Reading 1982 with permission).

These are mostly freshwater lakes but alkaline-saline lakes, whose chemistry is controlled by the nearby carbonatite volcanics, also occur along the rift (Reading 1986).

The Permo-Triassic Gondwana basins occupying three well-defined belts in the eastern part of the Indian Shield may be cited as Indian examples of intracontinental rift basins. The most well-studied of these Gondwana basins, the Jharia basin of Bihar in eastern India, was initiated by a general downwarping of the basin floor. At a later phase, the basin was intersected by a large number of parallel faults reaching the basement as a result of crustal tension. The smaller subbasins later merged into a single major depression, which subsided as a narrow intrabasinal graben bounded by contemporaneous normal faults. Sinking of the inselbergs along with rising of a basement ridge, reactivation of old faults and development of new post-depositional faults caused the basement to rise at either ends of the basin, thereby providing a source of sediments. Faulting, caused by crustal tension, continued until the deposition of the youngest sediments within the basin (Fig. 7.15).

Sedimentation within this fault-bounded graben took place by fluvial and fluvio-lacustrine processes. At a later stage the fluviatile clastics were intruded by dykes and sills of dolerites and lamprophyres. Emplacement of the lamprophyres, which are restricted to the basinal areas only, post-dates deposition of the major part of the Gondwana sediments. The tensional cracks, active since the early phase of basin formation, reached the mantle at an advanced stage of basin development, causing release of pressure, lowering of temperature and consequent intrusion of lamprophyre dykes in zones of maximum subsidence (Ghosh and Mukhopadhyay 1985).

Failed Rifts and Aulacogens: The 1000 km long and 100 km wide Benue Trough in the Gulf of Guinea in West Africa is generally cited as the best example of an aulacogen formed during drifting of Africa and South America. This trough is filled by more than 5 km of fluvial, deltaic and marine Cretaceous sediments. These are overlain by another sequence of fluvial, deltaic and submarine fan deposits extending into the Atlantic Ocean (Fig. 7.16). The Cretaceous aulacogen phase was separated from the post-Cretaceous sediments by a phase of folding, followed by erosion (Reading 1986).

Interior Basins

Interior basins occupy large circular or elliptical areas within cratons. They are generally symmetrical in profile. Their lithic fill includes clastics and carbonates.

The Michigan Basin, located in the east central part of the United States, is an example of an interior basin. This basin was initiated by thermal events. Deep crustal metamorphism leading to phase transformation (gabbro to eclogite) might have led to subsidence. This was followed by thermal sag

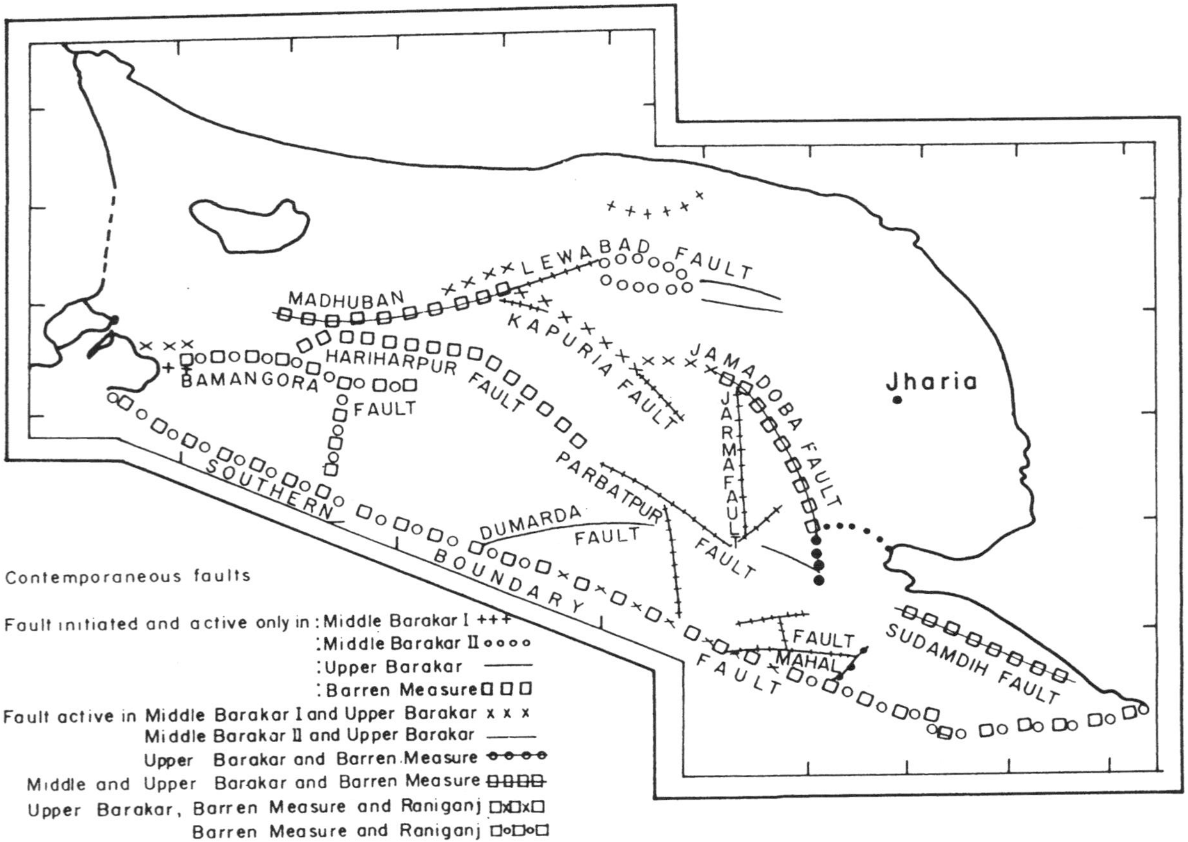

Fig. 7.15: Contemporaneous faults of the Jharia Basin, Bihar, India (after Ghosh and Mukhopadhyay 1985, reprinted with permission).

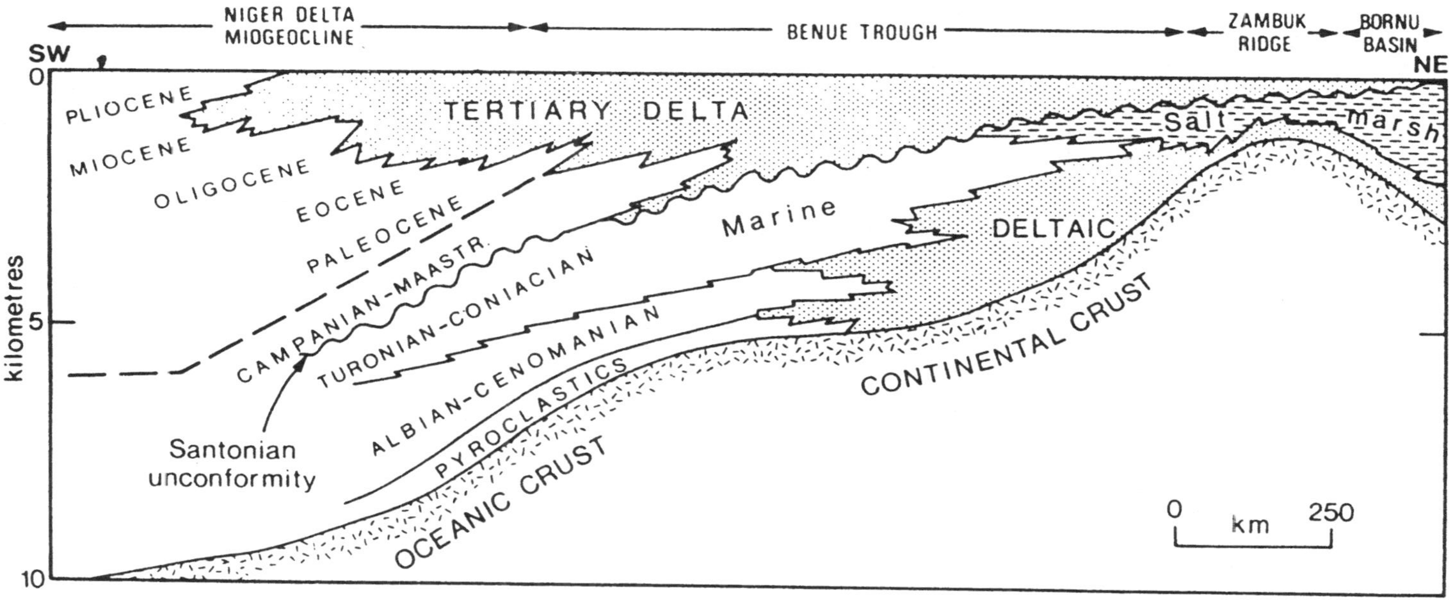

Fig. 7.16: Cross-section through Benue 'failed rift' (aulacogen) and the Tertiary Niger delta (modified by Reading after Burke et al. 1972; Petters 1978, reprinted from Reading 1982 with permission).

throughout the Lower Palaeozoic in a setting of low geothermal gradient. The sagging was effected by periodic movements along basement faults throughout the Palaeozoic. The basin contains clastic and non-clastic sediments ranging in age from Precambrian to Jurassic (Shannon and Naylor 1989).

The Jaisalmer and Bikaner-Nagaur basins of Rajasthan, India are interior basins. Bikaner-Nagaur, which was initially a stable shelf up to the Early Palaeozoic, later developed into a restricted ('silled') basin to accommodate about 2 km of Cambrian-Triassic sediments (Berger et al. 1983; see Fig. 7.11).

The Cuddapah Basin occupying a crescentic area of about 42,000 sq km in the south-eastern part of the Indian Shield is another example of an interior basin. A maximum thickness of about 10 km of clastic and non-clastic Proterozoic sediments is estimated to have been deposited in this basin. The total load of accumulated sediments and emplaced igneous material in this basin cannot account for a subsidence of 10 km. The main driving force for generation of this basin seems to be thermal heating and cooling, followed by isostatic loading. Three major unconformities within the sedimentary sequence of this basin indicate crustal uparching as a result of isostatic adjustment to rising temperature in the lithosphere. Rising temperature not only caused thermal expansion but also reduced the density and thickness of the lithosphere. A maximum crustal thickness of about 38 km below the basin, with a thinning of about 4 km towards the eastern fringe, is suggested by gravity data.

Step faults, tilting and local variations in the depth of the basin-floor revealed by Deep Seismic Sounding (DSS) profiles run across the basin were possibly caused by flexural loading. These deep-seated faults also provided passage for upward migration of mantle fluids (kimberlite-carbonatites) and other partial melt products, namely, calc alkaline, alkaline and felsic magmas. That the role of tensional faulting was diminishing during the later phase of depositional history is shown by dwindling of ingenous activity after Lower Cuddapah sedimentation. Compressional rather than tensional forces were active during the later (and post-depositional) phases of basin evolution. This is indicated by the presence of thrust faults in the eastern margin of the basin (Bhattacharji and Singh 1984).

Foreland Basins

Foreland basins generate at the toe of the overriding plates in the areas of continent-ocean or continent-continent collision. Subsidence in the foreland basin is caused largely by subduction. The lithosphere in front of the overriding plate is depressed also by the weight of the thrust slices. As the basin subsided, a relatively low but wide bulge developed on its cratonic side. The older surface, where not covered by deep water, is eroded in the form of an unconformity.

The western part of the West Canada Basin consisting of a folded and thrust belt on the western margin and a broad undeformed platform on the eastern side is described as a foreland basin (Cant 1989). The basin is elongate parallel to the orogenic belt. The preserved Mesozoic rocks occupy the southern part of this basin while Palaeozoic carbonates occupy the northern part. The basin continues to the south to merge with the Western Interior Basin of the United States. Reverse faults involving the basement divided the foreland basin in the latter area into a series of smaller basins during the Late Cretaceous Laramide orogeny.

The West Canada Basin initially developed as a downwarp to accommodate sediments ranging in age from the Cambrian to the Lower Jurassic. Apparently as a result of the Laramide orogeny, the same basin was converted to a foreland basin and received sediments ranging in age from the Middle Jurassic to the Lower Tertiary. Farther south, in South America, the Laramide basins in the foreland of the Andes resulted from shallow dipping subduction, causing drag at the base of the lithosphere (Cant 1989). A similar picture of basin development can be traced in the Ganga valley, India, where the downwarped Proterozoic Vindhyan Basin in the northern part of the Indian Shield gave way to a foreland basin, following the Himalayan orogeny, to accommodate Siwalik sediments of Miocene-Pleistocene age.

The Assam-Arakan Basin located between two thrust belts in north-eastern India is an example of a foreland basin. The Tertiary sediments of the Upper Assam belong to two distinct depositional facies—a relatively shallow-water shelf facies and a deep-water geosynclinal facies. Between the Himalayan and the Naga-Disang thrust complex occurs an autochthonous zone ('the foreland spur'; Fig. 7.17) containing sediments ranging in age from Eocene to Pleistocene. These are intersected by a number of gravity faults. Farther east, in the Patkai range, Tertiary sedimentation took place under deeper water conditions, punctuated by slight emergence at some places. The sea receded sourthwards towards the end of the Oligocene. Miocene sedimentation took place mostly under freshwater to brackish conditions. Oligo-Miocene was also the time when the deposits of the Assam-Arakan Basin were overthrust towards the north-west, over the north-eastern extension of the Indian Shield. The outermost of these thrusts, the Naga thrust belt, consists of a succession of six thrust sheets. In the thrust strip nearest to the foreland spur there are anticlinal structures, such as the one on which the Digboi oilfield is located (Evans 1932; Berger et al. 1983).

Subduction Basins

The California Basin located at the junction of the North American continental plate and the Pacific oceanic plate provides a well-documented ex-

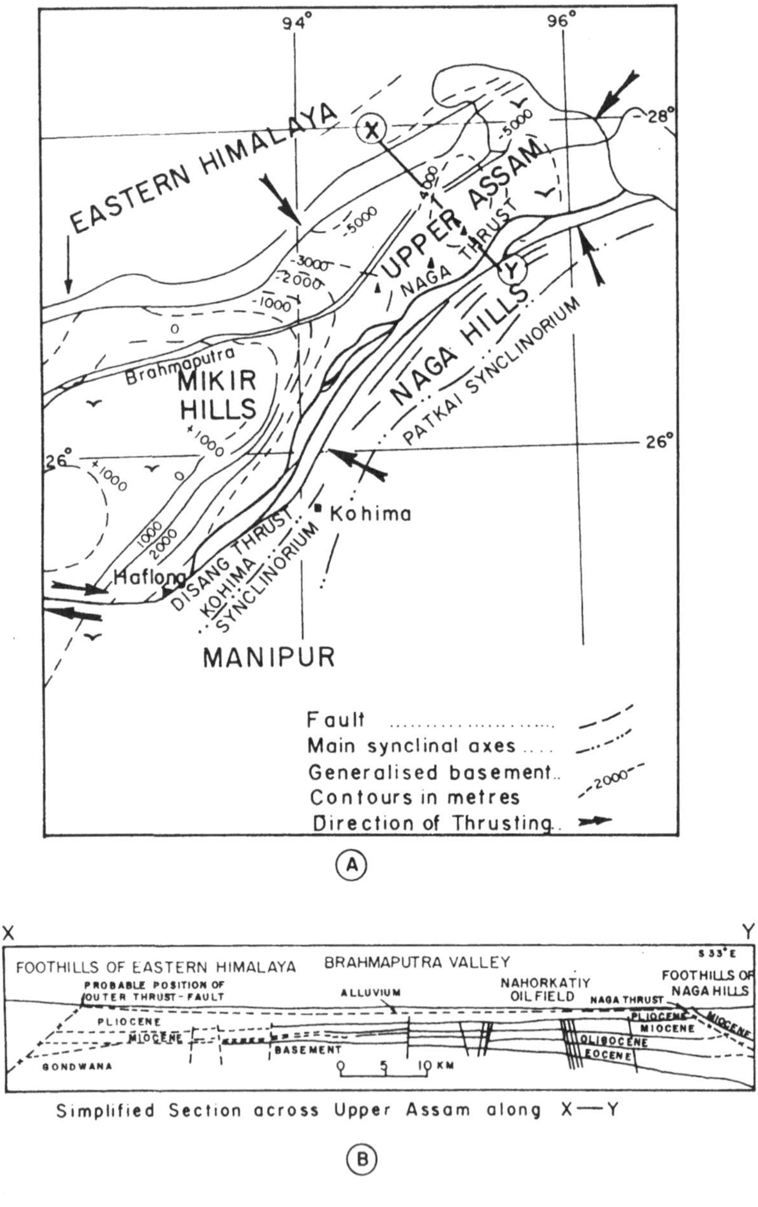

Fig. 7.17: (A) Foreland spur of north-eastern India, (B) NW-SE cross-section across the foreland spur shown in (A). (Redrawn from compilations by Berger et al. 1983, with permission of Schlumberger Technical Services, Inc.).

ample of a subduction basin which experienced compressive as well as wrench movements. Mesozoic sedimentation in California took place in forearc basins related to subduction movements, to be followed by sinistral, strike-slip basin development during the Neogene. The oceanic-continental plate collision and changing direction of plate movement brought complications to the area.

In complexity of character the California Basin is comparable to the deep-sea basin which runs for a length of about 3000 km from the offshore of Sumatra to the mainland of Burma via the western Andaman Sea (Fig. 7.18). The forearc basin off Sumatra contains about 4 km of sediments including deep-water turbidites, montmorillonitic clays and volcanic ash. As the Sumatran continental block subsided with time and the subduction zone migrated southwards, a gradual widening of the forearc basin took place. The basin shape is controlled by a number of strike-slip faults. In the Andaman Sea a central volcanic arc separates two troughs—a forearc basin on the western side and a system of rift valleys on the eastern side. The latter, associated with several strike-slip faults, is the spreading centre of several pull-apart basins (Rodolfo 1969). The total sediment thickness in the eastern Andaman basin is small (1.5 km), indicating that the Irrawaddy delta, the main source of the sediments in this basin, came into existence only in recent years (Reading 1982, 1986).

In the mainland of Burma a volcanic arc separates the forearc from the back-arc. The estimated sediment thickness is about 12 km in the former basin and 17 km in the latter. Both basins are possibly underlain by an oceanic crust.

The largest sediment cover on a presumably oceanic crust today is the Bengal deep-sea fan extending from the mouth of the Ganga-Brahmaputra delta to the Indian Ocean, covering an area of 3,000,000 km^2 of the Bay of Bengal floor. The average thickness of the fan sediments here is 7.5 km but the maximum estimated thickness is of the order of 12 to 15 km. Sedimentation in this fan has been going on since the beginning of the collision of the Indian and the Eurasian plates at a rate of 2–10 cm/1000 years. The phases of fan sedimentation are correlatable with the phases of upheaval of the Himalayas (Curray and Moore 1971; Moore et al. 1974).

Pull-apart Basins

Pull-apart basins developed at passive continental margins during rupture of the megacontinents and drifting away of the continents. Although they may have been initiated as rifts pull-apart basins are distinct in the sense that the sedimentary prisms of these basins prograded towards the ocean basins in the post-rift stage. Clastic sediments were deposited during the rift phase while evaporites, carbonates and turbidites dominated the spreading phase of pull-aparts.

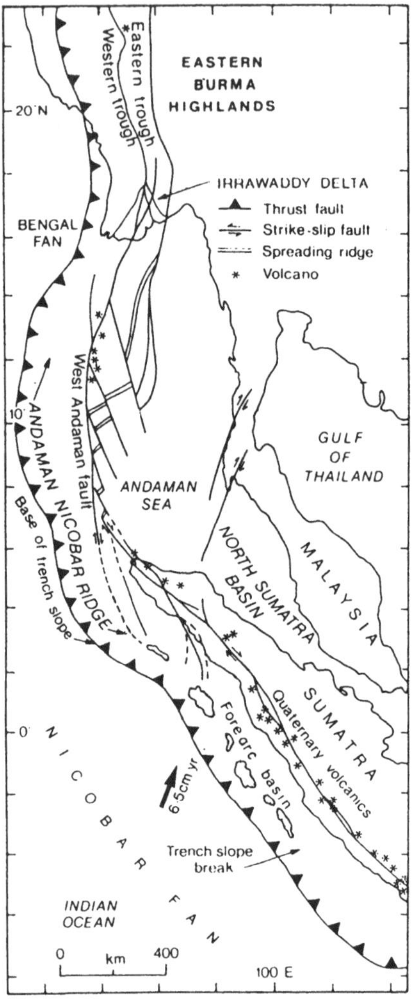

Fig. 7.18: Arc-trench system of Burma-Andaman-Indonesia showing strike-slip faults, volcanic arc, forearc and backarc areas (compiled by Reading 1982 after Curray et al. 1979; Kraig et al. 1980, reprinted with permission).

The Gabon Basin lying on the western coast of Africa, between the Niger delta and Angola, is a pull-apart basin (Fig. 7.19). The basin was initiated at the time of opening of the South Atlantic, following migration of South America from Africa. The onshore part of the basin has a maximum width of about 200 km. A Precambrian horst block separates the basin into two parts—a deeper basin opening to the Atlantic in the west, and a shallower,

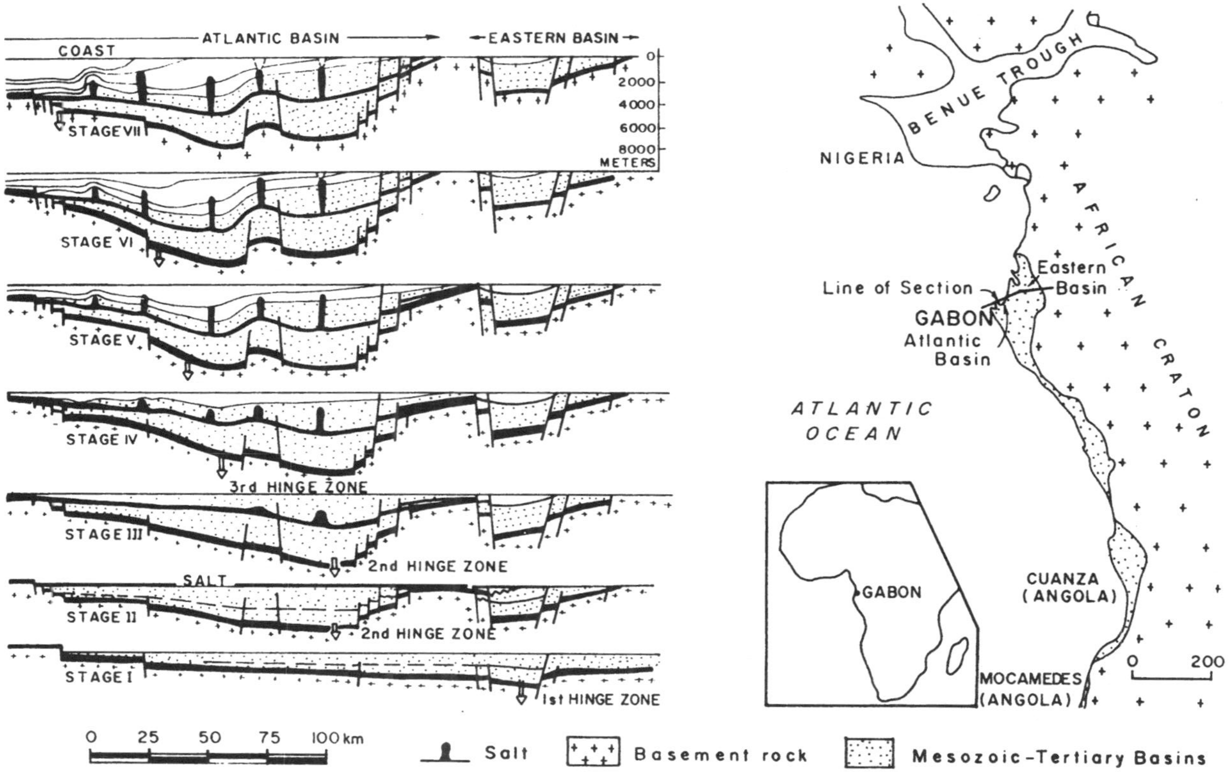

Fig. 7.19: Gabon 'pull-apart basin' of West Africa, and cross-sections showing stages of development of Gabon basin (compiled from Shannon and Naylor, 1989; Brink 1974, with permission of Graham & Trotman and American Association of Petroleum Geologists).

interior eastern basin. The Atlantic basin, an extensive monocline, contains more than 8000 m of sediments ranging in age from the Jurassic to the Miocene. A thick evaporite formation of Aptian age has resulted in structural discordance between the sequences above and below. In the eastern basin rocks ranging in age from the Jurassic to the Middle Cretaceous, up to a maximum thickness of 5000 m, occupy a faulted syncline.

The Late Jurassic-Early Cretaceous phase of rifting and basin development initiated by continental drifting, was followed in the Gabon Basin by a long post-rift phase of basin development. The rate of subsidence, which was high during the Jurassic to the Lower Cretaceous, gradually declined during the Upper Cretaceous and the Tertiary. The stages of basin development can be traced in the sections shown in Fig. 7.19. The three hinge lines, each representing a cluster of faults seen in the sections, had major roles to play in the development of the basin (Shannon and Naylor 1989).

Three sedimentary basins, at Kutch, Cambay and Narmada, occur in the western continental margin of India. The faults bordering these grabens open seawards (Fig. 7.20). The sediments within these basins, ranging in age from the Upper Bathonian/Callovian to the Neocomian (? Santonian), are essentially clastic, interspersed by calcareous or marly bands at some places in Kutch. A major marine transgression, followed by regression and deltaic environments with periodic marine influence, are the interpreted depositional environments in Kutch (Biswas 1981; Biswas and Deshpande 1983). These rift basins are supposed to have opened up successively as a result of the counterclockwise drifting of the Indian craton after its detachment from Gondwanaland in the Late Triassic or the Early Jurassic. Rifting is inferred to have developed from north to south in the following order: Kutch (Early Jurassic), Cambay (Early Cretaceous) and Narmada (Late Cretaceous), (Biswas 1982). These rift-bounded basins opening towards the sea satisfy the characters of the pull-apart basins of Shannon and Naylor (1989).

Delta type of Basins

In the Shannon and Naylor (1989) scheme (shown in Table 7.6), deltas have been classified as sedimentary basins. Deltas, as discussed in Chapter 6, represent a particular type of deposit produced by interaction of marine and non-marine agencies. They may develop whenever a sediment-water mixture is funnelled through a channel into a relatively quieter body of water provided that the rate of sediment input is higher than the rate of its dispersal. Deltas may develop on continental, transitional or oceanic crust, and on any type of plate margin, convergent or divergent. Although all deltas of the world have many broad similarities, their characters differ depending on the geological and tectonic setting and local climatic conditions, which control the rate of sediment supply.

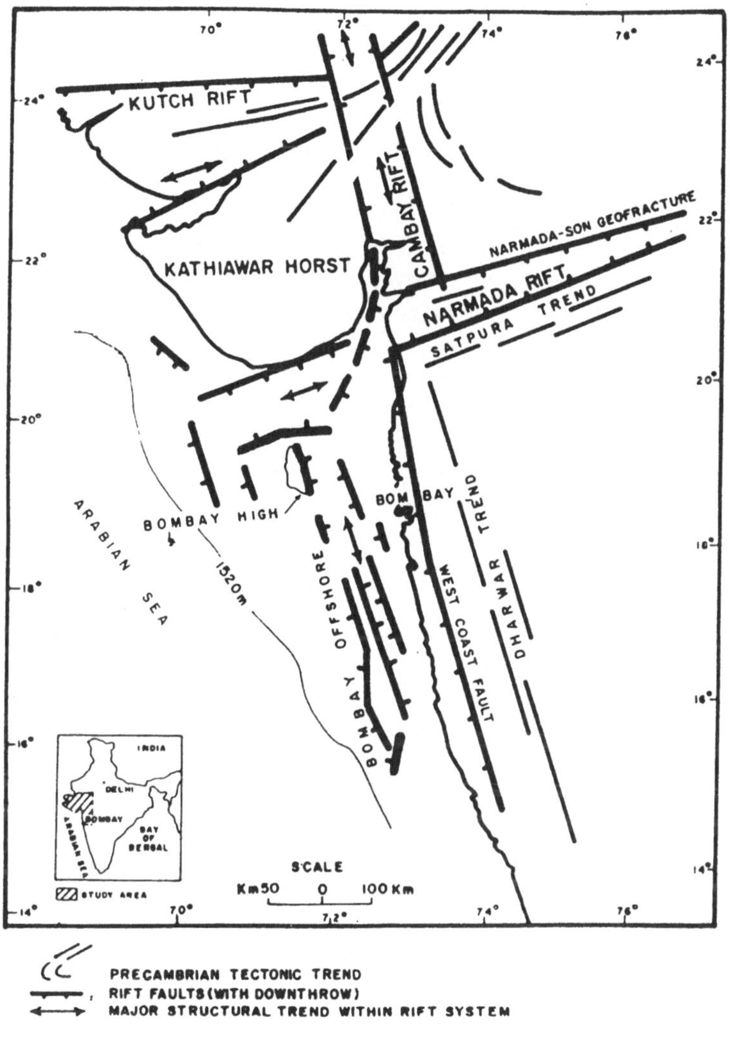

Fig. 7.20: Kutch, Narmada and Cambay rift basins on the western coast of India (modified after Biswas 1982, with permission of American Association of Petroleum Geologists).

In the Niger delta, cited by Shannon and Naylor (1989) as a typical delta-type of sedimentary basin, a mean annual supply of 2×10^7 m^3 of sediment is maintained through the Niger River and its tributary, the Benue River. These rivers flow through rift valleys believed to be the 'failed arm' (aulacogen) which extended into the continental interior from the 'tripple junction' formed during separation of South America from Africa. The rifting, initiated during the Late Jurassic, led to deposition of a wedge of sediment along the Benue trough until the Early-Middle Cretaceous. This was followed by deposition of a coastal prism of sediments during the Tertiary. The depositional centre of this wedge gradually shifted towards the sea as the pre-Tertiary depression was filled. The Oligo-Miocene delta complex, for example, is located wholly offshore on the rapidly subsiding oceanic crust (see Fig. 7.16).

Composite Basins with Polyhistory

The practice of compartmentalising sedimentary basins to specific associations of crust type and plate movement, followed in Table 7.6, is artificial to some extent. Many sedimentary basins have passed through a variety of lithospheric movements during their span of existence and hence are polygenetic in character. The Benue Basin, for example, developed into a delta, starting from an aulacogen. The West Canada Basin and the Vindhyan Basin, both of which were initiated as downwarps, experienced foreland depositional conditions following orogenic movements in the bordering areas. Some of the foreland basins again, like the one in north-eastern India, can be looked upon as a part of the adjacent subduction complex. Shannon and Naylor (1989) have cited the history of development of some of the complex basins in China as examples of composite basins. We discuss below the history of development of the Bengal Basin, located in eastern India and adjacent Bangladesh, as an example of a basin with a composite history.

The Bengal Basin occupies an area of more than 90,000 km^2, covering most of the present West Bengal and Bangladesh. The basin is open to the sea in the south. Extensive geophysical surveys and deep drilling in the alluvium-covered plains of Bengal have revealed a thick section of Cretaceous and Tertiary sediments on a basement of basaltic lava flows. The lava flows are of the same age as the Rajmahal volcanics ($\sim$ 100 m.y.) exposed on the north-western and north-eastern fringes of the basin. Geophysical surveys also suggested the presence of Gondwana sediments in several block-faulted troughs below the basalt flows. A series of fault-bounded buried basement ridges mark the western fringe of the basin. East of this marginal zone of fault lies the shelf area of the basin with a low, homoclinal dip towards the south-east. The total thickness of Mesozoic and Tertiary sediments on this shelf exceeds 9 km in the eastern part of West Bengal and is of the order of 16 km in the deepest part of the basin located in south-eastern Bangladesh (Fig. 7.21). Except for a few normal faults this sequence is structurally undisturbed.

The most prominent marker horizon within the whole sedimentary sequence of this basin is a bed of Nummulitic (Eocene) limestone. This bed dips uniformly at the rate of $1\frac{1}{2}^\circ$ over the whole of the Bengal shelf but in the subsurface of Calcutta exhibits a conspicuous basinward flexure traversing the entire basin in a NE-SW direction. Beyond this zone of flexure, which

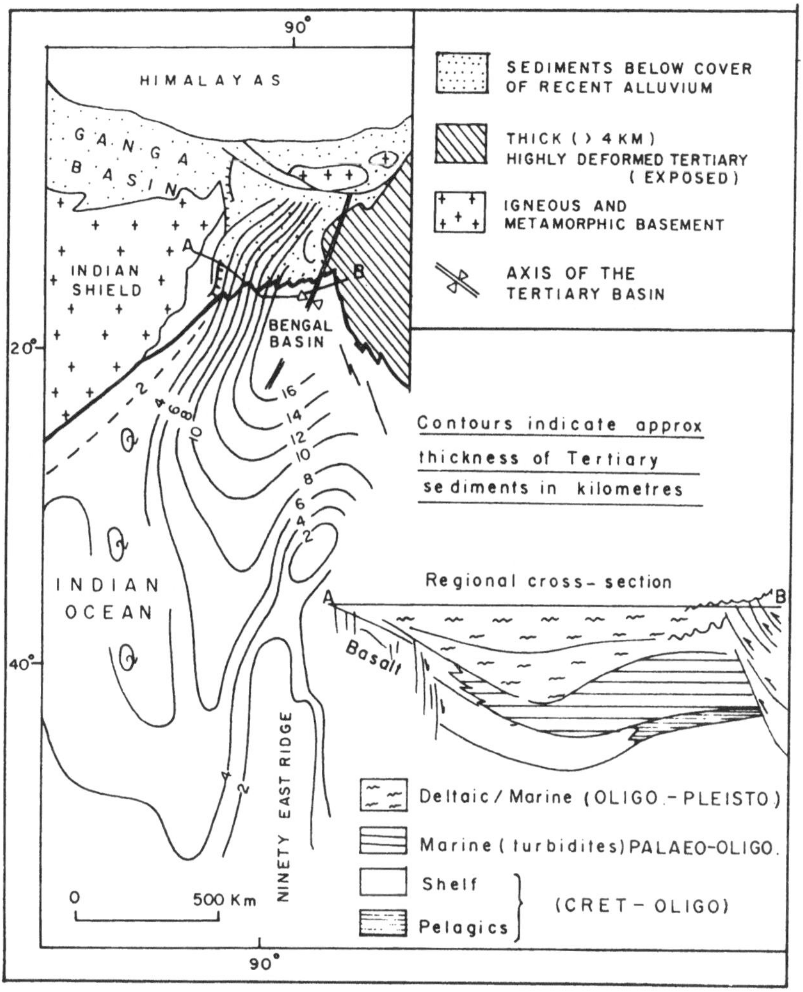

Fig. 7.21: Tertiary basins of Bengal and adjacent areas (compiled and redrawn after Paul and Lian 1975).

has been interpreted as the 'hinge zone' of the basin or as the Eocene shelf break, lies the deeper part of the basin with a greater rate of subsidence and a more argillaceous lithofacies. In the younger Teritiary sediments a similar change in lithofacies, from the arenaceous sediments of the stable shelf, to a dominantly argillaceous, deltaic sediment downdip, is noticed (Fig. 7.22). Except for phases of marine transgression occurring during the Upper Cretaceous (locally), Eocene (extensively) and Miocene (in the eastern parts

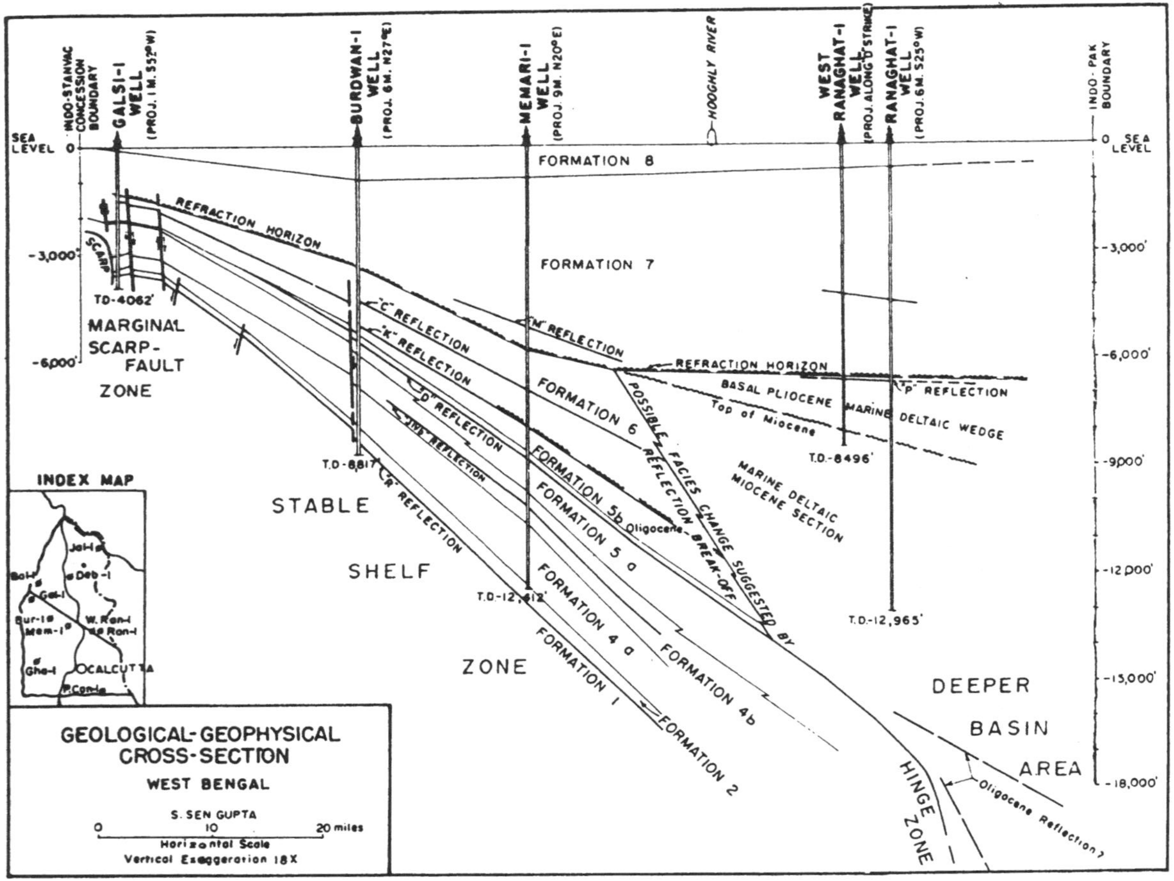

Fig. 7.22: Section across the western part of West Bengal, showing Tertiary sediments overlying basalt flow of Upper Jurassic age (the R-reflection) (after Sengupta 1966a, with permission of American Association of Petroleum Geologists).

only), the Tertiary sedimentation in this basin took place under freshwater, estuarine or deltaic conditions.

The evolution of the Bengal Basin took place in several stages. Block faulting within the Bengal plains and the adjacent areas was followed by deposition of freshwater Gondwana sediments of Permian to Early Jurassic age. During the Early Cretaceous a major rifting of the Bengal plains took place consequent to the break-up of Gondwanaland and the drifting away of Antarctica from India. The 'failed arm' of the aulacogen presumably rifted the entire Bengal plains while the other two arms mark the present boundaries of the eastern Indian coastline. The deep-seated faults located by seismic surveys between the Rajmahal and the Garo hills (Rao and Sengupta 1964) are perhaps the only detectable remnants of this 'failed rift' in the subsurface.

Rifting was associated with extensive eruption of flood basalts throughout the Bengal plains with the Kerguelen hotspot possibly serving as the heat source (Mahoney et al. 1983; Baksi et al. 1987). The eruption of basalt was followed by freshwater, brackish or estuarine sedimentation over the Bengal plains. This situation continued until the Lower Eocene when extensive marine transgression caused deposition of Nummulitic limestone over the whole Bengal plains extending north-east up to Assam. In the regressive phase which followed, most of the basin experienced deltaic sedimentation from the Miocene through the Pleistocene, when an enormous thickness of essentially argillaceous sediments was deposited, particularly in the south-eastern parts of the basin (Sengupta 1966a; Roybarman 1983). This phase of deltaic sedimentation is apparently continuing in the south and south-eastern parts of the present West Bengal and Bangladesh. Offshore explorations suggest continuation of the Bengal basin sediments southwards, into the present Bay of Bengal. The upper three sediment layers of the deep-sea fan in the Bay of Bengal are presumably the offshore continuation of the Miocene, Pliocene and Quaternary sediment layers of the Bengal basin inland (Curray and Moore 1971).

8

STRATIGRAPHY AND SEDIMENTATION

CLASSICAL CONCEPTS IN STRATIGRAPHY

Estimation of Geological Time

The ultimate aim of a sedimentological analysis is to unravel the history of depositional events leading to the development of particular stratigraphic sequences. Studies on sedimentology and stratigraphy are therefore inseparable. Evolution of the planet earth, both in the organic and inorganic sphere, through geologic time, has been the main concern of stratigraphy. Estimation of geological time, both in relative and absolute terms, has naturally been a matter of prime concern for stratigraphers as well as sedimentologists.

Steno's Law of Superposition, and Cuvier's studies on faunal assemblage made estimation of the relative ages of rock layers possible even in the early years of the nineteenth century. A layer of sediment occupying a higher position than the others, as also a layer containing an assemblage of organic remains with advanced anatomical features, is clearly younger than the rest. A novel method of age determination in absolute terms, introduced by Charles Lyell, was based on the principle that the younger rock layers should contain larger proportions of living organic species compared to extinct ones (see Chapter 1).

In an early method of age determination in terms of years, Lord Kelvin, a physicist, assumed that the earth had cooled at a uniform rate from a state of white heat to a condition wherein condensation of water vapour in the form of oceanic water was possible. The age of the earth obtained by this method in the late nineteenth century was a gross underestimate because the process of radioactivity, which replenishes much of the heat lost through radiation, was still unknown. This phenomenon, discovered jointly by H. Becquerel and Marie and Pierre Curie in 1896, and explained by Rutherford and Soddy in 1902 in terms of transmutation of atoms with a constant 'half-life', provided, for the first time, a dependable method for age determination of rocks in absolute terms. The radiometric method of dating is very much refined today and a host of radioisotopes, such as $U^{238} \rightarrow Pb^{207}$, $K^{40} \rightarrow Ar^{40}$, $Rb^{87} \rightarrow Sr^{87}$, are employed in radiometry. A similar method, using the ratio of radiogenic (C^{14}) and non-radiogenic carbon (C^{12} or C^{13}), can date woody matters which are less than 50,000 years old.

A more recent method of dating sediments uses the principle of magnetostratigraphy. Magnetically susceptible particles in sediments occurring in the deep ocean-floor reorient themselves, in a statistical sense, to the direction of the earth's magnetic field existing at the time of their deposition. By matching the magnetic polarity of an ancient sediment with the standard, radiometrically dated, magnetic stratigraphic column, some idea of the time of deposition of the sediment can be obtained. Problems are often created by bioturbation causing reorientation of sediment grains by deep-water burrowing organisms. On land the technique of magnetic stratigraphy can be used for dating ferruginous detrital (non-authigenic) particles laid down under oxidising conditions provided they have not been disturbed subsequent to deposition.

Dual Hierarchy in Stratigraphy

By the end of the nineteenth century it was recognised that the strata constituting the earth's crust could be grouped either as mappable rock units or on the basis of the time span they cover. This concept of 'dual hierarchy' in stratigraphy has continued to influence all our stratigraphic thinking since then. The problem of correlation between visible rock units and inferred time units was a difficult one. In the absence of a direct method for dating, the pioneers were obliged to use fossils as indicators of relative ages of rocks. This practice was later challenged on the ground that growth in the organic world is controlled not only by the trend of long-ranging organic evolution, but also by short-ranging local environmental conditions. Moreover, a parallel relationship between the rock and biostratigraphic units, an essential prerequisite for dating of rock layers by their fossil content, may not exist.

In an elegant solution to this problem, spans of time are recognised not by their fossil content, but by the deposits accumulated during specific time intervals (e.g. 'systems' deposited during 'periods' of time). The Schenck and Muller (1941) scheme of classification of stratigraphic units, reproduced in Table 8.1, is based on this principle. Although the geological time units and time-stratigraphic units are directly correlatable in this scheme, the rock unit terms have been deliberately written across to make direct correlation between rock- and time-stratigraphic units impossible. This trend of thinking eventually led to the classification of stratigraphic units into two distinct groups: 1) observable units consisting of litho- and biostratigraphic units and 2) inferential units consisting of chronostratigraphic units and geologic time units. This subdivision was the basis for creation of the Code of Stratigraphic Nomenclature. The categories and ranks of stratigraphic units defined in a modern version of this code, as proposed by the North American Commission for Stratigraphic Nomenclature in 1983, is reproduced in Table 8.2.

The 'material units' and the 'temporal and chronostratigraphic units' of Table 8.2 correspond respectively to the observable and inferential units of

Table 8.1. Schenck-Muller (1941) Scheme of stratigraphic classification

Geologic Time Units	Time Stratigraphic Units	Rock Stratigraphic Units			
		Group	Formation	Member	Bed
Era	—				
Period	System				
Epoch	Series				
Age	Stage				
—	Zone				

the conventional scheme. Of the material units, again the lithostratigraphic units are used for categorising and ranking of deposits of sediments in the field. The fundamental lithostratigraphic unit, called a formation, is defined as 'a body of rock identified by lithic characteristics and stratigraphic position; it is prevailingly, but not necessarily tabular and is mappable at the earth's surface or traceable in the subsurface.'

Subsurface explorations in large sedimentary basins conducted during the last few decades by deep drilling and geophysical techniques have provided a new perspective to stratigraphic concepts. The conventional as well as the newly developing ideas are briefly summarised below. Readers may refer to Dunbar and Rodgers (1957) for a masterly treatment of the traditional stratigraphic concepts. The newer concepts are summarised by Walker (1984) and Reading (1986).

VERTICAL AND LATERAL RELATIONSHIP

Vertical Relationship

Sediments accumulate within basins by aggradation, progradation and lateral accretion. Aggradation is the process of vertical filling, as in the flood basin of a river. Progradation is the filling from the margin, as in the case of an advancing delta front. Lateral accretion is caused by the process of accumulation against the margin of a channel, as in a river point bar (see Fig. 8.1). The vertical relationship between succeeding sedimentary units produced in this way may be either conformable or non-conformable. A conformable relationship may be either sharp or gradational. Two types of gradations are possible: a continuous gradation or an intercalated contact. A body of sand grading vertically up into a shale section through mixtures of sand and shale in various proportions, is an example of a continuous gradation. A sand body passing vertically up into a body of shale through an intermediate layer of closely spaced lenticles of sand and shale is an example of intercalation.

An unconformable relationship between two sedimentary bodies implies a break in the process of sedimentation on a large scale. This is generally

Table 8.2. Categories and ranks of units defined in the North American Stratigraphic Code (1983)*

A. MATERIAL UNITS

Lithostratigraphic	Lithodemic		Magnetopolarity	Biostratigraphic	Pedostratigraphic	Allostratigraphic
Supergroup Group	Supersuite Suite	Complex	Polarity Superzone			Allogroup
Formation	*Lithodeme*		*Polarity zone*	*Biozone* (Interval, Assem-blage or Abundance)	*Geosol*	*Alloformation*
Member (or Lens or Tongue) Bed(s) or flow(s)			Polarity Subzone	Subbiozone		Allomember

B. TEMPORAL AND RELATED CHRONOSTRATIGRAPHIC UNITS

Chronostratigraphic	Geochronologic Geochronometric	Polarity Chronostratigraphic	Polarity Chronologic		Diachronic
Eonothem	Eon	Polarity Superchronozone	Polarity Superchron		
Erathem (Supersystem)	Era (Superperiod)				
System (Subsystem)	*Period* (Superperiod)	*Polarity Chronozone*	*Polarity Chron*		*Episode*
Series	Epoch			Diachron	Phase
Stage (Substage)	Age (Subage)	Polarity Subchronozone	Polarity Subchron		Span
Chronozone	Chron				Cline

* Fundamental units are italicized.

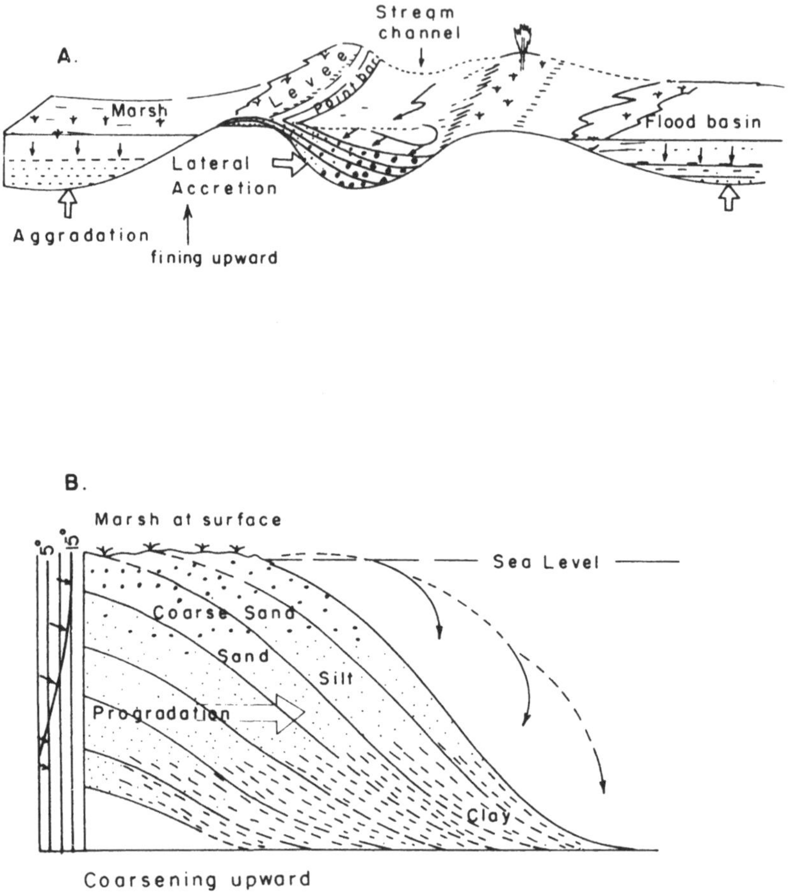

Fig. 8.1: Various mechanisms of sediment accumulation: (A) fining-upward sequences produced in a fluvial regime by lateral accretion within point bars and aggradation within flood basins, (B) coarsening-upward sequences produced by progradation of a delta (based on illustrations by Galloway and Hobday 1983; Scruton 1960 *in* Pettijohn 1975). Note 'blue pattern' in dipmeter log.

caused by tectonic uplift of the area of sedimentation leading to subaerial exposure and erosion of the previously deposited units. When the area subsides again, the newly deposited layers are said to be unconformably overlying the preceding ones. Various types of unconformable contacts are possible. When stratified rock layers overlie igneous or metamorphic bodies, the relationship between the two is said to be non-conformable. The unconformity in this case is readily recognisable. While mapping a non-conformity one should be certain that there is no intrusive relationship between the two sets of rocks. The possibility of contact metamorphism is also to be eliminated.

When the older beds within a sequence of sediments are tilted or deformed, before being buried by the younger layers, the resultant break is called an angular unconformity. Such unconformities are easily detectable during field mapping. When the older and younger layers lie parallel to each other with only an erosional break between the two, the relationship is designated a disconformity. This is often marked by a layer of conglomerate consisting of erosional products of the stratigraphically older unit. The term paraconformity was introduced by Dunbar and Rodgers (1957) to describe the situation wherein a time gap occurs without a detectable physical break between the upper and the lower units. Such time gaps between the two depositional units can be identified by the palaeontological or radiometric method (Fig. 8.2).

Recognition of the different types of unconformities can help in deciphering the depositional history of a sedimentary basin. A non-conformable relationship is expected at the basin fringe where layers of sediments overlie the igneous or metamorphic basement. Periodic withdrawal of the sea from the deeper parts of the basin or a temporary phase of subaerial exposure, may produce a disconformable relationship between the underlying and the overlying layers. Paraconformable relationships are to be expected at the centre of the basin where the sedimentary pile is very thick.

In the case of a gradual marine transgression over an unconformity, the younger layers overlap the older ones and are onlapped by the still younger sediments. The younger layers in this case are progressively displaced away from the centre of the basin. In a typical case of marine transgression on land, this results in a fining-upward sequence. In the event of a gradual marine regression, the sea is progressively displaced towards the basin centre and the younger layers are shifted away from the land area. The relation-

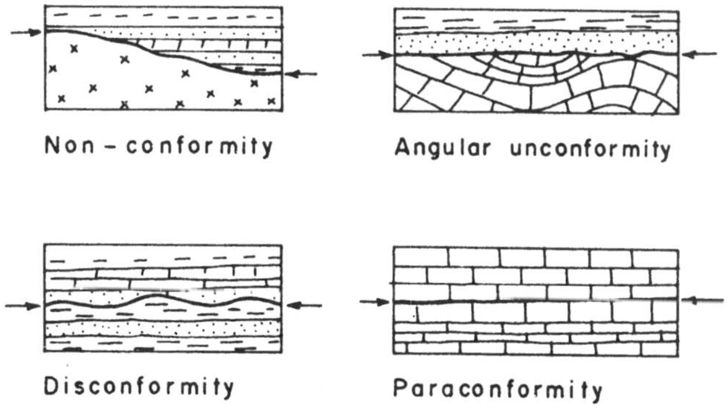

Fig. 8.2: Four types of unconformities (based on Dunbar and Rodgers 1957, fig. 57).

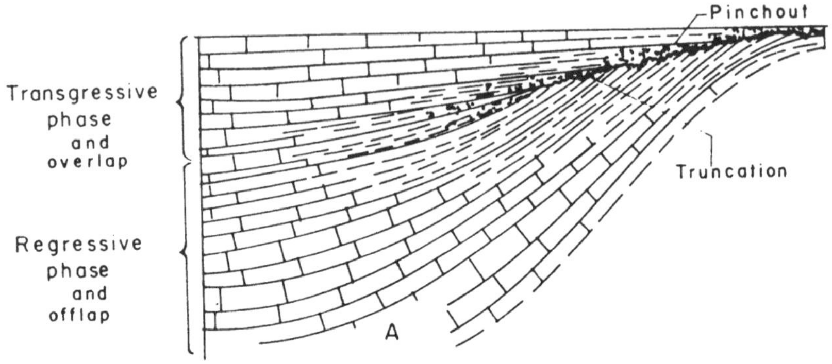

Fig. 8.3: Offlap and overlap relationship produced by marine regression and transgression. During a transgressive phase each unit onlaps the preceding one (modified after Krumbein and Sloss 1963, fig. 9-5).

ship between the successive sedimentary layers in this case is offlapping. Pinchouts are common in onlaps but truncations occur frequently in offlaps (Fig. 8.3).

While an unconformity represents a major break in sedimentation, on a regional scale over a protracted period of time many small breaks of short duration (diastems) may occur within a basin. Erosion and dispersal of loose sediments at sea bottom by sheets of moving water, a process termed 'sublevation' by Dunbar and Rodgers 1957 (Latin: *sublevare*—to lift up) is an example of such a break.

Lateral Relationship

Lateral changes in depositional conditions may cause different but contemporaneous rock types and faunas to be deposited in strike continuation of each other. One of the classical areas in which demonstration of such a change made the concept of facies variation acceptable in North America is the Devonian Catskill delta of New York state. Careful mapping in this area in 1933 indicated that the Catskill delta facies and the Chemung Group of rocks are of the same age, although the Catskill, resting on top of the Chemung, appears to be younger (Fig. 8.4). Similarly, in the Grand Canyon of Colorado, isochronous deposition of several Cambrian formations was demonstrated by McKee (1945, 1949). Several such classical examples of facies variation are to be found in Dunbar and Rodgers (1957). In the Permian Barakar Formation of the Southern Karanpura coalfield of Bihar, India, Banerjee (1960) demonstrated how a rhythmic pattern dominated alternately by coarse-grained, sandy channel facies and fine-grained (coal-shale) interchannel facies could develop by shifting of ancient river channels. Gradational and interfingering contacts between homotaxial channel

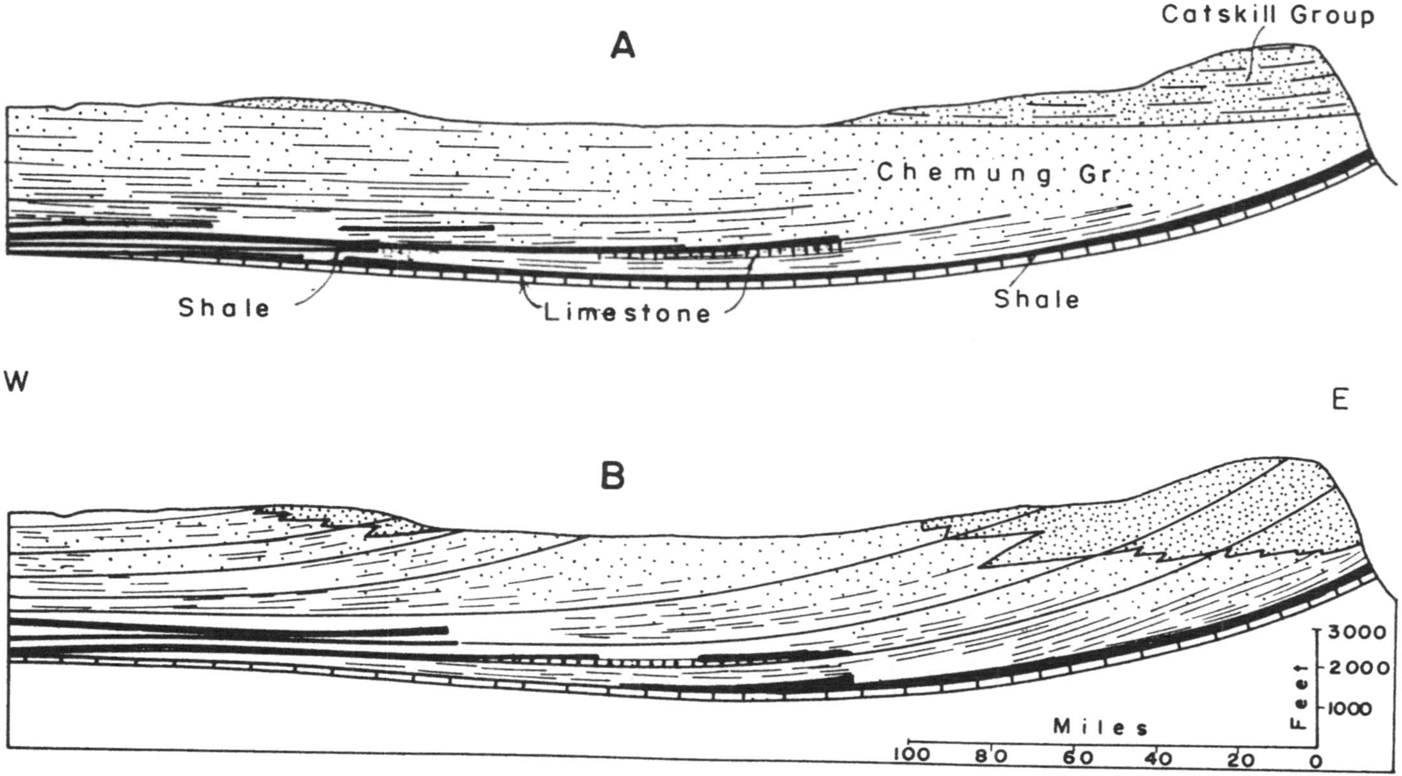

Fig. 8.4: Sections across the Devonian Catskill Delta, New York, showing stratigraphic interpretation: (A) before acceptance of the facies concept and (B) after introduction of the facies concept in the area by Chadwick and Cooper in 1933. The curved lines in B indicate time surfaces (simplified and redrawn after Dunbar and Rodgers 1957, fig. 71).

and interchannel permo-Triassic fluvial deposits were also established by Sengupta (1970) in the Godavari valley of southern India.

The rock units involved in the process of lateral (facies) change, like the situations described above, show up as three-dimensional masses of essentially uniform lithology having a gradational or interfingering relationship with adjacent masses of different lithology. Such a body of rock is designated a lithosome. Shapes of lithosomes, defined by external morphology and three-dimensional geometry (length, breadth and width), are related to specific tectonic and environmental settings and may provide clues to the interpretation of ancient environmental conditions. For this purpose facies associations, or groups of facies which occur together and are genetically connected, are more important than the geometry or character of an individual lithosome. Fig. 8.5 illustrates how the conventional lithostratigraphic terminology is to be applied during mapping of laterally or vertically occurring facies associations.

'A gradational transition from one facies to another implies that the two facies represent environments that were once adjacent laterally' is a concise expression of what was formulated by Johannes Walther in his Law of Correlation of Facies, usually referred to as Walther's Law of Succession of Facies (Middleton 1973). Continuous subsidence of the area of sedimentation is an essential prerequisite to such a preservation. The vertical stratigraphic record produced by laterally migrating deposits of a meandering stream channel may be cited here as an illustration of Walther's Law. At any point of time lenses of channel sand deposited by a meandering stream interfinger with silt and clay laid down in the interchannel flood basin. Continuous subsidence of this basin of deposition produces a succession of multistorey sand bodies, each of which grades laterally into the adjacent flood-basin deposits. The result is a vertical sequence of repetitive sedimentary cycles ('cyclothems') of channel sands overlying flood-basin silts and clays. An example of detrital sand lenses preserved due to subsidence of a hypothetical delta complex is given in Fig. 8.6.

Cyclicity

In the example cited above cycles of fining-upward sequences are repeated vertically by the process of channel migration within an environment without a major break in the process of sedimentation. Such repetitions are called autocyclic. Repetitions may also happen on a much larger areal and temporal scale, leading to superposition of disconformity-bounded deposits. These are termed allocycles. The North American Code (1983) provides for classification of these mappable allostratigraphic units into a hierarchical system of decreasing rank: allogroup, alloformation and allomember (Table 8.2).

The boundaries of these allocycles were produced by large-scale stratigraphic events connected with regional sea-level fluctuations. Such

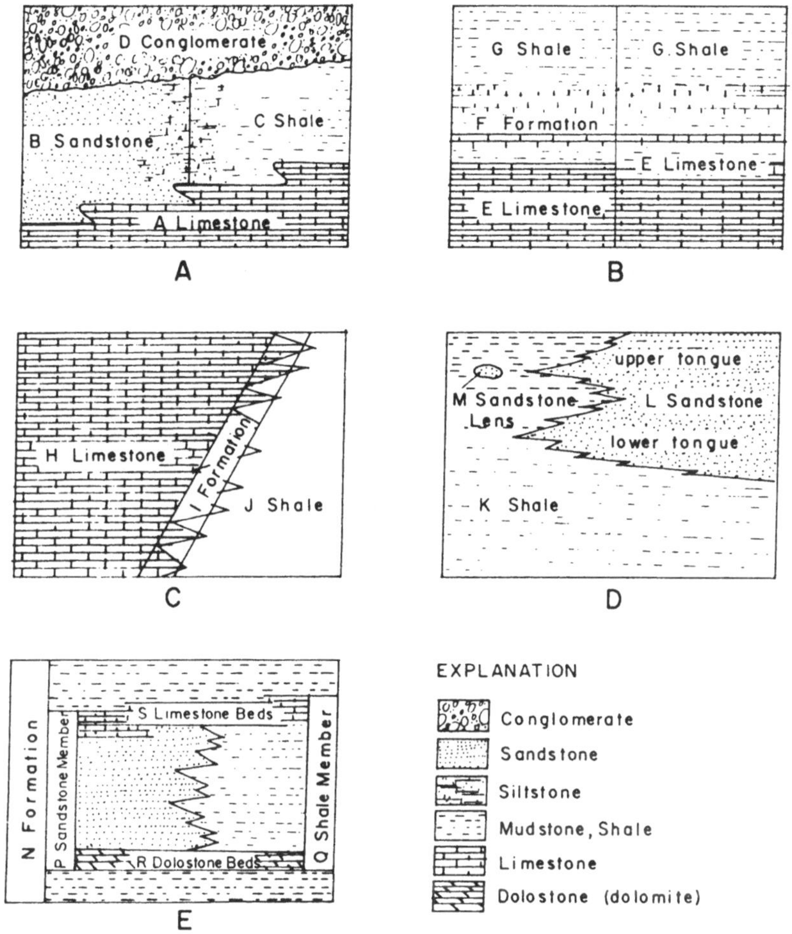

Fig. 8.5: Different types of lithostratigraphic boundaries in areas of facies change (redrawn from the Code of Stratigraphic Nomenclature proposed by the North American Commission 1983, fig. 2, with permission of American Association of Petroleum Geologists). (A) Boundaries at sharp lithological contacts and in laterally gradational sequence; (B) alternative boundaries in a vertically gradational or interlayered sequence; (C) possible boundaries for a laterally intertonguing sequence; (D) possible classification of parts of an intertonguing sequence; (E) key beds (R & S) are used as boundaries to distinguish Q-Shale Member from the other parts of N-Formation. A lateral change in composition between the key beds requires that another name, P-Sandstone Member, be applied. The key beds are part of each member.

allocycles have been discovered in many coal basins of the world. The Pennsylvanian cyclothem of Illinois and the Carboniferous cyclothem of Yorkshire, which are cited as classical examples, have been attributed to

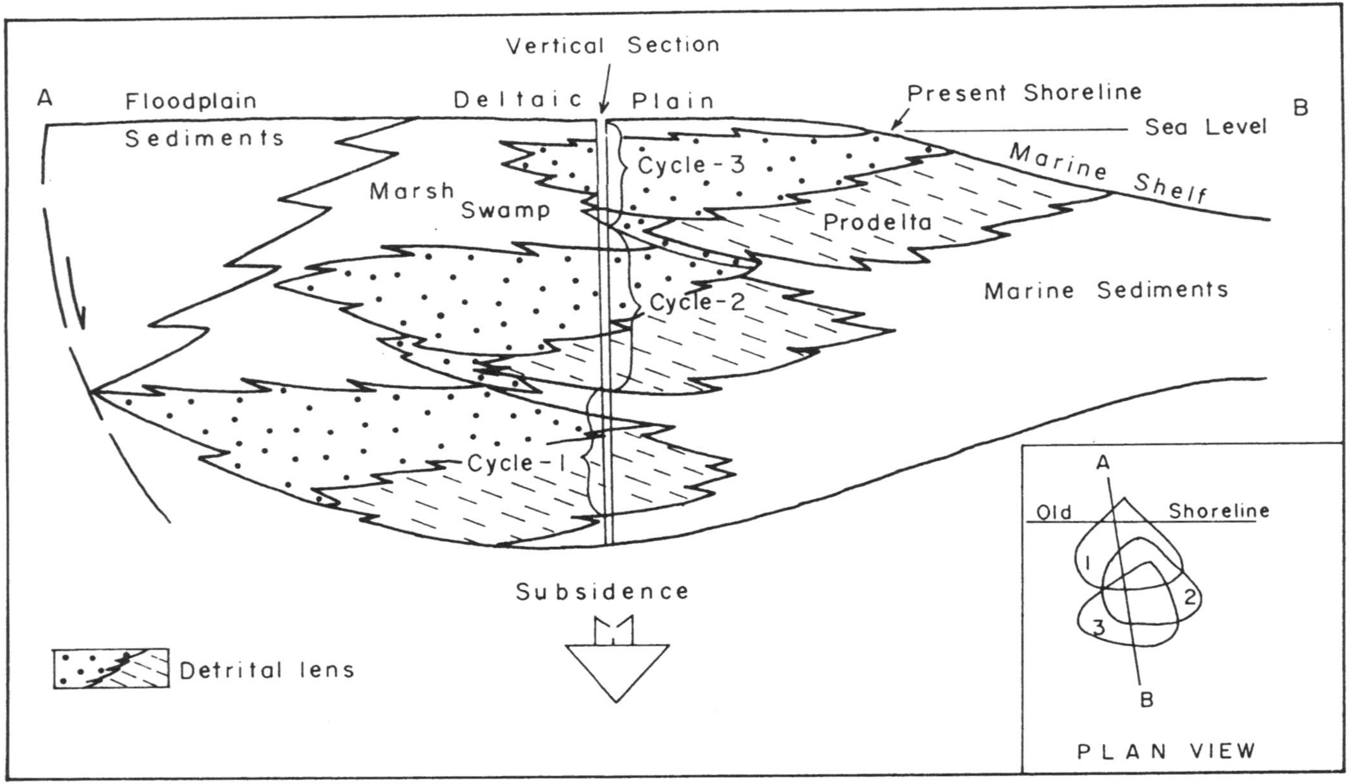

Fig. 8.6: Overlapping cycles produced in a hypothetical delta complex by subsidence (after Coleman and Gagliano 1964, as modified by Weimer 1976, fig. 11.18, modified and redrawn with permission of American Association of Petroleum Geologists).

sea-level fluctuations, either due to periodic diastrophism or to waxing and waning of land glaciers. Some authors have also attempted to correlate these periodic fluctuations to the orbital cycles of the earth's precession (31,000 year cycles), obliquity (43,000 year cycles) and eccentricity (100,000 year cycles), discovered by Milancovitch in 1941 and subsequently named after him.

During the last two decades earth cycles of various magnitudes have been discovered. Thirty-two million year cycles involving changes in the distribution of marine unconformities, patterns of faunal diversity, occurrence of black- and red-clays have been recorded. These have been attributed to cyclic changes in global climatic conditions caused by variation in the rates of volcanic activity at oceanic ridges (Fischer and Arthur 1977). These 32 m.y. cycles are superimposed on much larger cycles of about 200 m.y. duration.

Seismic surveys have led to the discovery of major cyclic erosional breaks of different magnitude during the Phanerozoic (Fig. 8.7). Of these, the major (first-order) breaks have been correlated to global sea-level fluctuations caused by changes in ocean-floor spreading rates (Vail, Mitchum and Thompson III 1977). These have some correspondence to the 200 m.y. cycles of Fischer and Arthur. The rise in sea level could have been caused by buoying up of sea-water due to heating near the spreading midoceanic ridges (Pitman 1978). The lower order sea-level changes have been attributed to glacio-eustatic changes (Vail, Mitchum and Thompson III 1977) or, alternatively, to flexures during basin formation on an elastic crust (Watts et al. 1982).

Freezing and locking up of sea-water in the form of land ice would cause a general lowering of the sea level, but whether this would cause a transgression or a regression of the shoreline would depend on the rates of sediment input, tectonism and isotatic adjustments. A regression may take place even when the sea level is high, if the sediment input is also high at this stage. Similarly, a transgression may occur even when the sea level is low, if the rate of subsidence of the land area is fast, either due to tectonic or due to isostatic causes. The local sea-level changes therefore, are the net product of many factors of which eustacy, isostasy, tectonism and sediment input are more important (see Reading 1986 for a more detailed discussion).

Two orders of transgressive-regressive cycles (called T-R cycles), the larger one corresponding to the third-order cycle of Vail et al. (1977) and the smaller (named fourth-order cycle) coinciding with the transgressive-regressive maxima of Vail's third-order cycle, have been identified in the coal-bearing Upper Cretaceous strata of Utah (Reyer 1983). Discovery of short (metre scale) basinward discontinuities within the Helderberg Group of rocks (Devonian) in New York state, led some modern stratigraphers to suggest that the stratigraphic record, barring minor exceptions, consists of thin, shallowing-up cycles bounded by geologically instantaneous rise

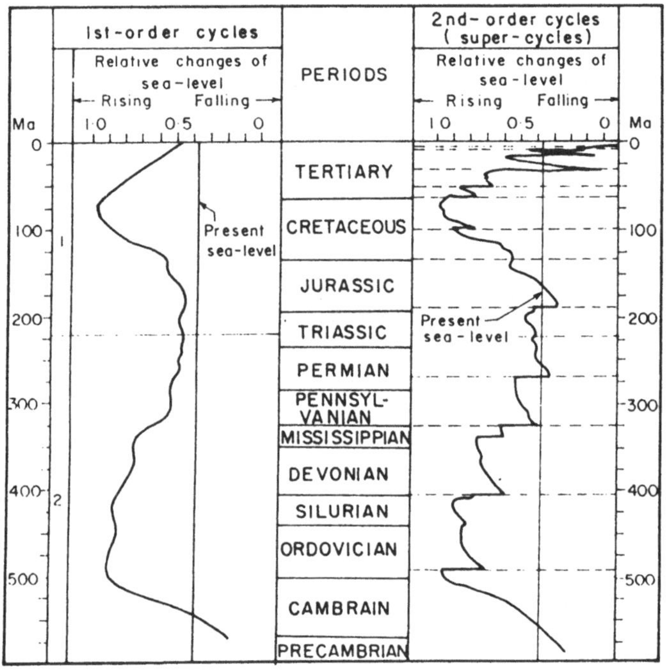

Fig. 8.7: First- and second-order global cycles of sea-level fluctuations during the Phanerozoic Eon (after Vail, Mitchum and Thompson III 1977, with permission of American Association of Petroleum Geologists).

of base level. The basic assumption behind this model is that the mode of stratigraphic accumulation is episodic rather than gradual, as traditionally believed. Goodwin and Anderson (1985), the advocates of this model, have demonstrated how episodic deposits might be produced by eustatic responses to orbital perturbations superimposed on continuous basin subsidence (Fig. 8.8). They have also worked out in detail how (geologically) instantaneous sea-level rise followed by aggradation might lead to the development of what they termed Punctuated Aggradational Cycles (PAC). Such

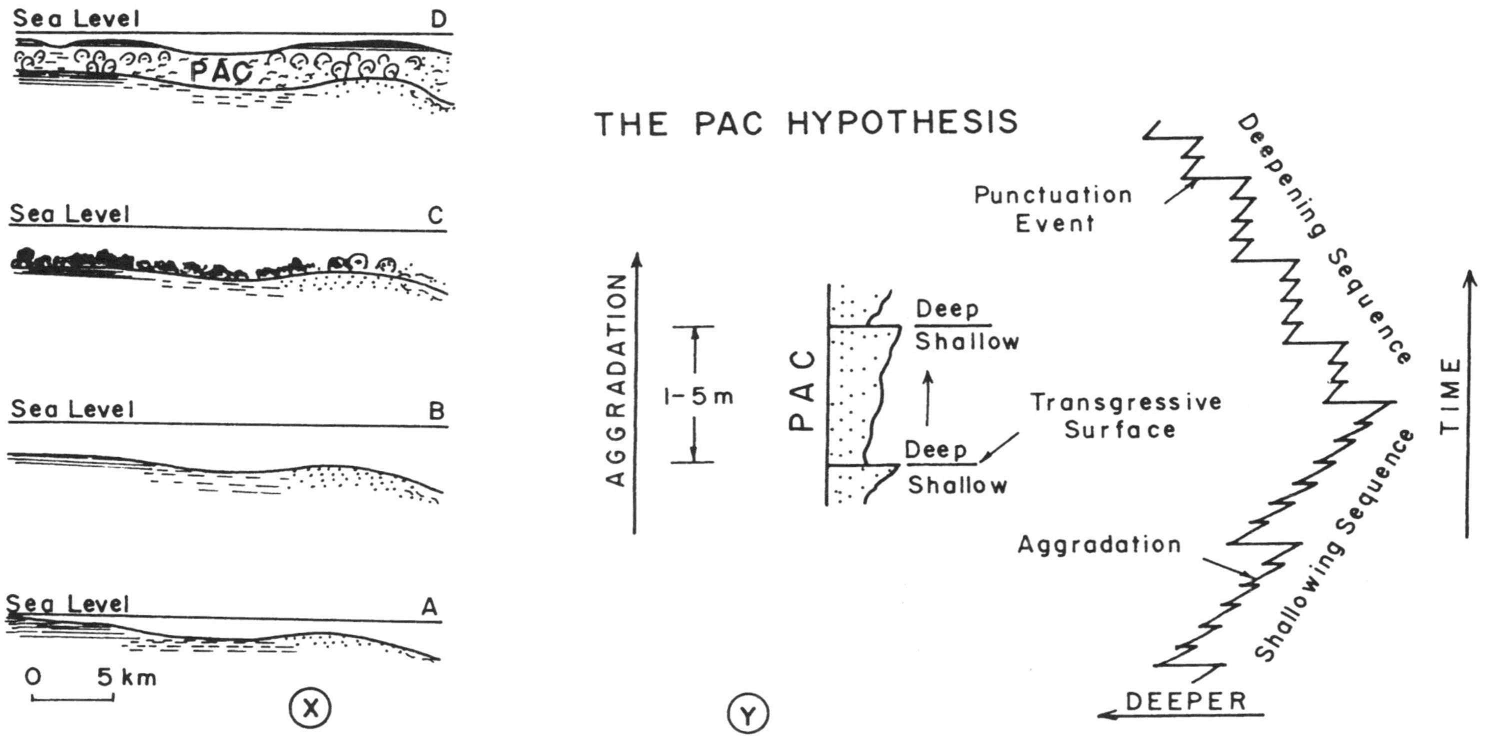

Fig. 8.8: (Y) Sequence of events leading to 'Punctuated Aggradational Cycles' (PAC). Each PAC is bounded by surface of abrupt change to deeper water facies (X) illustrates how this change may take place by sea-level fluctuation (redrawn from Goodwin and Anderson 1985, figs. 1 & 2, with permission of University of Chicago Press).

a model of sedimentation, based on inferred, short-lived episodes, precludes the application of Walther's Law except within cycles. The advocates of PAC believe that the PAC boundaries, being short-ranging stratigraphic units, will provide the tool for detailed intrabasinal correlation.

A model for occurrence of major depositional episodes, each consisting of many small depositional events, is shown in Fig. 8.9. This model, formulated on the basis of observations in the Quaternary of the Gulf Coast Basin (Frazier 1974) is applicable to major marine and lacustrine deposits. The major depositional episodes of this model are bounded by large-scale hiatal surfaces of regional extent. Periodic subsidence of the land area or eustatic changes of the sea level during glacial cycles cause marine transgressions and flooding to recur within a depositional episode. During the progradational phase of sedimentation, which is represented by a time line extending from the shore to the centre of the basin, basinward thickening of sediments takes place. This is followed by an aggrading phase when sedimentation occurs on top of those already deposited. The aggrading deposits thin towards the basin.

The depositional episodes of Frazier and the conventional stratigraphic units may coincide. The transgressive facies are generally given the status of formations. The deposits of a progradational and an aggradational phase together constitute a facies sequence. The large-scale hiatal surfaces are indicated by hard grounds and continuous thin deposits of marl, limestone, glauconitic or phosphatic beds. The small-scale hiatal surfaces are represented by soil, peat, or simply by local erosional surfaces. Recognition and documentation of these depositional episodes constitute an essential part of facies analysis.

Analysis of Cyclicity

Rock units, arranged apparently randomly in vertical stacks, may contain a hidden regularity. This regularity, if deciphered, may provide clues to the inherent genetic relationship between different facies. In a simple method of studying this relationship, de Raff, Reading and Walker (1965) worked out the frequency of facies which are in contact with each other in a vertical succession. This led to a 'spider diagram' depicting the facies relationship. Nowadays, more sophisticated methods are used for finding the 'hidden motif' within cycles (Miall 1973; Cant and Walker 1976).

The procedure for quantitative analysis of cyclicity, based on Gingerich's (1969) technique for analysis of cyclicity in fluvial sequences, determines the probability with which a certain rock type will follow another in the vertical column. If the probable sequence of transition, worked out in this way, does not give a complete picture, the 'hidden elements' may be detected by a second stage of analysis when the 'difference matrices' are computed. Finally, a test for the Markovian property is carried out in order to know if each

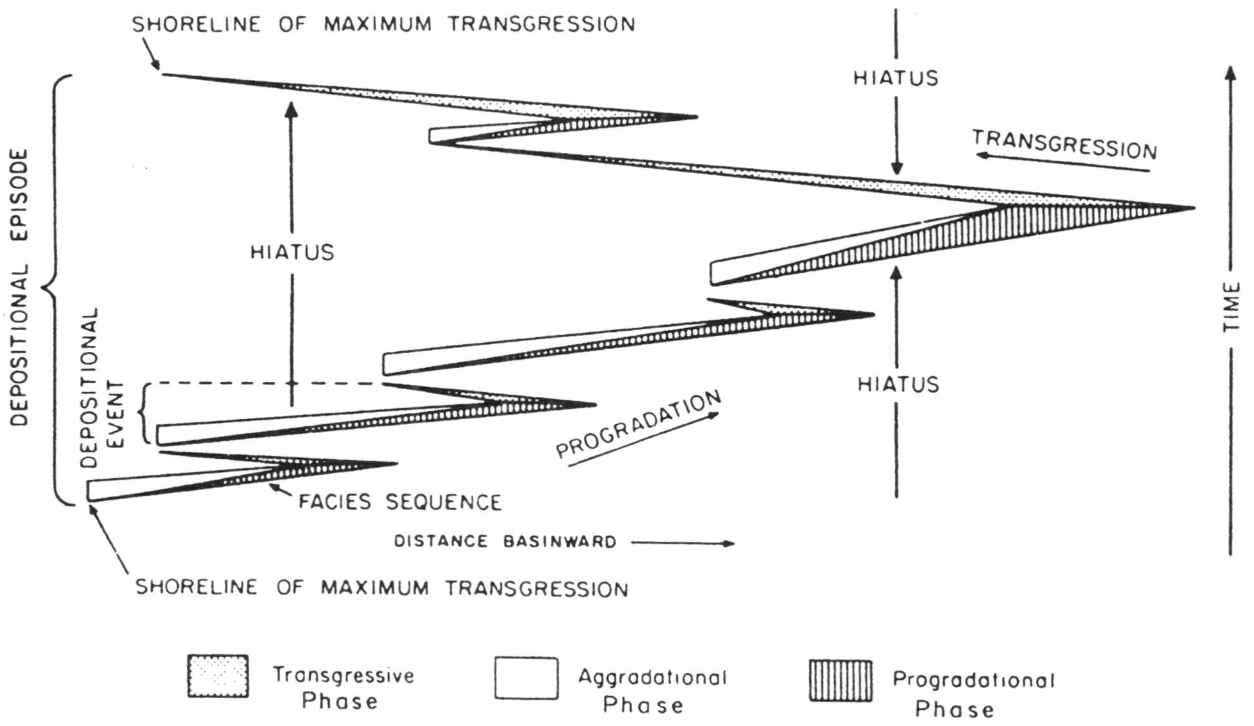

Fig. 8.9: Temporal-spatial relationship of an ideal depositional episode. Hatched areas represent intervals of active terrigeneous deposition. The depositional episode is bounded by external hiatal surfaces (modified by Galloway and Hobday 1983 after Frazier 1974; reprinted with permission of Springer-Verlag, Heidelberg).

depositional event of the system was influenced by the event immediately preceding it, i.e., whether the depositional processes have followed a continuous sequence.

The method outlined above has been shown to be statistically incorrect (Walker 1984a), but the alternative procedure of analysis, claimed to be dependable (Harper, Jr. 1984), is too sophisticated to be discussed here. Experience, however, has shown that the system of analysis proposed by Gingerich (1969) gives fairly dependable clues to genetic interpretation of facies in many cases.

A mechanical procedure of analysis of cyclicity may fail to take into account important geological features such as the nature of contact between facies, and the conclusions which prove to be statistically significant, may influence interpretations even if they are geologically unimportant. Moreover, every method of mathematical analysis based on summarisation of the original observations, runs the risk of losing valuable geological information in the process. The answer to this problem should be to return to the raw data at every stage of analysis. Only the geologically meaningful solutions should be accepted (Reading 1986).

SUBSURFACE CORRELATION

Correlation is demonstration of time equivalence of particular stratigraphic units from the type sections to the areas of their geographic extension. The criteria used for correlation should ideally be independent of lithology, fossil content or other material bases of stratigraphic division, that is, some criteria for determination of time, which is independent of all these, should be used (North American Commission on Stratigraphic Nomenclature 1983). But in practice, stratigraphers often use mineralogical and biological contents of stratigraphic units for tracing their geographic extension. Since the advent of geophysical tools, it has also become customary to trace the contacts of stratigraphic units having contrasting physical properties for correlating them. There seems to be no particular objection to this method of correlation as long as the specific stratigraphic character being correlated is appreciated.

In the early days of geological exploration heavy mineral assemblages were used for the purpose of stratigraphic correlation. The technique is based on the assumption that the heavy minerals are incorporated into layers of sediments by progressive deroofing of a source rock. When such a process takes place at different geographic locations, but within a short span of time, the layers of sediments are taken to be time equivalent. The method has worked well in many cases. The proportions of epidote, hornblende and staurolite, for example, could be successfully used for correlating the Tertiary stratigraphic units of Assam in India. The technique reached such a refinement that local stratigraphic irregularities could be detected by studying the heavy mineral assemblages of only a few odd samples. Some new techniques for analysis and representation of heavy mineral data were also developed by the gèologists of the Assam Oil Company (Evans et al. 1934).

When the heavy minerals occur within a stratigraphic column in reverse order of their stability (i.e., only the more stable minerals are found in the older horizons), one may suspect that the less stable ones have been removed by circulating interstratal fluids. Any attempt at time stratigraphic correlation by heavy minerals is of little value in such an instance (Pettijohn 1975).

Borehole Logging

The best possible way of obtaining subsurface information is by drilling. Well cuttings continuously brought to the surface by circulating drilling fluid ('mud') do not provide precise information on the nature of stratigraphic contacts in the subsurface. To obtain this information core samples must be collected continuously during drilling. As this is an extremely expensive process, a continuous record of the electrical, sonic and radioactive properties of the sections drilled is generally maintained by wireline logging. Logging is conducted by introducing into the borehole a sonde containing an energy source and a detector. The source is a combined current and potential electrode for electrical logging, an ultrasonic transducer for sonic logging, a neutron emission source for neutron logging or a gamma ray (Co^{60}, Cs^{137}) source for density logging. Recording is done continuously as the sonde is moved up the borehole. The geological use of different types of logs is shown in Table 8.3.

The conventional electrical logs record spontaneous potential (S.P.) and resistivity. The former is a measure of the electrical current generated at the contact of rock layers by interaction of the drilling mud with the formation water. The latter measures the voltage drop across potential electrodes for known currents passed through current electrode(s) (Fig. 8.10A).

Contrasting salinity of the pore fluid and the drilling mud is essential for the generation of spontaneous potential. As the mud comes into contact with sand or a shale layer filled with salt-water, the electrochemical potentials add up to produce S.P. currents across the sand-shale sequence. The smooth S.P. curve that generates against a shale layer, when it is thick, is called the shale line or base line. Against porous sand beds, the S.P. curve shifts to the left. The spontaneous potential generated is related to resistivity of the formation water (R_w) and the resistivity of the mud filtrate (R_{mf}). Hence the S.P. log can be used not only for delineating the bed boundary and bed thickness, but also for estimating the resistivity of the formation water. Under suitable conditions the S.P. curve can also be used as an indicator of the

Table 8.3. Log types, properties measured, and geologic uses (modified after Cant 1984)

Log	Property measured	Units	Geologic uses
Spontaneous Potential	Natural electric potential (compared to drilling mud)	Millivolts	Lithology (in some cases), correlation, curve shape analysis, identification of porous zones
Resistivity	Resistance to electric current flow	Ohm-metres	Identification of coals, bentonites, fluid evaluation
Gamma-ray	Natural radioactivity of K, Th, U	API units	Lithology (shaliness), correlation, curve shape analysis
Sonic	Velocity of propagation of sound wave	Microseconds/metre	Identification of porous zones, coal, tightly cemented zones
Caliper	Diameter of hole	Centimetres	Evaluate hole conditions and reliability of other logs
Neutron	Concentrations of hydrogen (water and hydrocarbons) in pores	Per cent porosity	Identification of porous zones
Density	Bulk density (electron density) includes pore fluid in measurement	Kilograms per cubic metre (kg/m^3)	Identification of hydrite, halite, non-porous carbonates, etc.
Dipmeter	Resistivity changes indicating dips	Degrees (and direction)	Structural and stratigraphic analysis, environmental interpretation.

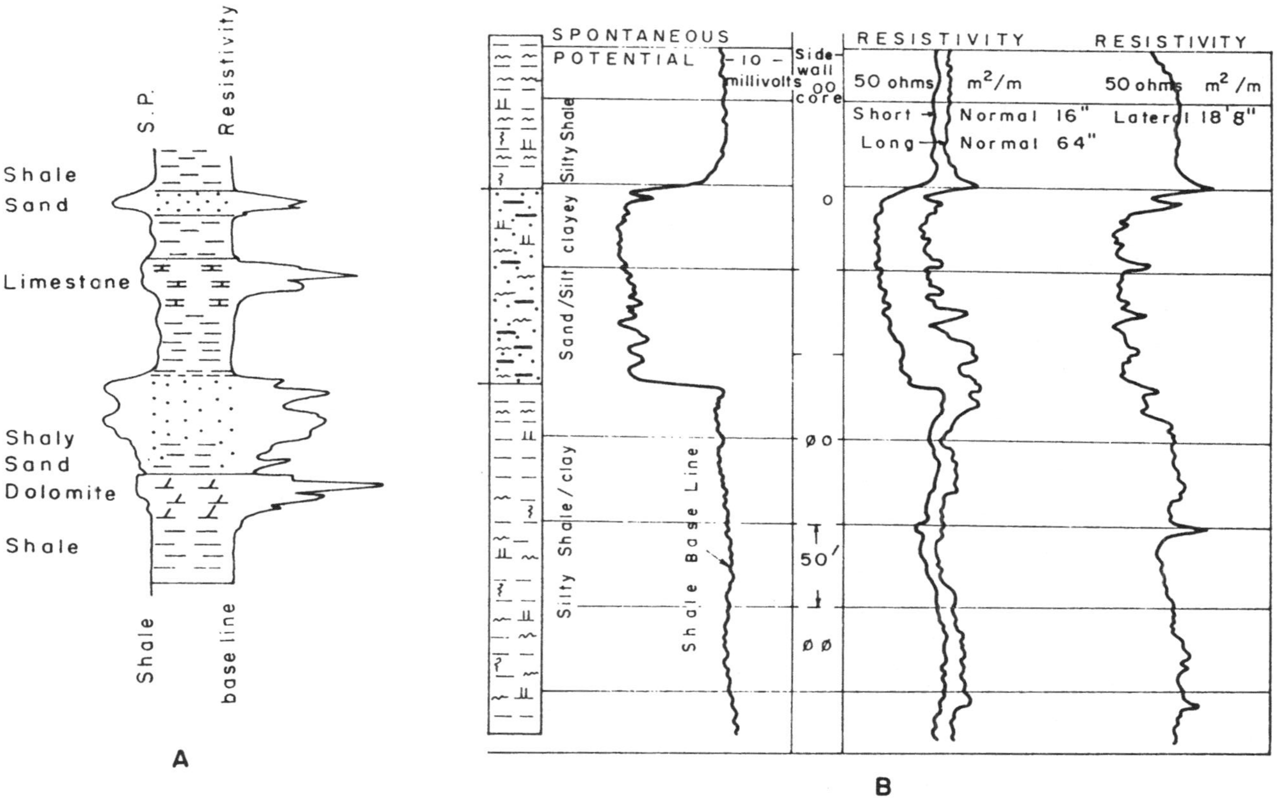

Fig. 8.10: A. Typical Spontaneous Potential (S.P.) and Resistivity patterns for sand, shale and limestone. B. An electrolog showing S.P. and Resistivity (Short and Long Normal, and Lateral) patterns for different lithological units within a borehole.

shale content of a sand layer. The higher the shale concentration, the lower the amplitude of S.P. (Fig. 8.10B).

Since S.P. is generated by the difference in concentration of electrolytes in the mud and the formation water, there will be no S.P. if the salt concentration is of the same order in the mud and the formation in contact. In such a case a gamma-ray log may be used for determination of the sand-shale boundary. Shales containing K^{40} isotope provide a natural source of gamma-ray flux. In fact, a combination of a number of logs—S.P., sonic, density and neutron logs—is used for accurate determination of the proportion of shale in a sandy shale. The combination of gamma-ray and sonic logs is perhaps the best for the purpose of subsurface correlation (Fig. 8.11). The sonic log is effective in detecting very high velocity rocks (e.g., carbonates or well-cemented calcareous sandstones) and also very low velocity rocks (e.g., coal, porous or fractured rocks with interstitial fluid, unconsolidated sediments). The gamma-ray log, by detecting the shaliness of a rock, may indicate whether a sequence is 'coarsening' or 'fining-upward' although this log does not measure grain-size.

The equipment for resistivity measurement in boreholes consists of a current and a potential electrode. An electric current is sent through the current electrodes and the potential difference between the two potential electrodes is measured. The greater the spacing of the electrodes, the deeper the penetration of the current into the formation. By using electrodes at different spacings (Normal, 16″, Long Normal, 64″, Lateral, 18′8″) resistivities of the mud-invaded zone and also the true formation resistivity beyond the zone of flushing can be measured (Fig. 8.10B). Resistivity measurement can aid particularly in correlating the carbonate beds, which have high resistivity and low porosity. A well-cemented calcareous sandstone, which is more resistive than a porous, non-calcareous one, may be detected by resistivity logging. Both sonic and resistivity logs may be used for detecting the fluid content of a rock.

Resistivity measurements using direct current work only when the drilling mud is conducting. In empty wells, as also in wells filled with oil-based drilling fluids or gas, resistivity is measured by transmitting alternating current into the formation. The amount of current induced into the system is a measure of the formation resistivity. This is the method of induction logging.

Resistivity logs may also be used for measurement of the dip of a formation within a borehole. For this purpose microresistivity is measured simultaneously by four (five in the case of high resolution measurements) identical microresistivity electrodes mounted on four equally spaced pads of the dipmeter tool inserted into the borehole. A dipping bed produces a resistivity record of comparable magnitude at different depths on different walls of the borehole. Mathematical correlation methods are then applied for measurement of these displacements. Individual features or short intervals are

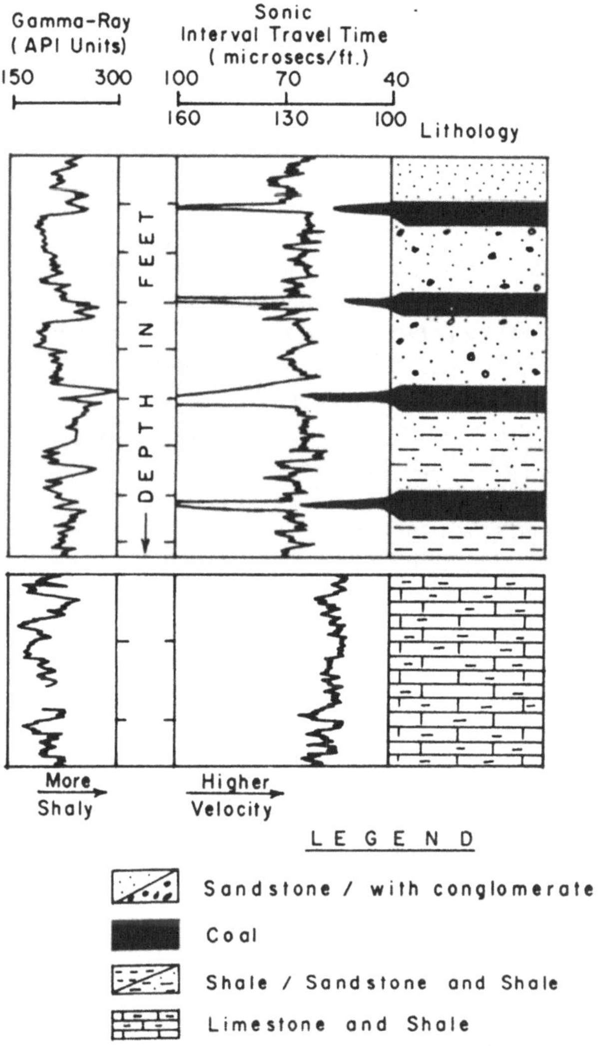

Fig. 8.11: Gamma-ray and sonic logs against typical siliciclastic and carbonate sections (modified and redrawn after Cant 1984).

matched together for computation of the dip and azimuth of the bedding. These computations, as well as correction for the effect of deviation of the borehole, are done by a microcomputer built into the system.

The dipmeter output is in the form of an arrow plot. The combined dipmeter-inclinometer plot, which shows dip as well as azimuth is called a 'tadpole plot'. Vertical variations in dips recorded in a dipmeter log help in long- and short-interval correlation. Structural and stratigraphic features can also be detected in the subsurface by dipmeter logging.

The methods commonly used for correlation of borehole logs are: tracing of marker beds, sequence analysis and slicing (Cant 1984). A bed having a distinctive log character may be used as a marker bed for the purpose of correlation when it has an appreciable areal extent. The deposits in low-energy environments, such as lakes and oceans, which generally extend over large areas, make good markers. Those in high-energy environments, such as rivers and sea beaches, are impersistent and lack distinctive characters. Such beds may be correlated for short distances with reference to the markers that may occur immediately above or below. In the Nahorkatiya oilfield in Assam, for example, a number of multistorey sand bodies, each of small areal extent, were mapped between two well-established marker horizons by S.P. and Short Normal logs (Fig. 8.12).

Sequence analysis in wireline logs involves matching of readily recognisable vertical patterns such as coarsening or fining-upward sequences by their log characters. Such sequences, called 'events', may match palaeontologically established time lines but they often cut across lithological or facies boundaries. When all other techniques of correlation fail, a log interval may be subdivided arbitrarily into slices for studying the nature of variation within small intervals. Thick deltaic deposits, for example, may be correlated over large areas with the help of the overlying and underlying markers but the intermediate thick pile of sediment is 'sliced' to study the nature of facies variation within short vertical distances.

Interpretation of facies or depositional environment from the shape of log signatures alone is risky because no log pattern is uniquely related to any particular environment. Fig. 8.13 gives the idealised gamma-ray or S.P. patterns found in some of the common depositional environments. Each log pattern shown in this diagram can generate in any of the depositional conditions mentioned. In choosing between them one must utilise some other independent line of evidence connecting it to a genetic (facies) model.

For example, a 'cylindrical' log pattern can develop from a fluvial sand within a distributary channel in a delta sequence, a carbonate shelf or a reefal growth in a shallow marine environment. Therefore, before interpreting the log signature it is important to know whether one is dealing with a marine or non-marine sequence. Direct information on lithology, granulometry and sedimentary structure, as may be obtained from core samples, is invaluable for this purpose. In the absence of borehole cores, side-wall cores may be extracted from desired locations of the borehole by special techniques.

Seismic Correlation

Seismic surveys are conducted by detonating a charge near the surface and by detecting the waves reflected back to the surface from subsurface interfaces. By converting the two-way travel times of elastic waves into depths

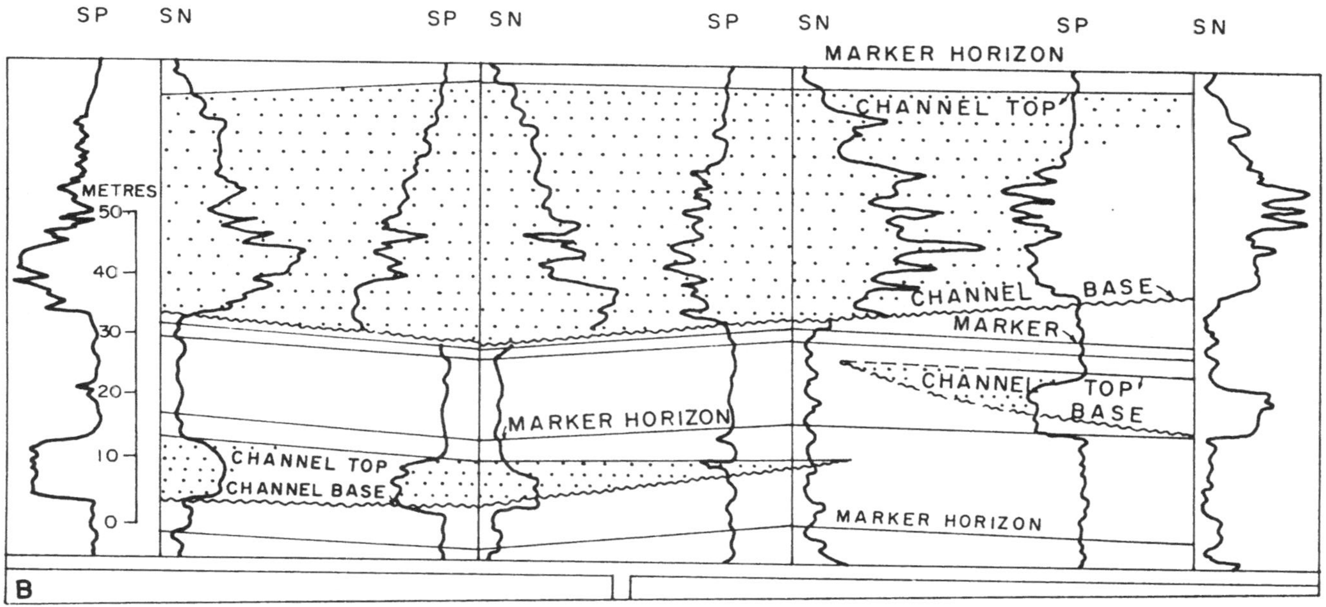

Fig. 8.12: Correlation of fluvial channel sands in the subsurface of Nahorkatiya, Assam, India by Spontaneous Potential (SP) and Short Normal (SN) resistivity logs (after Azad et al. 1971). (redrawn with permission of Schlumberger Technical Services, Inc.).

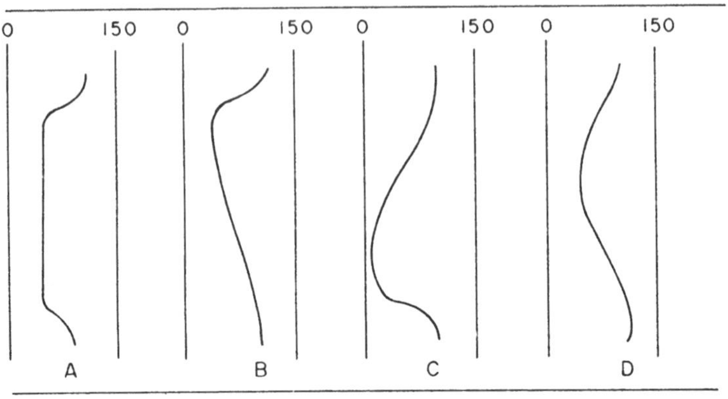

Fig. 8.13: Idealised gamma-ray and SP log patterns in deposits of different types. Note that similar curves may be produced in different environments (simplified and redrawn after Cant 1984).

(A) Cylindrical - clean sand of eolian, braided fluvial, carbonate shelf, reef or submarine canyon-fill origin.
(B) Funnel-shaped, coarsening-upward. Likely depositional environment: crevasse splay, fluvial distributary bar, barrier island, submarine fan lobe.
(C) Bell-shaped, fining-upward. Point bar, deep sea channel, transgressive shelf sand.
(D) Symmetrical. Sandy offshore bar or transgressive shelf sand.

with the help of interval velocities, accurate information about subsurface disposition of the reflecting horizons can be obtained by this method. The data obtained are utilised for subsurface structural mapping, correlation and facies analysis. Modern techniques of seismic recording and display allow visualisation of the subsurface picture over large areas.

For the purpose of geological interpretation composite cross-sections running across the basin under study are constructed using seismic reflections. One such composite profile prepared for the western part of the Bengal Basin by means of seven seismic reflectors is shown in Fig. 7.22. Two of these reflectors, one emanating from the top of an extensive basalt flow underlying the Tertiary sediments ('R'-reflector) and the other corresponding to the interface of a limestone and a shale bed ('C'-reflector), were used as markers for geological interpretation. The seismic signatures of both the reflectors could be correlated over large areas. Five other reflections from shallower horizons terminated at different distances basinward. This phenomenon was interpreted as a basinward change of lithofacies. The type of regional correlation achieved in the Bengal Basin by co-ordinated electro logging and seismic survey is illustrated in Fig. 8.14.

The conventional technique of seismic interpretation expects the reflections to correspond to lithological interfaces characterised by sharp contrasts

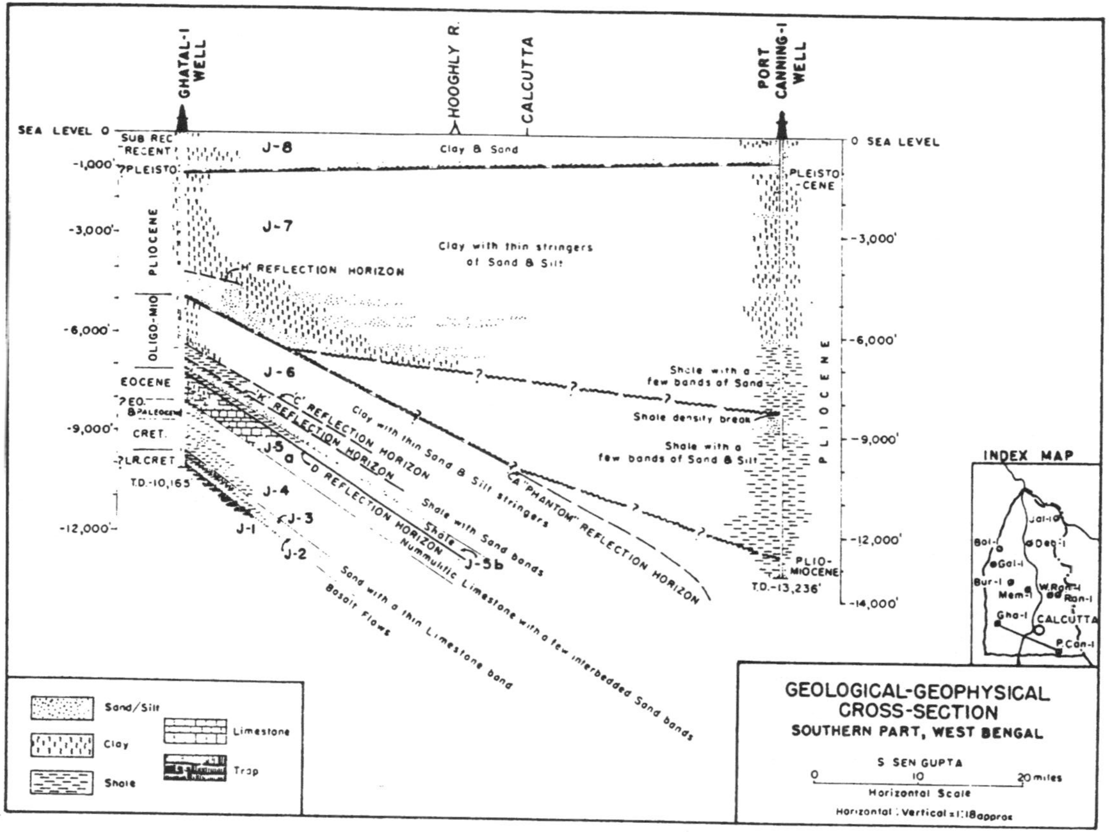

Fig. 8.14: Cross-section across southern part of West Bengal showing correlation of subsurface rock units by seismic reflections and electrical logs (after Sengupta 1966, with permission of American Association of Petroleum Geologists).

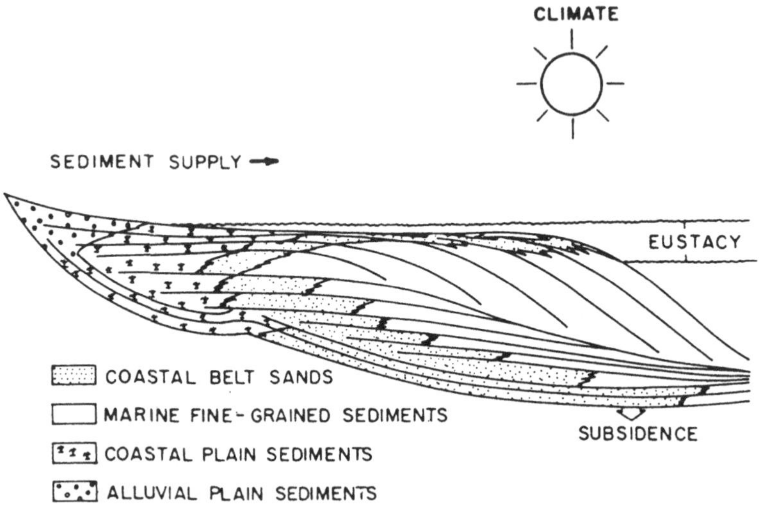

Fig. 8.15: Diagram illustrating development of stratigraphy as a function of sediment supply, subsidence, sea-level change and climate. Thick black lines indicate seismic reflections (redrawn from Vail, in Belley 1987, fig. 1, with permission of American Association of Petroleum Geologists).

in acoustic impedance (product of velocity and density). Marker reflections generated from such interfaces of regional extent are characterised by their coherence over a number of seismograms. The 'C'- and the 'R'-reflections in the Bengal Basin are examples of such reflections. Apart from the marker reflections a number of minor ones, apparently not connected with strong lithological interfaces, are also found in every sedimentary basin. Such a situation is represented in Fig. 8.15. Seismic reflections, represented by thick black lines in this diagram, follow depositional (stratal) surfaces which are ancient depositional planes, corresponding to time lines. These reflections cross rock formation (lithofacies) boundaries.

Studies by Gardner, Gardner and Gregory (1974) indicate that the wave propagation velocities for shales, sandstones and limestones overlap at depth, thereby making their interfaces unsuitable for seismic reflections. Constructive and destructive interferences (called 'Fresnel Zone Effect') between reflections from the top and bottom of interfaces also place a limit on seismic resolution. Although some rock units may be physically persistent over large areas, their reflections may go undetected because they are too thin. Many reflections traced from one seismogram to another have been shown to be parallel not to any lithological interface but to isochronal planes ('time significant surfaces') terminating at different distances during a phase of onlap (Cross and Lessenger 1988).

Unconformities are often the best seismic reflectors. Angular unconformities involving sharp changes in dip in the overlying and underlying horizons are the easiest to detect. Even when angular changes are not involved, an unconformity may be detected by the contrast in acoustic impedance created by the presence of a weathered zone at the plane of erosion. Unconformities correlatable on a global scale by seismic reflections have been interpreted to be the effect of global sea-level fluctuations.

MODERN CONCEPTS IN STRATIGRAPHY

Seismic Stratigraphy

High-resolution seismic techniques now make detection of the three-dimensional external morphology of subsurface bodies possible. The nature of facies ('seismic facies') constituting these bodies can also be inferred from reflection configuration, continuity, amplitude, frequency and interval velocity. Packages containing sets of internally consistent seismic reflections, bounded by unconformities (abrupt termination of reflections), have been termed seismic sequences. Seismic stratigraphy reconstructs the geologic history from these seismic sequences (Mitchum et al. 1977; Sheriff 1982).

Stratal patterns and lithofacies distributions, in the seismic stratigraphic concept are controlled by four variables: rate of tectonic subsidence, rate of eustatic change, volume of sediment input and climate. The first three control the rate of sediment input and the pattern of sediment dispersal. The nature of the sediments involved—siliciclastic, carbonate or evaporite—to some extent depend on temperature and rainfall (Fig. 8.15). The sequence boundary in the seismic stratigraphic model, as defined by the geologists of the EXXON Production Research Company, is an erosional unconformity caused by a relative fall in sea-level, exposing large parts of continental shelves to subaerial action. During this subaerial exposure rivers incised deep valleys on the continental shelves, shelf-edge deltas were formed, submarine canyons were rejuvenated and submarine fans, which underwent rapid sedimentation due to sediment feeding through canyons, onlapped on the continental slopes. These phenomena comprise what has been termed a Lowstand Systems Tract or LST (Fig. 8.16A).

With a rise in sea-level, the sea transgressed on the shelf. Two situations arose depending on the relative rates of subsidence and sedimentation. When the rate of sea-level rise was slow, sediments blanketed the shelf, covering the previously incised river valleys. The terrigenous sediment load was deposited mainly in the coastal deltas and sediment supply to submarine fans slowed down. This stage has been named the Transgressive Systems Tract (TST). When the sea level was high and the sediment supply was also large, the coastal plain prograded across the shelf. The thick alluvial and deltaic complex developed at this stage constituted what has been named a Highstand Systems Tract or HST (see Fig. 8.16B). The well-known post-glacial delta lobes of the Mississippi River (Fig. 6.31) are cited

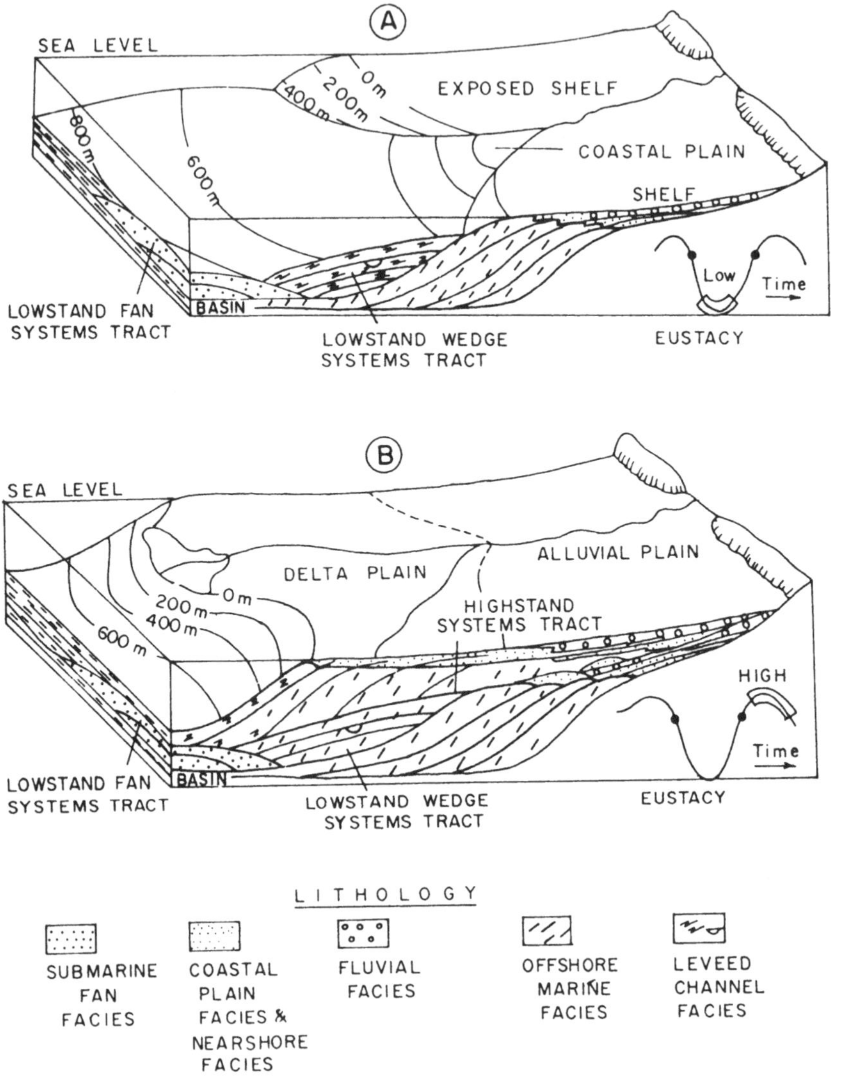

Fig. 8.16: (A) Lowstand Systems Tract (LST) and (B) Highstand Systems Tract (HST) produced respectively by sea-level fall and rise. Note the timing of the model with respect to the sea-level sine curve (modified and redrawn after Posamentier et al. 1988).

as examples of High- and Lowstand Systems Tracts. The first three of these lobes represent the transgressive systems tract. The others are highstand systems tracts (Kolb and Van Lopik 1966, see also Fig. 6.31). Each of these lobes represent a parasequence (Miall 1984), which, as defined by seismic stratigraphers, is 'a relatively conformable succession of genetically related beds or bedsets bounded by marine flooding surfaces and their correlative surfaces. Parasequences are progradational and therefore the beds within parasequences shoal upward' (Van Wagoner et al. 1990).

Sequence boundaries, downlap surfaces, and systems tracts are shown in an idealised cross-section in Fig. 8.17. A disconformity characterised by a regional onlap above and a surface of truncation below, is quite likely a sequence boundary. Reflections dipping at sharp angles are the downlap surfaces. Such facies interpretations, applicable to marine and coastal deposits, cannot be readily continued landwards because the depositional surfaces are levelled there by erosional processes leaving only parallel to subparallel reflections.

Sequence Stratigraphy

Identification of genetically related reflection packages bounded above and below by regional onlap or truncation led to a new concept in stratigraphy named sequence stratigraphy. The basic definitions of sequence stratigraphy are quoted below from publications by the EXXON Research Group.

'A sequence is defined as a relatively conformable, genetically related succession of strata bounded by unconformities or their correlative conformities' (Mitchum 1977). The building blocks of sequences which form in deep water are parasequence and parasequence sets.

A parasequence is 'a relatively conformable, genetically related succession of beds or bedsets bounded by marine flooding surfaces or their correl-

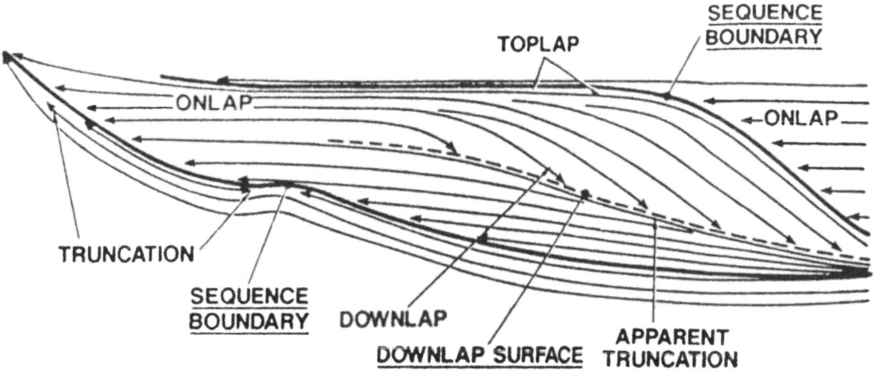

Fig. 8.17: Depositional discontinuities and seismic reflection termination patterns produced in a situation comparable to that shown in Fig. 8.15 (after Vail 1987, fig. 3, with permission of American Association of Petroleum Geologists).

ative surfaces.' A parasequence set is 'a succession of genetically related parasequences that form a distinctive stacking pattern, bounded, in many cases, by major marine flooding surfaces and their correlative surfaces. Parasequence and parasequence set boundaries form in response to an increase in water depth. Under certain depositional conditions parasequence and parasequence set boundaries may coincide with sequence boundaries' (Van Wagoner et al. 1990).

Parasequences are composed of bedsets, beds, laminasets, and laminae. 'Each stratal unit, with the exception of the lamina, is a genetically related succession of strata bounded by chronostratigraphically significant surfaces, and each surface is a single, physical boundary that everywhere separates all of the strata above from all of the strata below over the extent of the surface... Because of these properties, bounding surfaces that are correlated using well logs, cores, or outcrops, provide a high-resolution chronostratigraphic framework for facies analysis' (Van Wagoner et al. 1990).

Time-range, thickness and areal extent of the stratal units in sequence hierarchy are shown in Fig. 8.18. The characteristics of lamina, laminaset, bed, and bedset are given in Table 8.4.

The concept of stratigraphic sequences produced by unconformity-bounded strata is not new. The idea, conceived by Sloss in 1941, was elaborated by him in subsequent writings (see Sloss 1988 for a review of these developments). Recognition of regional unconformities caused by sea-level fluctuations on a global scale and identification of eustacy as the main driving mechanism for generation of unconformity bound sequences (Vail 1987; Haq et al. 1987) provided the ground for revival of the concept of sequence stratigraphy. With the development of high-resolution seismic techniques in the 1980s, the 'systems tracts' constituting sequences could be delineated. Documentation of the hierarchical units of stratigraphic sequences in well logs, cores and outcrops, firstly in the siliciclastic rocks by the geologists of the EXXON Production Research Company, and later, by others in the carbonate rocks, provided the framework for high-resolution sequence analysis. Attempts are now being made to develop quantitative models connecting sequence stratigraphy to crustal behaviour during basin subsidence and sea-level changes both on regional and global scales (see Miall 1984 for details).

The concepts of seismic and sequance stratigraphy have no doubt provided new perspectives to sedimentary basin analysis but the tools necessary for putting these concepts into practice are not readily available to those outside the major oil exploration agencies. Many practising field geologists do not have the type of familiarity with borehole logs and reflection seismograms needed for applying the concept of sequence stratigraphy to practice.

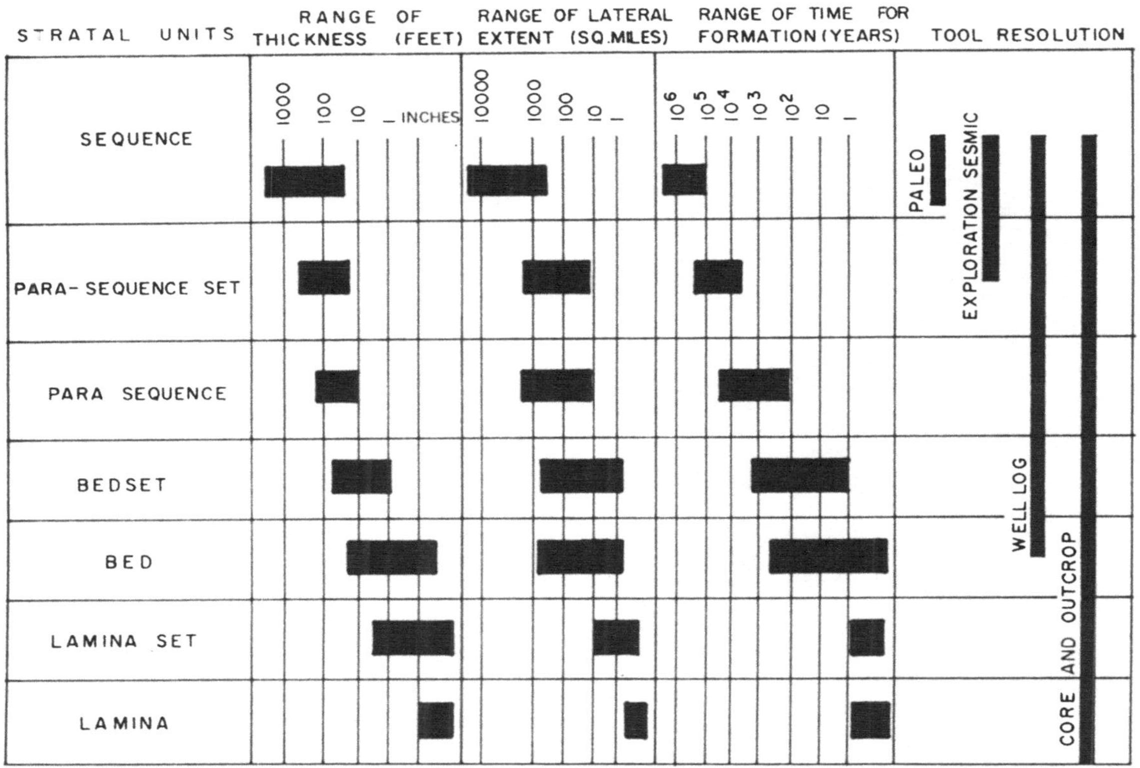

Fig. 8.18: Hierarchy of stratal units in sequence stratigraphy (modified after Van Wagoner et al. 1990, Table 1, with permission of American Association of Petroleum Geologists).

Table 8.4. Detailed characteristics of lamina, laminaset, bed and bedset (from Campbell 1967)

Stratal unit	Definition	Characteristics of constituent stratal units	Depositional processes	Characteristics of bounding surfaces
BEDSET	A relatively conformable succession of genetically related beds bounded by surfaces (called bedset surfaces) of erosion, non-deposition, or their correlative conformities	Beds above and below bedsets always differ in composition, texture, or sedimentary structure from those composing the bedset	Episodic or periodic (same as bed below)	(Same as Bed below) plus • Bedsets and bedset surfaces form over a longer period of time than beds • Commonly have a greater lateral extent than bedding surfaces
BED	A relatively conformable succession of genetically related laminae or laminasets bounded by surfaces (called bedding surfaces) of erosion, non-deposition or their correlative conformities	Not all beds contain laminasets	Episodic or periodic Episodic deposition includes deposition from storms, floods, debris flows, turbidity currents Periodic deposition includes deposition from seasonal or climatic changes	• Form rapidly, minutes to years • Separate all younger strata from all older strata over the extent of the surfaces • Facies changes are bounded by bedding surfaces • Useful for chronostratigraphy under certain circumstances • Time represented by bedding surfaces probably greater than time represented by beds • Areal extents vary widely from square feet to 1000s square miles
LAMINASET	A relatively conformable succession of genetically related laminae bounded by surfaces (called laminaset surface) of erosion, non-deposition or their correlative conformities	Consists of a group or set of conformable laminae that compose distinctive structures in a bed	Episodic, commonly found in wave- or current-rippled beds, turbidites, wave rippled intervals in hummocky bedsets, or cross-beds as reverse flow ripples or rippled toes of foresets	• Form rapidly, minutes to days • Smaller areal extent than encompassing bed
LAMINA	The smallest megascopic layer	Uniform in composition/texture Never internally layered	Episodic	• Form very rapidly, minutes to hours • Smaller areal extent than encompassing bed

The techniques of identification of sequence boundaries and their component systems tracts in outcrops are still being perfected by the oil company research groups. Meanwhile, practising geologists may not find it easy to match the building blocks of a sequence with the units of the Stratigraphic Code with which they have had long familiarity. Those working on intracratonic, non-marine sequences may like to have more distinctive, local evidence of transgression and regression than global eustatic changes (see for example, the Genetic Stratigraphic Sequence model of Galloway 1989). A number of problems concerning the global sea-level changes also remain to be solved. For example, it is not clear why the proposed model of global sea-level fluctuation satisfies the sedimentary record in only a few (but not all) basins (see de V. Klein 1987). Such confusions and controversies will continue until the concept of sequence stratigraphy has been widely applied and its merits and demerits evaluated on a global scale.

9

BASIN ANALYSIS: A SYNTHESIS

Basin analysis, the ultimate goal of a sedimentological study, unravels the complete depositional history of a sedimentary basin. This involves synthesis of observations on the framework of the basin, its sediment fill and also the biota associated with the sediments. The stages of basin evolution and the patterns of sediment dispersal are also to be worked out in detail. In short, basin analysis calls for complete knowledge of all aspects covering the processes and products of sedimentation within the sedimentary prism which constitutes a basin. This also demands a command over all the techniques and methods of sedimentology and related disciplines on the part of the analyser.

Until about the fifties of the present century basin analysis was a relatively simple affair. It involved, broadly speaking, erection of local stratigraphic columns within the sedimentary basin and correlation of these units with the components of the conventional geosynclinal model which was popular at that time. Basin analysis today is far more complicated. Mapping the basement configuration of a sedimentary basin now calls for intensive application of various geophysical techniques. The geometry of the sediment fill has to be worked out in three dimensions using subsurface data, the nature of the fill has to be interpreted in terms of facies models, and the dispersal patterns have to be determined in quantitative terms, using the modern techniques of palaeocurrent analysis. To complete the story of basin evolution, one would like to have radiometric dates not only for the layers of sediments, but also for the intrusives occurring within the basin.

Finally, today's students of sedimentology would like to interpret the stages of basin evolution in terms of lithospheric plate movements, following the currently popular model of plate tectonics. Constant feedback of information from field to laboratory and vice versa is essential. The ultimate aim is to interpret the products of sedimentation in terms of the processes involved. Co-ordinated effort on the part of specialists in various branches of earth science is required to achieve this goal. In most of the exploratory organisations this is done by a team designated as 'the basin study group'.

A simple flow chart for basin analysis is presented in Table 9.1. The procedure to be followed is divided into three broad categories—geological, geophysical and geochemical. The sequence of analysis to be followed is

Table 9.1. Flow sheet for basin analysis

			Work to be undertaken	Phase I		Phase II		Phase III	
				Field	Lab.	Field	Lab.	Field	Lab.
Nature of Investigation	Geological	Field studies	Mapping of the basin boundary and the lithic fill. Establishing lithostratigraphic units, unconformities etc.	I					
			Preparation of measured stratigraphic columns from surface and subsurface data. Study of sedimentary structures, biota.			II			
		Laboratory	Thin section petrography, modal, textural and heavy mineral analysis.		I				
			Clay mineral studies by XRD, SEM etc. Palaeontological, biometric and palynological studies.				II		
	Geo-chemical		Determination of major and trace elements, organic content, maturation etc. Radiometric dating.						III
	Geophysical	Field	Magnetostratigraphy						III
			Aeromagnetic and gravity surveys.	I					
			Reflection seismic survey, DSS.			II			
			Drilling and borehole logging.			II		III	
Purpose			Determination of:	Basin configuration, nature of sediment fill, facies variation.		Basin floor pattern, palaeocurrent, palaeohydraulics, palaeoecology, depositional and diagenetic conditions.		Seismic facies, and sequences. Regional correlation. Pattern of sedimentation in relation to tectonics. Maturation of sediments.	
Conclusions				Nature of the basin, its lithic fill and provenance of sediments.		Basin architecture. Sediment dispersal pattern, depositional enviornment and diagenetic trends.		Basin evolution with respect to geodynamics and global sea-level changes. Hydrocarbon prospects.	

briefly discussed below. For details of the analytical techniques the reader should refer to the chapters cited.

The study of a virgin sedimentary basin commences with mapping the depositional boundary and obtaining as much information as possible on the basin architecture. The approach is necessarily different for relatively small, intracratonic basins (e.g. pockets of Gondwana deposits within the Indian craton) and large sedimentary basins of the type of the Bengal Basin, which run across cratonic or continental margins and continue on the oceanic crust. While conventional geological mapping followed by exploratory drilling provides answers to most of the questions in the former case, a well co-ordinated geological-geophysical approach is essential for investigation of basins of the latter type.

An aeromagnetic survey, which provides the quickest method for obtaining the depth of the basin-floor by detecting the boundary between the magnetically susceptible, presedimentary, often metamorphosed basement and the non-magnetic sediments overlying it, is useful in detecting the gross thickness of sediments. The overall direction of thickening and thinning can also be determined by the same method. An aeromagnetic survey is generally followed by ground gravitymeter survey for obtaining information on density anomalies in the subsurface. The data obtained are sometimes useful although the inherent ambiguity in the interpretation of gravity data may not always provide a unique solution to problems.

Basin and its Lithic Fill

Geological mapping of the lithic fill should be started simultaneously with the aeromagnetic survey. The purpose of mapping is to establish the lithostratigraphic units following the Code of Stratigraphic Nomenclature. Mapping begins at the level of formations. The lower and the higher ranking units in the hierarchy (member and group respectively) can be demarcated within an area once the formations have been established. The nature of unconformity below and within the stratigraphic sequence should also be worked out in the course of mapping (Chapter 8). Three-dimensional disposition of the rock units within a basin is represented by a variety of composite maps prepared from measured stratigraphic sections. In the absence of suitable exposures, information on rock units is obtained either from boreholes or from two-way seismic travel times from reflecting interfaces in the subsurface. The reflection times, converted to depths by means of interval velocities, are used for preparation of structural maps, isopachs, palaeogeologic and palaeotopographic maps.

Structural maps show relief of the surface of interest with respect to a datum which is commonly the mean sea level. These are comparable to the topographic maps prepared at the surface. Palinspastic adjustments for folding, faulting and regional tilt may be incorporated into these maps.

Isopach maps depict variation in thickness of the stratigraphic unit of interest. This variation is represented by contours; zero contour, indicative of total merger of the upper and lower surfaces of the unit, defines the limit of pinchout. The shape of the unit thus defined often helps to decipher the origin and depositional environment of the concerned rock unit.

Lithofacies maps depict aerial variations in rock types (e.g., sand, shale, carbonates), their proportions, volumes and ratios (e.g., sand/shale, shale/carbonate) within particular stratigraphic units. Isolith maps show distributions as well as thickness variation of individual facies. Aerial distributions of cumulative thickness of particular rock types is shown by net rock-type maps.

Having obtained a broad picture of the basin architecture and the overall nature of the basin fill, one should undertake a petrographic study of the rock samples collected from the field. Modal analysis of sandstone samples by the point-counting method provides quantitative information on the mineral proportions. The sandstones are broadly classified into arenites and wackes on the basis of matrix content. The proportions of quartz, feldspars and rock fragments are then plotted on a QFR diagram for further subdivision of the sandstone type (see Fig. 4.3). Proportions of undulose, non-undulose and polycrystalline varieties of quartz, plotted on the Indiana diamond diagram (see Fig. 2.3), give clues to the nature of the non-sedimentary source rock. The heavy mineral assemblage (see Table 2.2), degree of sorting and rounding (see Fig. 3.9) of the clastic particles provide insight not only into the possible provenance, but also into the transportational history of the sediments. Grain-size data may be utilised in a limited way for interpreting palaeo-environment and palaeohydraulics (Chapter 3).

The carbonate rocks may be classified either by their depositional texture, as proposed by Dunham (1962), or by identification of the allo- and orthochemical components under the microscope (Folk 1959). The type of carbonate (calcite/aragonite/dolomite) is deciphered by staining (Friedman 1959). Replacement textures, if any, may also be identified at this stage. Identification of mineralogy of the carbonates and the fine-grained argillaceous sediments, by more sophisticated methods, are undertaken at an advanced stage of investigation. The mineralogical assemblage within terrigenous or chemical sediments are often indicative of the chemical conditions (Eh and pH) prevailing at the site of deposition (see Fig. 4.18).

In the second phase of field work the primary sedimentary structures are studied in detail. These can aid in stratigraphic studies in many different ways. Graded bedding and ripple marks, for example, provide clues to the attitude (top and bottom) of the strata. For this purpose, wave ripples are of greater help than current ripples. Sharp crests and indented troughs of wave ripples can be easily identified even when they occur in reversed position at the sole of beds (see Fig. 5.23). In areas of structural complexity, when the original sequence is inverted, the depositional top and bottom of the sequence can be determined from the wave ripples. The external morphology of a current ripple is of little help in the determination of the depositional top of the sequence but the internal cross-lamination might be useful in such a case (see Fig. 5.23). Some of the common scour marks, such as flutes and grooves, have been traditionally used for determination of the depositional top. Such interpretations are based on the belief that the groove and the flute marks are the counterparts of the original casts formed as scours on the bed surface. However, beds having positive reliefs resembling flutes have been reported from the upper surface of non-cohesive sediment beds also (Friedman and Sanders 1974). Thus the technique of determination of the depositional top from scour marks may not always provide dependable results.

Palaeoslope and Palaeocurrent

The initial subaqueous dip or slope of a basin-floor at the time of sedimentation is called the palaeoslope. The term is related to certain subaqueous deposits only. Eolian deposits, controlled wholly by wind direction, have very little relationship with the slope of the depositional surface. For fluvial environment the palaeoslope is indicated by cross-bedding foreset azimuths, averaged over a large area. This is not necessarily so in marine or tidal environments. Before attempting an interpretation of the palaeoslope therefore, the nature of the depositing medium (water or wind) and the environment of sedimentation must be ascertained by criteria which are independent of palaeocurrent direction.

Palaeocurrent analysis involves the study of ancient sediment dispersal patterns. An asymmetric ripple mark is one of the most common linear structures used for palaeocurrent study. Crest lines of asymmetric ripples are generally normal to the direction of sediment transport. Palaeocurrent directions can also be obtained from scour marks such as flutes and current crescents. The direction of flare of these structures indicates the direction of sediment transportation. Orientation of sole marks and other current oriented linear features, such as parting lineation and grain orientation, may also be used as palaeocurrent indicators.

Cross-beddings of various shapes and sizes are the commonest tools used for palaeocurrent analysis. The reliability of different types of cross-bedding for palaeocurrent interpretation has been discussed by High and Picard (1974). For a planar-tabular cross-bedding, palaeocurrent is indicated by the direction of inclination of the foreset, but random measurement of dips on the flanks, in the case of a trough type cross-bedding, might give misleading results. Such observations must be corrected, taking into consideration the plunge of the trough axis, to obtain the trough axial azimuth (Slingerland and Williams 1979).

When the area of study is structurally tilted or folded, a correction has to be applied to restore the bed containing the directional elements to its original position. In the case of cross-bedding, this is done by rotating the plane of the cross-bedded unit around the strike of the true bedding through the angle of the bedding (see Potter and Pettijohn 1977, for the technique involved). Tilts in linear structures may also be similarly treated. Empirical studies have shown that tilt correction is negligible when the structural dip is less than 25° and also when the difference between the strike directions of bedding and cross-bedding is small (<10°). In all other cases tilt correction is necessary. Computer programs for tilt correction are now available (Saha and Chakraborty 1990).

Conclusions on regional palaeocurrent trends should be based, as far as possible, on a large number of observations on cross-bedding azimuth of the same type, because they are likely to be indicative of the same depositional environment. At the same time, when the geological formation under study contains a large number of directional features of a particular type, it is necessary to know the minimum number of observations that would give the mean direction for the whole formation with a specified precision. This is a sampling problem. The technique of sampling developed by Rao and Sengupta (1970, 1972), answers the following questions: 1) the minimum sample size required for estimating, with a desired precision, the mean palaeocurrent direction of a formation and 2) the optimum allocation of samples between and within the outcrops that would allow efficient sampling at minimum cost. The answers to these questions are naturally related to the amount of variation between the cross-bedding azimuths. In the scheme of sampling proposed by Rao and Sengupta (1970), the between- and within-outcrop variations in cross-bedding azimuths are determined by a pilot survey before the actual sampling plan is applied.

Summation and representation of palaeocurrent directions also pose special problems (see discussions in Rao and Sengupta 1970; Potter and Pettijohn 1977). The current directions are generally represented graphically, in the form of a circular histogram ('rose diagram'). Circular histograms plotted on linear scales may give a wrong impression about the nature of distribution because the area of the 'pie' in such a case is not proportional to the frequency (density). Circular histograms with areas proportional to density may be constructed by using the specially designed, non-linear frequency net devised by Nemec (1988).

For the purpose of representation of palaeocurrent directions, the basin under study is generally divided into a number of sectors of equal dimensions (Fig. 9.1). The mean direction of sediment transportation (vector mean given by $\hat{\gamma} = \tan^{-1} W/V$, where W and V are respectively the sums of sines and cosines of current directions within each sector) is plotted by an arrow at the centre of each sector. The length of the arrow, in each case, is made

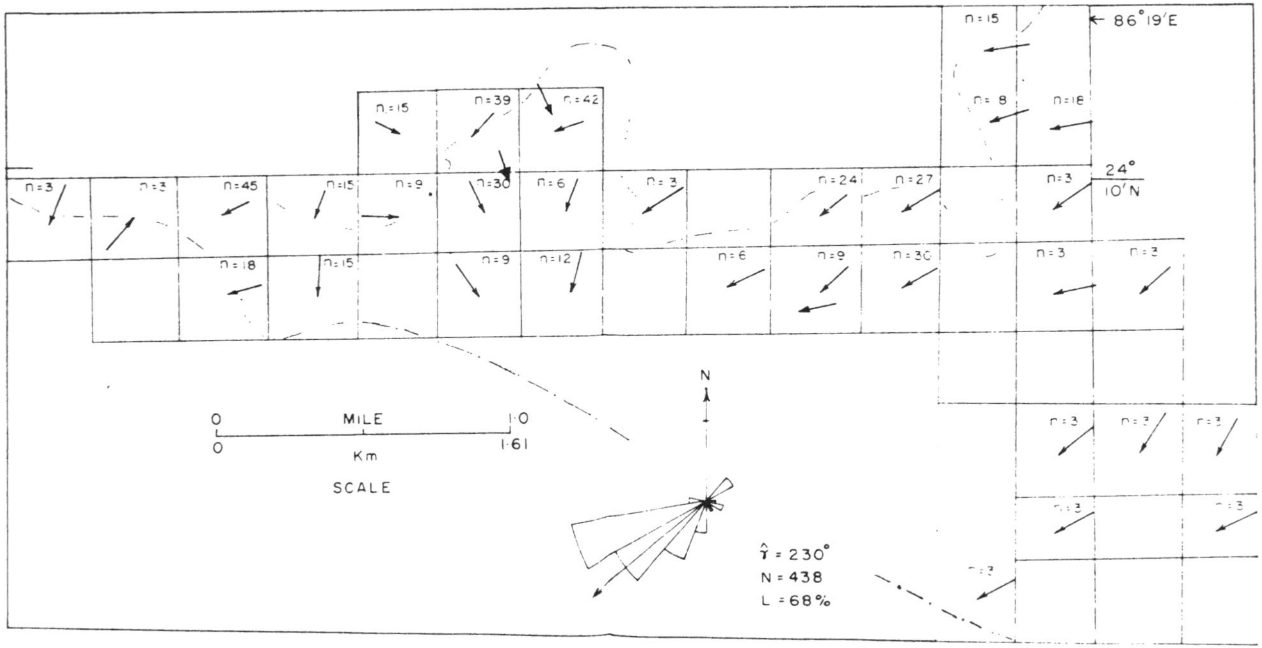

Fig. 9.1: Palaeocurrent directions worked out from cross-bedding foreset azimuths in the Giridih Basin, Bihar, India (from Sengupta et al. 1988).

proportional to the strength of the vector resultant or consistency ratio, given by $L = R/N$, where R and N represent the strength of the resultant vector and the number of observations respectively. L is represented in percentage by multiplying it by 100. R is given by $R = \sqrt{W^2 + V^2}$.

Palaeocurrent directions can also be obtained with the help of scalar attributes such as grain-size, grain roundness and bed thickness. Mean grain-size of a sediment population decreases while roundness increases towards the direction of the current flow. Both bedding and cross-bedding thickness also show overall decrease towards the direction of current. Systematic mapping, taking into consideration observations on as many points as possible, is needed for working out the direction of maximum change of the scalar property under consideration. Unlike directional properties, a single observation on a scalar property provides no clue to the palaeocurrent. In the case of subsurface data, however, scalar elements have a distinct advantage over the directional ('vectorial') ones. Palaeocurrent maps utilising scalar elements can be prepared from unoriented core samples but oriented cores are essential when directional properties are involved.

Depositional Environment

It is not advisable to draw a conclusion on depositional environment from a single line of evidence because a particular sedimentary structure or palaeocurrent pattern may occur in several environments in which similar hydraulic conditions prevail. An integrated facies model approach, taking into consideration the information drawn from various lines of evidence—lithofacies, sedimentary structure, texture, geometry of the sand bodies, nature of coarsening or fining-upward sequences—should be put together for drawing a meaningful conclusion.

Most of the common sedimentary structures may originate in more than one depositional environment but particular combinations of sedimentary structures are often diagnostic of certain depositional conditions (see Chapter 6 for details). The dispersion pattern of the palaeocurrent may, under certain circumstances, be indicative of a specific depositional environment (Selley 1968). For example, the palaeocurrent patterns within fluvial systems may be of two types—radiating and converging. The former is indicative of alluvial fans and the latter is characteristic of coal basins. The dispersion of current directions within piedmont fans and braided stream channels is naturally high. For meandering channels, on the other hand, the dispersion is particularly low (Fig. 9.2, a–d).

The palaeocurrent patterns for deep-sea fans are unimodal, as in fluvial systems. Considered on a regional scale, however, two different patterns are possible. For small-scale turbidity currents, as in the outlet of submarine canyons or on slopes of delta lobes, the pattern is unimodal with small dispersion (Fig. 9.2, a). In turbidite-filled troughs the general flow direction

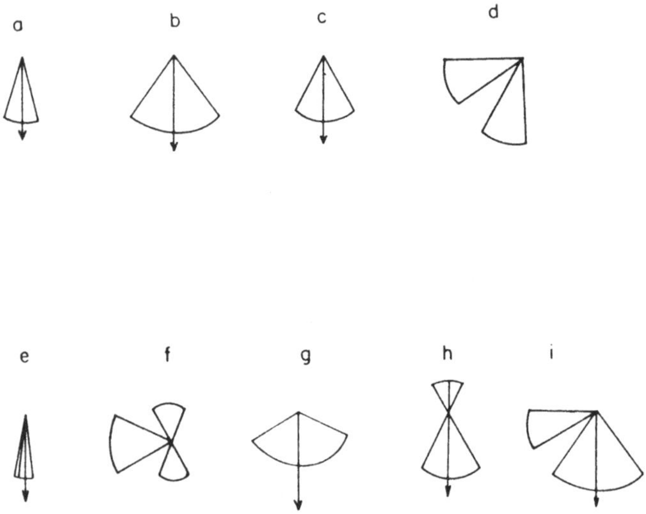

Fig. 9.2: Idealised palaeocurrent dispersion patterns in various depositional environments: (a) meandering stream channels or turbidity flows, (b) alluvial or submarine fans, (c) braided streams, (d) converging fluvial channels as in coal basins, (e) tidal inlets, (f) turbidite filled troughs, (g) radiating distributaries in deltas, (h) tidal flats, (i) longshore drifts superposed on fluvial currents (based on Selley 1968, Allen 1968a).

remains longitudinal and parallel to the trough axis although sedimentation may take place laterally from the margins. This leads to an essentially unimodal pattern, with two subordinate modes lying nearly at right angles to the main flow direction (Fig. 9.2, f).

In tidal flats bimodal and bipolar current directions, normal to the shoreline, are common (Fig. 9.2, h). Bipolarity develops due to landward flow of current during flood tide and flow in the opposite direction during ebb tide. A perpendicular arrangement of current directions may also be produced when fluvial currents flowing seaward alternate with longshore drifts (Fig. 9.2, i). In tidal inlets, on the other hand, the currents are unimodal, with small dispersion, being guided either by ebb or by flood (Fig. 9.2, e). Sea beaches may show a variety of palaeocurrent patterns, depending on whether the agencies responsible for sediment transportation are wholly marine or have both marine and tidal influence. Deltas produced by fluvial systems debouching into sea or lake may produce radiating palaeocurrent patterns (Fig. 9.2, g).

Subtle changes within the sedimentary structures may also be used as indicators of variation in physical condition within a depositional environment. Cross-strata sets, for example, are sometimes interrupted by sloping surfaces called reactivation surfaces (Fig. 9.3B). These may be caused by changes in flow stages (a lowering of water level followed by rising) of the depositing medium. In the case of reversal of flow direction due to tidal effect,

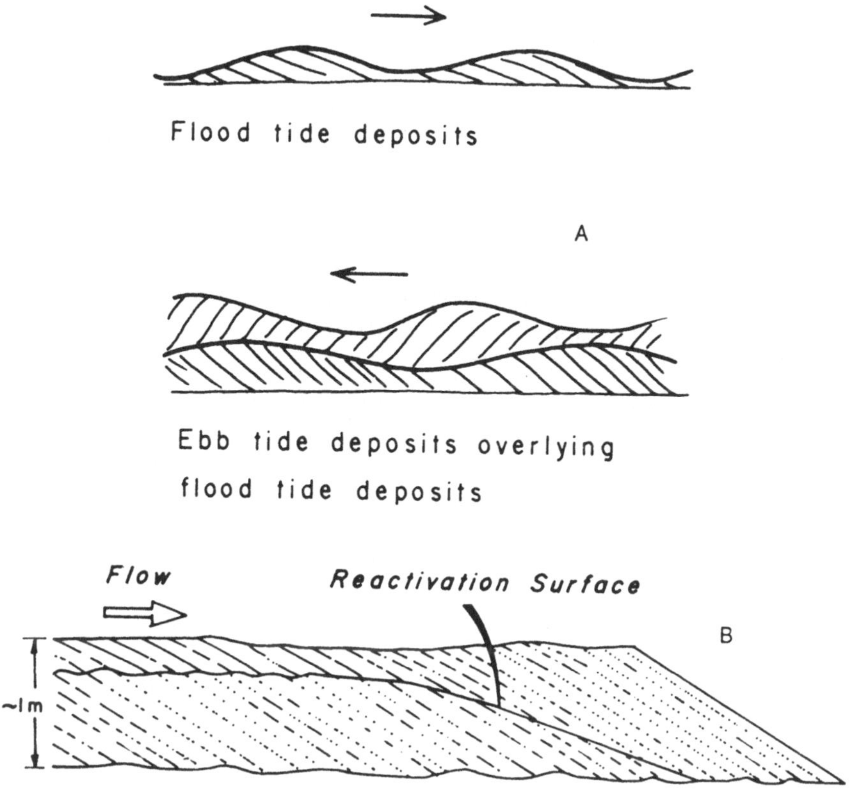

Fig. 9.3: (A) Herringbone structures produced by reversal of current directions in tidal environment (modified and redrawn after Fritz and Moore 1988, fig. 7.25). (B) Reactivation surface produced by changes in flow stage (after Harms et al. 1975, fig. 3-4).

the cross-strata above and below the reactivation surface may dip in opposite directions, giving the impression of herringbone structures (Fig. 9.3A).

The interpretation of depositional environment is aided also by the logs of vertical stratigraphic columns, generally prepared during an advanced phase of field work. These may be constructed either from the exposed sections or, preferably, from borehole data when these are available. The field (or borehole) data may be utilised for studying the vertical variation in lithofacies and also for preparing maps. Correlation of borehole data and preparation of stratigraphic sections also provide clues to the three-dimensional geometry of sand bodies, which are useful for environmental interpretation. The choice of an appropriate datum is important for this purpose. For example, the same sand body may look like a distributary channel sand having a convex bottom or a sand bar with a convex top, depending on whether the upper or the lower surface is used as the datum (Conybeare 1979).

Wireline logs, run within the boreholes, help not only in regional correlation, but also in defining the facies boundaries. A bed having a distinctive log character may be used as a marker bed for the purpose of correlation. To some extent the log characters are indicative of depositional environments. The deposits in low-energy environments, as in lakes for example, are good markers but those in high-energy environments, such as rivers, are impersistent. No log pattern, however, can be uniquely related to any particular depositional environment. For example, the gamma-ray or S.P. patterns shown in Fig. 8.13 can be found in any of the depositional environments listed in this figure. A line of evidence, independent of the log pattern, must be used for choosing the most probable depositional environment.

Under suitable conditions dipmeter logs may be also used for interpreting the depositional environment. Four different dipmeter log patterns are generally recognised (Fig. 9.4): 1) the 'green pattern' showing vertically constant dips, 2) the 'blue pattern' showing downward decrease of dip, 3) the 'red pattern' showing downward increase of dip and 4) the 'random pattern' showing no systematic variation of dip. Although each of these four patterns can generate in a number of depositional environments, with some prior information on the type of environment to be expected, one can make an intelligent guess about the type of environment involved. In the case of an alluvial channel, for example, the thickest and the steepest cross-beds are expected near the basal scour ('the red pattern') (see Fig. 6.19), but in a prograding delta the dips will decrease downwards ('the blue pattern') (see Fig. 8.1B). The Schlumberger (1970) dipmeter manual discusses in detail the technique of interpretation.

Other uses of stratigraphic logs include analysis of the nature of cyclicity, i.e., identification of auto- and allocycles. The vertically fining- or coarsening-upward cycles may also be worked out following the techniques suggested in Chapter 8. These also have an important bearing on interpretation of depositional environment.

Palaeohydraulic Interpretation in Fluvial Channels

The dimensions and types of sedimentary structures are intimately related to the hydraulic conditions responsible for their origin. Sedimentary structures may thus be used as clues to the geometry and hydraulics of ancient flow channels. In one such attempt Bridge (1978) used the following relationship:

$$V_m = \sqrt{\frac{8 g y_m \cdot S_m}{f}},$$

where V_m is the mean velocity for a cross-section at the position of mean depth (y_m), S_m is longitudinal water-surface slope along the locus of mean depth (≈ channel centre line) of a curved channel, and f is the Darcy-

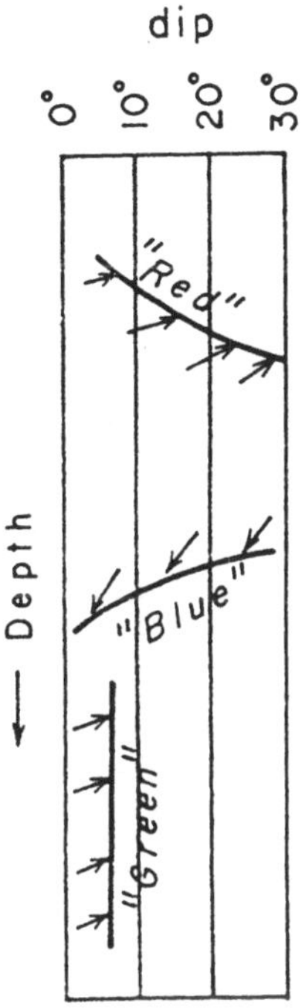

Fig. 9.4: Typical dipmeter log patterns.

Weisbach friction factor, estimated to be 0.02 for horizontal lamination in upper flow regime plane beds, 0.04 to 0.06 for large-scale trough cross-stratification in fully developed dunes and 0.03 for small-scale (ripple) cross-stratification, or tabular cross-stratification from diminished dunes (Bridge 1978, p. 736).

Using a series of hydraulic relationships which are valid for meandering stream channels, Casshyap and Khan (1982) worked out the palaeohydraulics of the Permian streams responsible for Gondwana sedimentation in the Bokaro Basin of India. Using similar techniques, Gardner (1983) worked out the palaeohydrology and palaeomorphology of a Carboniferous meandering stream in eastern Kentucky. That such interpretations can provide a consistent picture of the regional palaeohydraulic changes was shown by contouring the palaeohydraulic parameters of the Karharbari and the Barakar Formations (Permian) in the Giridih Basin, India (Sengupta et al. 1988; Fig. 9.5). Several well-known hydraulic relationships were used at Giridih in the following sequence to estimate palaeohydraulic parameters from dimensions of sedimentary structures:

$H = 0.086\, d^{1.19}$ (Allen 1968)

$W = 42\, d^{1.11}$ (Allen 1968)

$$L_m = 10.9\ W^{1.01}$$ *(Leopold, Wolman and Miller 1964)

$$L_m = 106.1\ \bar{Q}^{-0.46}$$ *(Carlston 1965)

$$V = \bar{Q}_{ma}/A$$ (Schumm 1972)

*These equations, where the units are in f.p.s. system, are used only as intermediate steps for estimating the water discharge.

where H is the mean dune height in metres, d is the bankfull depth in metres, W is the channel width in metres, L_m is the meander wavelength in feet, $\bar{Q}$ is the mean annual discharge in cubic feet per second, V is the flow velocity in metres per second, $\bar{Q}_{ma}$ is the mean annual flood in cubic metres per second = bankfull discharge (Q), A is the channel cross-sectional area in square metres. This technique of palaeohydraulic interpretation uses dune height as the starting point for computations. The complete height of the original dune is rarely preserved in geological record and hence a correction is often necessary. Jopling (1980) suggested a method of correction for the height lost on the assumption that about four-fifths of the actual dune heights are preserved in the cross-beddings from which the dune height is estimated.

In the Giridih Gondwana Basin, where channel geometry and palaeohydraulic parameters were estimated by the above-mentioned method, the contours of channel depth, width, and discharge run normal to the palaeoflow direction obtained from cross-bedding azimuths. Moreover, the trends of width, depth and discharge parallel each other and all these parameters progressively increase in the downstream direction, as in modern rivers (see Leopold, Wolman and Miller 1964, p. 240), thereby confirming the dependability of the use of sedimentary structures in palaeohydraulic interpretations.

Diagenesis and Maturation

Detailed laboratory investigation of the core and surface samples is recommended towards the end of the second phase of study. While optical microscopy is adequate for identification of the framework minerals, XRD or SEM techniques are needed for determination of the clay mineral species. Such studies often provide clues to the diagenetic changes which the sediments have undergone. The pattern of diagenesis undergone by sediments in turn reflects the burial, thermal and fluid circulation events to which the sediments were subjected after deposition. Under suitable conditions the trend of diagenesis can also be used as a clue to geodynamics and sedimentary basin evolution. For example, precipitation of high-temperature zeolite cements within sandstones in the back-arc basins of the Western Pacific was interpreted by Lee and Klein (1986) as the effect of upwelling thermal convective circulation followed by a lower temperature downwelling phase. These reactions were preserved because the sediments underwent a slow burial rate during the active rifting phase of basin formation.

Maturation of the organic materials within sediments during burial in deep basins is of particular interest to petroleum geologists. During pro-

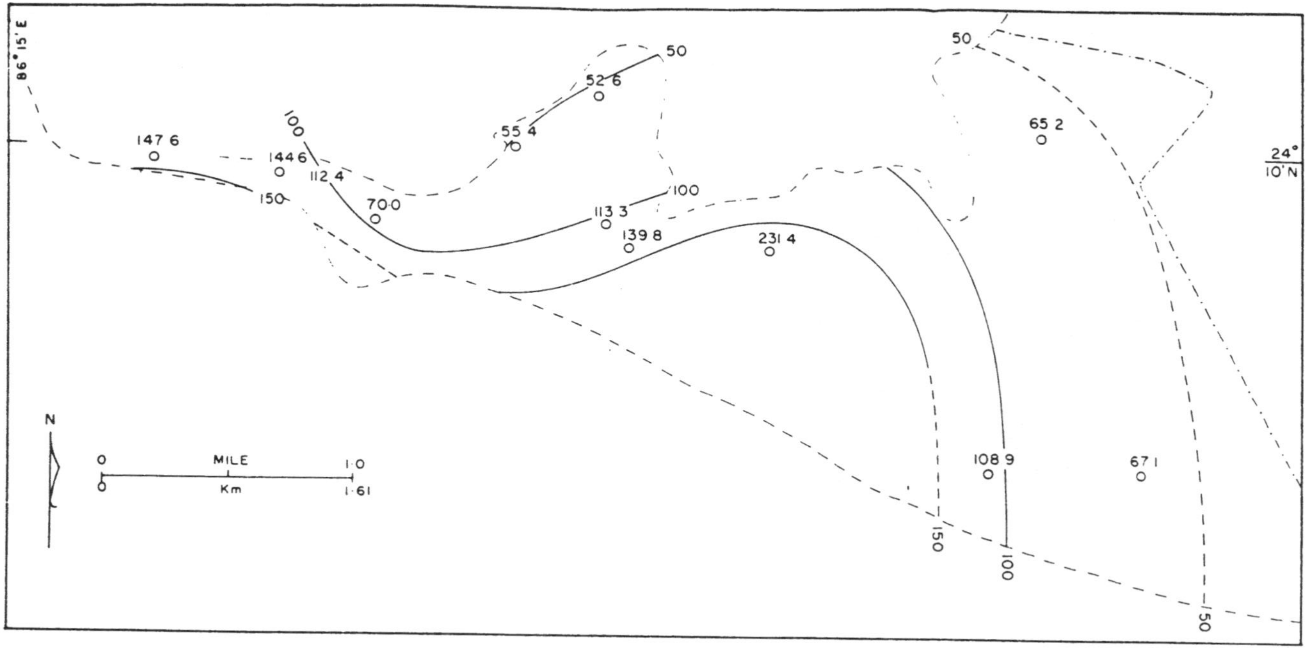

Fig. 9.5: Barakar (Permian) stream discharge (m^3/s) in the Giridih basin worked out from magnitudes of bedforms using hydraulic equations (from Sengupta et. al. 1988).

gressive burial of sediments the organic components are converted to liquid or gaseous hydrocarbons, depending on the pressure and temperature to which the sediment has been subjected. Liquid petroleum being preserved only within a narrow range of pressure and temperature, called the 'liquid window', it is of utmost importance for petroleum geologists to obtain clues to the burial history of the sediments.

Porosity and permeability of the reservoir rocks, factors of importance in prospecting for natural hydrocarbon, are affected during diagenesis and cementation. Reduction of porosity, either due to compaction during burial or due to chemical precipitation, takes place during diagenesis. Although some idea of the packing density may be obtained by counting under a microscope the number of grain-to-grain contacts in randomly cut rock sections, *in situ* determination of porosity is best done by using various types of wireline logs. Combined SP-resistivity log, sonic log, neutron log and density log are particularly helpful for *in situ* determination of porosity. Sonic log measures the time taken by acoustic waves to travel through one foot of rock formation along the borehole. Porosity (ϕ_s) is then determined by the formula:

$$\phi_s = (\Delta t - \Delta t_{ma})/(\Delta t_f - \Delta t_{ma}),$$

where Δt_f = transit time through the pore fluid,

Δt_{ma} = transit time through the rock matrix,

Δt = interval transit time

(in modern logging manuals Δt is often replaced by t).

The sonic log, which responds to primary porosity, is valid only for clean compact formations with intergranular pores filled by liquids. The presence of hydrocarbon in a liquid or gaseous state upsets measurements.

Sediment Chemistry

Analyses of the major and trace elements of the sediment samples are undertaken at an advanced stage of basin analysis. Ratios of various oxides, $(Fe_2O_3 + MgO)/TiO_2$, Al_2O_3/SiO_2, K_2O/Na_2O, $Al_2O_3/(CaO + NaO)$, may help decipher the tectonic setting of the sedimentary basin (see Table 7.2). A similar analysis may also be attempted by plotting the framework mineraology of sandstones on Dickinson-Suczek diagrams with QFL or QmFLt as end-members (see Fig. 7.10). Analysis of the immobile trace elements may also be used for working out the tectonic settings of the basins. A progressive increase in the proportions of light rare earth elements and decrease in U and Sc are expected from oceanic island arcs to passive margin settings passing through continental island arcs and active continental margins (Chapter 7).

Basin Evolution and Tectonics

Most of the large sedimentary basins are defined by a series of step faults bordering the cratonic margin, a sharp increase in sediment thickness at the basin centre, and highly deformed strata along the borderland. While the mobile belt at the borderland can be studied by conventional geological techniques, geophysical methods have to be applied for investigating the first two features because they generally lie concealed in the subsurface.

Subsidence of the basin margin concomitant with sedimentation allows a thick pile of sediment to accumulate. Simultaneously, a series of growth faults develop in the areas of rapid subsidence (see Dailly 1976 for details). The beds located above these growth faults at shallower levels show flexures. Another zone of faulting generally marks the belt of adjustment between the stable shelf of the basin and the deeper trough of relatively rapid subsidence. This was formerly designated as the 'hinge zone' of the basin (Weeks 1952). In the Bengal Basin a conspicuous flexure in the Nummulitic (Eocene) limestone horizon, detected by seismic survey in the subsurface of Calcutta, was designated as the 'hinge zone' (Sengupta 1966). This is also the area where the seismic reflections from the shallower sections above the 'hinge' terminate sharply. The zone is also marked by an abrupt increase in thickness of sediments downdip (see Fig. 7.22). This zone, therefore, is not a hinge in the tectonic sense but possibly a stratigraphic feature resulting from increase in thickness of the sediment section and consequent compaction of layers in the lower part (see Conybeare 1979). Such an interpretation suggests that the so-called hinge line of the Bengal Basin is in reality the depositional limit of the shelf sediments. On the other hand, the later discovery of continuation of the Nummulitic limestone into deeper horizons in the form of a flexure across the 'hinge line' suggests that the limestone hinge is only the shallower expression of a deep-seated system of faults which has remained undetected. The challenge of basin analysis lies in searching for the most plausible answer from a host of such possible solutions.

A sedimentary basin occupies only a thin layer in the outermost part of the earth's crust. Thus our knowledge of a sedimentary basin remains incomplete without information on the lithospheric plate below the basin. While conventional reflection seismic survey can provide information on most of the major tectonic features within the basin, deep seismic surveys (DSS) are required to obtain information from deeper parts of the lithosphere. The importance of detection of thinning of the crustal layer (e.g., the North Sea graben) below a basin, or of a shear zone traversing the whole lithospheric plate (as in the Rocky Mountains) in tracing the stages of evolution of a sedimentary basin, can hardly be overemphasised. A basin can be classified according to its genesis only when its location with respect to the plate, the type of the underlying crust (continental or oceanic) and the major structural movements which have affected it are known (see Chapter 7 and Table 7.6). Well co-ordinated geological, geophysical and geochemical explorations are needed for this purpose.

The ultimate aim of basin analysis should be to trace the history of evolution of a sedimentary basin in the context of global dynamics. 'Correlation

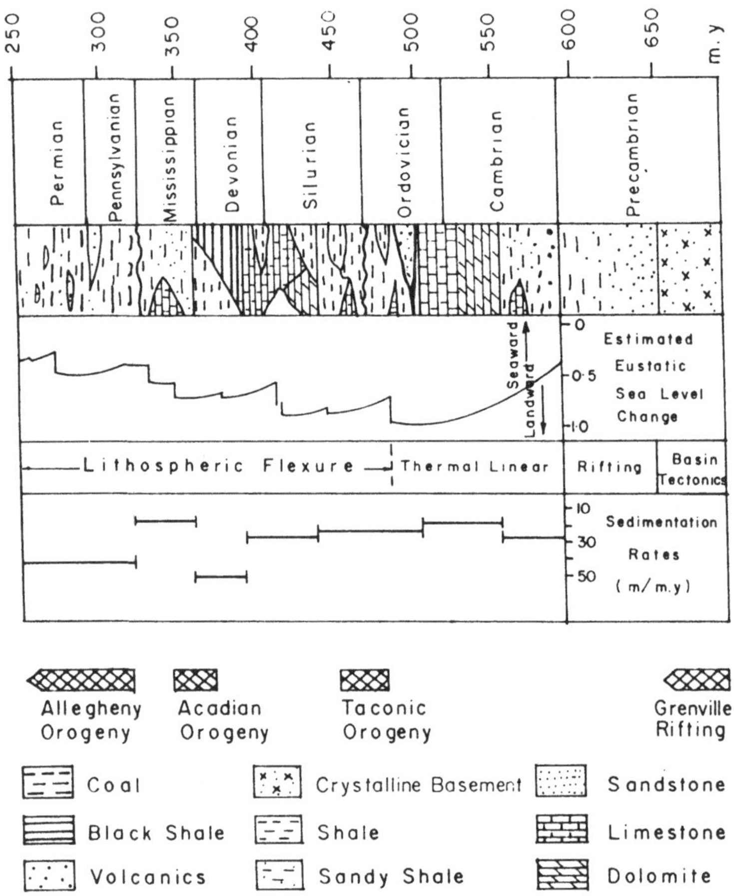

Fig. 9.6: Correlation diagram of the Appalachian Basin combining information on sediment accumulation rates, sea-level changes, orogenic events and stages of basin evolution against time scale. The ultimate aim of basin analysis is to prepare such correlation diagrams for sedimentary basins (redrawn after de V. Klein 1987, with permission of Elsevier Science Publishers B.V.).

of the local sections into a composite time scheme', as suggested by Dunbar and Rodgers (1957), is essential for this purpose. Measurement of the time span involved in basin evolution and sedimentation by radiometric, time-stratigraphic or other similar methods is needed. All these are represented in composite correlation diagrams of the type shown in Fig. 9.6. Recognition of seismic stratigraphic 'sequences' and 'systems tracts' and correlation of the local events with global phenomena such as plate movements and sea-level changes are taken to be the goal of basin analysis today.

REFERENCES

Adhikari, A.K., Roy, J. and Sengupta, S. 1981. Estimates of parameters of grain-size distribution from weight frequencies. *Mathematical Geology*, vol. 12, pp. 247-265.

Ahlbrandt, T.S., and Fryberger, S.G. 1982. Introduction to eolian deposits. *In* P.A. Scholle and D. Spearing (eds.), *Sandstone Depositional Environments*, pp. 11–47. American Association of Petroleum Geologists, Tulsa, OK., 410 pp.

Albritton, C.C. Jr. (ed.), 1963. *The Fabric of Geology*. Addisson-Wesley Publication Co. Inc., Reading, 372 pp.

Allen, J.R.L. 1963. A classification of cross-stratified units with notes on their origin. *Sedimentology*, vol. 2, pp. 93–114.

Allen, J.R.L. 1964. Studies in fluviatile sedimentation: six cyclothems from the Lower Old Red Sandstone, Anglo-Welsh basin. *Sedimentology*, vol. 3, pp. 163–198.

Allen, J.R.L. 1965. A review of the origin and characteristics of Recent alluvial sediments. *Sedimentology* (special issue), vol. 5, 2:89–191.

Allen, J.R.L. 1965a. Upper Old Red Sandstone palaeogeography (Farlovian) in South Wales and the Welsh Borderland. *Journal Sedimentary Petrology*, vol. 35, pp. 167–195.

Allen, J.R.L. 1968. The nature and origin of bedform hierarchies. *Sedimentology*, vol. 10, pp. 161–182.

Allen, J.R.L. 1968a. *Current Ripples*. North Holland, Amsterdam, 433 pp.

Allen, J.R.L. 1970. *Physical Processes of Sedimentation. An Introduction*. George Allen and Unwin Ltd., London, 248 pp.

Allen, J.R.L. 1970a. Studies in fluviatile sedimentation: a comparison of fining-upwards cyclothems, with special reference to coarse-member composition and interpretation. *Journal Sedimentary Petrology*, vol. 40, pp. 298–323.

Allen, J.R.L. 1973. Development of flute mark assemblages, I. Evolution of pairs of defects. *Sedimentary Geology*, vol. 10, pp. 157–177.

Allen, J.R.L. 1975. Development of flute mark assemblages, 2. Evolution of trios of defects. *Sedimentary Geology*, vol. 13, pp. 1-26.

Allen, J.R.L. 1977. The possible mechanism of convolute lamination in graded sand beds. *Journal Geological Society London*, vol. 134, pp. 19–31.

American Society of Civil Engineers (Task Force on Bedforms in Alluvial Channels). 1966. Nomenclature for bedforms in alluvial channels. *Journal Hydraulics Division*, ASCE, 92 (HY3):51–64.

Ashley, G.M. 1990. Classification of large-scale subaqueous bedforms: a new look at an old problem (report of the SEPM bedforms and bedding structures symposium). *Journal Sedimentary Petrology*, vol. 60, pp. 160–172.

Aubouin, J. 1965. *Geosynclines: Developments in Geotectonics*. Elsevier, Amsterdam, 335 pp.

Auden, J.B. 1933. Vindhyan sedimentation in the Son Valley, Mirzapur district. *Memoir, Geological Survey India*, 62(2), pp. 141–256.

Azad, J., Bhattacharyya, S., Datta, B.D. and Stevens, T.E. 1971. Hydrocarbon accumulation in the Nahorkatiya oilfield. Assam. *Proceedings VIII World Petroleum Congress, Moscow*, vol. II, pp. 1-10. Applied Science Publishers Ltd., U.K.

Bagati, T.N. 1979. Geochemistry of carbonate rocks of the Krol Formation of the Simla Himalaya, India. *Himalayan Geology*, vol. 9, pt. II, pp. 462–497.

Bagnold, R.A. 1941. *The Physics of Blown Sand and Desert Dunes*. Methuen, London (reprinted 1954), 265 pp.

Bagnold, R.A. 1954. Experiments on gravity-free dispersion of large solid spheres in a Newtonian fluid under shear. *Roy. Soc. London. Phil Trans.* Ser.A 225, pp. 49–63.

Baksi, A.K., Ray Barman, T., Paul, D.K. and Ferrad, Edward. 1987. Widespread Early Cretaceous flood basalt volcanism in eastern India: Geochemical data from the Rajmahal-Bengal-Sylhet traps. *Chemical Geology*, vol. 63, pp. 133–141.

Banerjee, D.M. 1986. Proterozoic and Cambrian phosphorites—regional review: Indian subcontinent. *In* P.J. Cook and J.H. Shergold (eds.), *Phosphatic Deposits of the World*, vol. I. *Proterozoic and Cambrian Phosphorites*, pp. 70–90, Cambridge University Press, Cambridge.

Banerjee, Indranil. 1960. Stratigraphy and sedimentation in the South Karanpura Coalfield, Bihar. *Quarterly Journal Geological, Mining & Metallurgical Society India*, vol. 32, pp. 189–203.

Banerjee, Indranil, 1964. Size-roundness relation in the Barakar sandstones of the South Karanpura Coalfield, India. *Sedimentology*, vol. 3, pp. 22–28.

Banerjee, Indranil. 1966. Turbidites in a glacial sequence: a study from the Talchir Formation, Raniganj Coalfield, India. *Journal Geology*, vol. 74, pp. 593–606.

Banerjee, Indranil. 1973. Sedimentology of Pleistocene glacial varves in Ontario, Canada. *Geological Survey Canada Bulletin* 226.

Banerjee, Indranil. 1974. Barrier coastilline sedimentation model and the Vindhyan example. *Geological, Mining & Metallurgical Society India.* Golden Jubilee Volume, (vol. 46), pp. 101–127.

Banerjee, Indranil. 1980. The Vindhyan tidal sea. *In* K.S. Valdiya, et al. (eds.), *Geology of Vindhyanchal*, pp. 80–87, Hindusthan Publishing Corporation (India), New Delhi, 231 pp.

Banerjee, Indranil and Sengupta, Supriya. 1963. The Vindhyan Basin—a regional reconnaissance of the eastern part. *Quarterly Journal Geological, Mining & Metallurgical Society India*, vol. 35, pp. 141–149.

Banerjee, Indranil and McDonald, B.C. 1975. Nature of esker sedimentation. *In.* A.V. Jopling and B.C. McDonald (eds.), *Glaciofluvial and Glaciolacustrine Sedimentation*, pp. 132–154. Society of Economic Paleontologists and Mineralogists, Spl. Pub. No. 23.

Banerji, A.K. 1977. On the Precambrian Banded Iron Formations and the managanese ores of the Singhbhum region, Eastern India. *Economic Geology*, vol. 72, pp. 90–98.

Barrell, J. 1912. Criteria for recognition of ancient delta deposits. *Geological Society America Bulletin*, vol. 23, pp. 377–446.

Basu, Abhijit, Young, S.W., Suttner, L.J., James, W.C. and Mack, G.H. 1975. Re-evaluation of the use of undulatory extinction and polycrystallinity in detrital quartz for provenance interpretation. *Journal Sedimentary Petrology*, vol. 45, pp. 873–882.

Basu, Abhijit, 1976. Petrology of the Holocene fluvial sand derived from plutonic source rocks: implications to paleoclimatic interpretations. *Journal Sedimentary Petrology*, vol. 46, pp. 694–709.

Basu, Abhijit, 1977. Provenance and AlSi order-disorder of detrital alkali feldspars. *Journal Geological Society India*, vol. 18, pp. 477–492.

Basu, Abhijit, 1985. Reading provenance from detrital quartz. *In* G.G. Zuffa (ed.), *Provenance of Arenites*, pp. 231-247. NATO-ASI. V.C-148. Reidel, Holland.

Basu, Abhijit and Suttner, L.J. 1975. Use of structural state of alkali feldspars in provenance interpretations. *Proceedings IXth International Sedimentological Congress*, Nice, Theme 3, pp. 1-9.

Basu, P.K., Biswas, S. and Acharyya, S.K. 1987. Late Quaternary ash beds from the Son and Narmada Basins, Madhya Pradesh. *Indian Minerals*, vol. 41, pp. 66–72.

Basumallick, Sudhir, 1962. Petrology and zonation of the Rewa sandstone in Maihar, Madhya Pradesh. *Quarterly Journal Geological, Mining & Metallurgical Society India*, vol. 34, pp. 11–28.

Bass-Becking, L.G.M., Kaplan, I.R. and Moore, D. 1960. Limits of natural environment in terms of pH and oxidation-reduction potentials. *Journal Geology*, vol. 68, pp. 243–284.

Bates, C.C. 1953. Rational theory of delta formation. *American Association Petroleum Geologists Bulletin*, vol. 37, pp. 2119–2162.

Bathurst, R.G.C. 1975. *Carbonate Sediments and Their Diagenesis*. Elsevier, Amsterdam, 658 pp.

Berger, P. et al. 1983. Schlumberger Well Evaluation Conference, India. Schlumberger Technical Services, Inc., 263 pp.

Bernard, H.A., LeBlanc, R.J. and Major, C.F. 1962. Recent and Pleistocene geology of southeast Texas. *In* E.H. Rainwater and R.P. Zingula (eds.), *Geology of Gulf Coast and Guidebook of Excursions*, pp. 175–205. Houston Geological Society and Geological Society of America Annual meeting, Houston, Texas.

Berner, R.A. 1981. A new geochemical classification of sedimentary environments. *Journal Sedimentary Petrology*, vol. 51, pp. 359–365.

Bhatia, M.R. 1983. Plate tectonics and geochemical composition of sandstones. *Journal Geology*, vol. 91, pp. 611–627.

Bhatia, M.R. and Crook, A.W. 1986. Trace element characteristics of graywackes and tectonic setting discrimination of sedimentary basins. *Contribution to Mineralogy & Petrology*, vol. 92, pp. 181–193.

Bhattacharji, S. and Singh, R.N. 1984. Thermo-mechanical structure at the southern part of the Indian shield and its relevance to Precambrian basin evolution. *Tectonophysics*, vol. 105, pp. 103–120.

Bhattacharyya, A. and Morad, S. (1993). Proterozoic braided ephemeral fluvial deposits: an example from the Dhandraul Sandstone Formation of the Kaimur Group. Son Valley, Central India. *Sedimentary Geology*, vol. 84, pp. 101–114.

Biswas, B. 1959. Sub-surface geology of West Bengal, India. *Proceedings Symposium Development Petroleum Resources*. ECAFE Mineral Resources Development Series, No. 10, pp. 159–161.

Biswas, B. 1963. Results of exploration for petroleum in the western part of the Bengal basin, India. *Proceedings 2nd Symposium Development Petroleum Resources*. ECAFE, Mineral Resources Development Series, No. 18, pt. 1, pp. 241–250.

Biswas, S.K. 1981. Basin framework, Palaeo-environment and depositional history of the Mesozoic sediments of Kutch basin, Western India. *Quarterly Journal Geological, Mining & Metallurgical Society India*, vol. 53, pp. 56–85.

Biswas, S.K. 1982. Rift basins in western margin of India and their hydrocarbon prospects with special reference to Kutch basin. *American Association Petroleum Geologists Bulletin*, vol. 66, pp. 1497–1513.

Biswas, S.K. and Deshpande, S.V. 1983. Geology and hydrocarbon prospects of Kutch, Saurashtra and Narmada basins. *In* L.L. Bhandari et al. (eds.), *Petroliferous Basins of India*, vol. I, pp. 111–126. *Petroleum Asia Journal*, V–VI, No. 4. 189 pp.

Blatt, H. and Christie, J.M. 1963. Undulatory extinction in quartz of igneous and metamorphic rocks and its significance in provenance studies of sedimentary rocks. *Journal Sedimentary Petrology*, vol. 33, pp. 559–579.

Blatt, H., Middleton, G. and Murray, R. 1972. *Origin of Sedimentary Rocks*. Prentice-Hall, Englewood Cliffs, N.J., 634 pp.

Boersma, J.R. 1967. Remarkable types of mega cross-stratification in the fluviatile sequence of a subrecent distributary of the Rhine Amerongen, the Netherlands. *Geol. Mijnbouw*, vol. 46, pp. 217–235.

Borchert, H. 1960. Genesis of marine sedimentary iron ores. *Trans. Inst. Min. Met.*, vol. 69, pp. 261–279.

Bose, Abhijit and Sengupta, Supriya. 1993. 'Infra-Kamthi' of the central Godavari Valley—petrological evidence of marine influence during the Permian. *National Acad. Sci. India, Proc.*, vol. 63, Sec.A Pt.1 (Spl. Pub. on Petrology), pp. 149–166.

Bose, P.K. and Das, N.G. 1986. A transgressive strom- and fair-weather wave-dominated shelf sequence: Cretaceous Nimar Formation, Chakrud, Madhya Pradesh, India. *Sedimentary Geology*, vol. 46, pp. 147–167.

Bose, P.K., Chaudhuri, A.K. and Seth, Ashoke. 1988. Facies, flow and bedform patterns across a storm-dominated inner continental shelf: Proterozoic Kaimur Formation, Rajathan, India. *Sedimentary Geology*, vol. 59, pp. 275–293.

Bouma, A.H. 1962. *Sedimentology of Some Flysch Deposits: A Graphic Approach to Facies Interpretation*. Elsevier, Amsterdam, 168 pp.

Bouma, A.H. 1969. *Methods for the Study of Sedimentary Structures*. Wiley Interscience, New York, 458 pp.

Bridge, J.S. 1978. Palaeohydraulic interpretation using mathematical models of contemporary flow and sedimentation in meandearing channels. *In* A.D. Miall (ed.), *Fluvial Sedimentology*, pp. 723-742, Canadian Society of Petroleum Geologists Memoir 5, Calgary, Alberta, 859 pp.

Bridge, J.S. 1987. Descriptive classification of fluvial bedforms: SEPM Classification of Large-Scale Flow-Transverse Bedforms Symposium (Unpub.expanded abstract cited by Ashley, 1990).

Brink, A.H. 1974. Petroleum Geology of Gabon basin. *American Assoc. Petroleum Geologists Bulletin*, vol. 58, pp. 216–235.

Brookfield, M.E. 1984. Eolian Facies. *In* R.G. Walker (ed.), *Facies Models*, 2nd ed., pp. 91–104. Geological Association of Canada, Toronto, Ont. 317 pp.

Burger, Heinz and Skala, Wolfdietrich. 1976. Comparison of sieve and thin-section technique by Monte-Carlo model. *Computers and Geoscience*, vol. 2, pp. 123–139.

Burke, K., Dessauvagie, T.F.J. and Whiteman, A.J. 1972. Geological history of the Benue valley and adjacent areas. *In* T.F.J. Dessauvagie and A.J. Whiteman (eds.), *African Geology*. Ibadan, Nigeria, pp. 187–205.

Burley, S.D., Kantorowicz, J.D. and Waugh, B. 1985. Clastic diagenesis. *In* P.J. Brenchley and B.P.J. Williams (eds.), *Sedimentology, Recent Developments and Applied Aspects*, pp. 189–226. Published for the Geological Society by Blackwell Scientific Publications, Oxford.

Busch, D.A. 1974. *Stratigraphic Traps in Sandstones—Exploration Techniques*. Memoir 21, American Association of Petroleum Geologists, Tulsa, Oklahoma, 174 pp.

Cant, D.J. 1984. Subsurface facies analysis. *In* R.G. Walker (ed.), *Facies Models*, 2nd ed. pp. 277-310. Geol. Assoc. of Canada, Toronto, Ont. 317 pp.

Cant, D.J. 1989. Zuni Sequence: the foreland basin. Lower Zuni sequence: Middle Jurassic to Middle Cretaceous. Chapter 11. *In* B.D. Rickets (ed.) *Western Canada Sedimentary Basin: A Case History*. pp. 251-254. Canadian Society of Petroleum Geologists. Calgary.

Cant, D.J. and Walker, R.G. 1976. Development of a braided-fluvial facies model for the Devonian Battery Point Sandstone, Quebec. *Canadian Jour. Earth Sc.*, vol. 13, pp. 102–119.

Carlston, C.W. 1965. The relation of free meander geometry to stream discharge and its geomorphic implications. *American Journal Science*, vol. 34, pp. 864–885.

Carozzi, A. 1960. *Microscopic Sedimentary Petrography*. John Wiley, New York, 485 pp.

Carver, R.W. (ed.). 1971. *Procedures in Sedimentary Petrology*. John Wiley, New York, 458 pp.

Casshyap, S.M. 1970. Sedimentary cycles and environment of deposition of the Barakar Coal Measures of Lower Gondwana, India. *Journal Sedimentary Petrology*, vol. 40, pp. 1302–1317.

Casshyap, S.M. 1973. Paleocurrents and paleogeographic reconstruction in the Barakar (Lower Gondwana) sandstones of Peninsular India. *Sedimentary Geology*, vol. 9, pp. 283–303.

Casshyap, S.M. and Khan, Z.A. 1982. Paleohydrology of Permian Gondwana seams in Bokaro Basin, Bihar. *Journal Geological Society India*, vol. 23, pp. 419–430.

Casshyap, S.M. and Tewari, R.C. 1982. Facies analysis and paleogeographic implications of a late Paleozoic glacial outwash deposit, Bihar, India. *Journal Sedimentary Petrology*, vol. 52, pp. 1243-1256.

Chakrabarti, Amitabha. 1977. Upward flow and convolute laminations. *Senckenbergiana Marit.*, vol. 9, pp. 285–305.

Chakraborty, Tapan. 1991. Sedimentology of a Proterozoic erg: the Venkatpur Sandstone, Pranhita-Godavari Valley, South India. *Sedimentology*, vol. 38, pp. 301–322.

Chanda, S.K. 1963. Cementation and diagenesis of the Lameta Beds, Lametaghat, M.P. India. *Journal Sedimentary Petrology*, vol. 33, pp. 728–738.

Chanda, S.K. and Bhattacharyya, Ajit. 1974. Ripple-drift cross-lamination in tidal deposits: example from the Precambrian Bhander Formation of Maihar, Satna District, Madhya Pradesh, India. *Geological Society America Bulletin*, vol. 85, pp. 1117-1122.

Chepil, W.S. 1959. Equilibrium of soil grains at the threshold of movement by wind. *Soil Science Society America, Proceedings*, vol. 23, 6, pp. 422–428.

Chow, Ven Te. 1954. *Open-Channel Hydraulics* (International Student Edition). McGraw-Hill Kogakusha Ltd., Tokyo, 680 pp.

Christiansen, C., Blæsild, P. and Dalsgaard, K. 1984. Re-interpreting 'segmented' grain-size curves. *Geological Magazine*, vol. 121, pp. 47-51.

Colby, B.R. and Hembree, C.H. 1955. Compositions of total sediment discharge Niobrare River near Cody, Nebraska. *U.S. Geological Survey Water Supply* Paper No. 1357. 187 pp.

Coleman, J.M. 1969. Brahmaputra River, channel processes and sedimentation. *Sedimentary Geology*, vol. 3, pp. 131-239.

Coleman, J.M. and Gagliano, S.M. 1964. Cyclic sedimentation in the Mississippi river deltaic plain. *Trans. Gulf Coast Association of Geological Societies*, vol. 14, pp. 67–80.

Conybeare, C.E.B. 1979. *Lithostratigraphic Analysis of Sedimentary Basins*. Academic Press, New York, 555 pp.

Cook, H.E., Field, M.E. and Gardner, J.V. 1982. Characteristics of sediments on modern and ancient continental slopes. *In* P.A. Scholle and O. Spearing (eds.), *Sandstone Depositional Environments*, pp. 329–364. AAPG Memoir No. 31, 410 pp.

Crook, K.A.W. 1974. Lithogenesis and geotectonics: the significance of compositional variation in flysch arenites (graywackes). *In* R.H. Dott and R.H. Shaver (eds.), *Modern and Ancient Geosynclinal Sedimentation*, pp. 304–310. Society of Economic Paleontologists and Mineralogists, Spl. Pub. no. 19, 380 pp.

Cross, T.A. and Lessenger, M.A. 1988. Seismic Stratigraphy. *Annual Review of Earth and Planetary Sciences*, vol. 16, pp. 319-353.

Curray, J.R. and Moore, D.G. 1971. Growth of the Bengal deep-sea fan and denudation in the Himalayas. *Geological Society America Bulletin*, vol. 82, pp. 563-572.

Curray, J.R., et al. 1979. Tectonics of the Andaman Sea and Burma. *In* J.S. Watkins, L. Montadert and P.W. Dickerson (eds.). *Geological and Geophysical Investigations of Continental Margins*. pp. 189-198, American Association of Petroleum Geologists, Memoir No. 29, Tulsa, OK.

Curtis, C.D. 1976. Stability of minerals in surface weathering reactions: a thermochemical approach. *Earth Surf. Proc.*, vol. 1, pp. 63-70.

Dailly, G.S. 1976. A possible mechanism regarding progradation, growth faulting, clay diapirism and overthrusting in a regressive sequence of sediments. *Bulletin, Canadian Petroleum Geology*, vol. 24, pp. 92-116.

Dasgupta, H.C. and Batabyal, A.K. 1988. Genesis of the manganese and iron ore deposits of Noamundi basin, Eastern India: a dilemma (abs.). *Intl. Conf. on Metallogeny Related to Tectonics of the Proterozoic Mobile Belts*. Calcutta.

Demaison, G.J. and Moore, G.T. 1980. Anoxic environments and oil source bed genesis. *American Association Petroleum Geologists Bulletin*, vol. 64, pp. 1179–1209.

de V. Klein, G. 1987. Current aspects of basin analysis. *Sedimentary Geology*, vol. 50, pp. 95–118.

de Raff, J.F.M., Reading, H.G. and Walker, R.G. 1965. Cyclic sedimentation in the Lower Westphalian of North Devon, England. *Sedimentology*, vol. 4, pp. 1-52.

Dickinson, K.A., Berryhill, H.L. Jr. and Holmes, C.W. 1972. Criteria for recognizing ancient barrier coastlines. *In* J.K. Rigby and W.K. Hamblin (eds.), *Recognition of Ancient Sedimentary Environments*, pp. 192-214. Society of Economic Paleontologists and Mineralogists, Spl. Pub. no. 16, 340 pp.

Dickinson, K.A. and Suczek, C.A. 1979. Plate tectonics and sandstone composition. *American Association Petroleum Geologists Bulletin*, vol. 63, pp. 2164-2182.

Dickinson, W.R. 1970. Interpreting detrital modes of graywacke and arkose. *Journal Sedimentary Petrology*, vol. 40, pp. 695–707.

Dickinson, W.R. 1974. Plate tectonics and sedimentation. *In* W.R. Dickinson (ed.), *Tectonics and Sedimentation*, pp. 1-27. Society of Economic Paleontologists and Mineralogists, Spl. Pub. no. 22, 204 pp.

Dietz, R.S. 1963. Collapsing continental rises: an actualistic concept of geosynclines and mountain building. *Journal Geology*, vol. 71, pp. 314–333.

Dietz, R.S. and Holden, J.C. 1974. Collasping continental rises: actualistic concept of geosynclines—a review. *In* R.H. Dott and R.H. Shaver (eds.), *Modern and Ancient Geosynclinal Sedimentation*, pp. 14-25. Society of Economic Paleontogists and Mineralogists, Spl. Pub. no. 19, 380 pp.

Dobkins, J.E. Jr. and Folk, R.L. 1970. Shape development on Tahiti-Nui. *Journal Sedimentary Petrology*, vol. 40, pp. 1167-1203.

Dott, R.H. Jr. 1964. Wacke, graywacke and matrix—what approach to immature sandstone classification? *Journal Sedimentary Petrology*, vol. 34, pp. 625–632.

Dott, R.H. Jr. and Bourgeois, Joanne. 1982. Hummocky stratification: significance of its variable bedding sequences. *Geological Society America Bulletin*, vol. 93, pp. 663–680.

Drake, C.L., Ewing, M. and Sutton, G.H. 1959. Continental margin and geosynclines: the east coast of North America, North of Cape Hatteras. *In* L.H. Ahrens et al. (eds.), *Physics and Chemistry of the Earth*, vol. 3, pp. 110-198, Pergamon Press, London.

Dunbar, C.O. and Rodgers, John, 1957. *Principles of Stratigraphy*. John Wiley & Sons. Inc., New York, 356 pp.

Dunham, R.J. 1962. Classification of carbonate rocks according to depositional texture. *In* W.E. Ham (ed.), *Classification of Carbonate Rocks*, pp. 108-121, AAPG Memoir no. 1, American Association of Petroleum Geologists, Tulsa, Oklahoma.

Dzulynski, S. and Walton, E.K. 1962. *Sedimentary Features of Flysch and Graywackes*. Elsevier, Amsterdam, 300 pp.

Edwards, M. 1986. Glacial environments. *In* R.G. Reading (ed.), *Sedimentary Environments and Facies*, pp. 445–470. Blackwell Scientific Publications, Oxford, 2nd ed. 615 pp.

Einstein, H.A. 1950. The bed-load function for sediment transportation in open channel flows. United States Department of Agriculture, Soil Conservation Service, Washington, D.C., Technical Bulletin no. 1026, pp. 1-71.

Evans, P. 1932. Tertiary succession in Assam. *Mining & Geological Institute India, Transactions*, vol. 27(3).

Evans, P., Hayman, R.J. and Majeed, M.A. 1934. The graphical representation of heavy mineral analyses. *Quarterly Journal Geological, Mining & Metallurgical Society India*, vol. 6, pp. 27–48.

Eyles, N. and Miall, A.D. 1984. Glacial facies. *In* R.G. Walker (ed.), *Facies Models*, 2nd ed. pp. 15–38. Geological Association of Canada, Toronto, Ont. 317 pp.

Feo-Codecido, Gustavo, 1956. Heavy-mineral techniques and their application to Venezuelan stratigraphy. *American Association Petroleum Geologists Bulletin*, vol. 40, pp. 984–1000.

Fermor, L.L. 1918. Preliminary note on the burning of coal seem at the outcrop. *Min. Geol. Inst. India, Transactions*, vol. 12, pp. 50–63.

Fischer, A.G. and Arthur, M.A. 1977. Secular variations in the pelagic realm. *In* H.E. Cook and P. Enos (eds.), *Deep Water Carbonate Environments*, pp. 19–50. Soc. Econ. Palaeont. Min. Spl. Pub. no. 25.

Fischer, A.G. and Judson, S. 1975 (eds.). *Petroleum and Global Tectonics*, Princeton University Press, New Jersey, 322 pp.

Fisk, H.N. 1944. Geological investigation of the alluvial valley of the lower Mississippi River. Mississippi River Commission, Vicksburg, Mississippi, 78 pp.

Flügel, Erik. 1982. *Microfacies Analysis of Limestones* (trans. K. Christenson). Springer-Verlag, Berlin, 633 pp.

Folk, R.L. 1959. Practical petrographic classification of limestones. *American Association Petroleum Geologists Bulletin*, vol. 43, pp. 1–38.

Folk, R.L. 1968. *Petrology of Sedimentary Rocks*. Hemphill's Book Store, Austin, Texas, 170 pp. 2nd ed. (1974), 182 pp.

Folk, R.L. 1974. The natural history of crystalline calcium carbonate: effects of magnesium content and salinity. *Journal Sedimentary Petrology*, vol. 44, pp. 44–53.

Folk, R.L. and Ward, W.C. 1957. Brazos River bar, a study in significance of grain-size parameters. *Journal Sedimentary Petrology*, vol. 27, pp. 3–27.

Folk, R.L. and Land, L.S. 1975. Mg/Ca ratio and salinity: two controls over crystallization of dolomite. *American Association Petroleum Geologists Bulletin*, vol. 59, pp. 60–68.

Fouch, T.D. and Dean, W.E. 1982. Lacustrine and associated clastic depositional environments. *In* P.A. Scholle and D. Spearing (eds.) *Sandstone Depositional Environments*, pp. 87–114. American Association Petroleum Geologists, Tulsa, Ok, 410 pp.

Francis, J.R.D. 1973. Experiments on the motion of solitary grains along the bed of a water-stream. *Proc. Royal Society London*. Ser. A. vol. 332, pp. 443–471.

Frazier, D.E. 1974. Depositional episodes: their relationship to the Quaternary stratigraphic framework in the north-western portion of the Gulf Basin. *Bull. Econ. Geol. Univ. Texas*, Austin Geol. Circle, 71-1.

Friedman, G.M. 1958. Determination of sieve-size distribution from thin section data for sedimentary petrological studies. *Journal Geology*, vol. 66, pp. 394–416.

Friedman, G.M. 1959. Identification of carbonate minerals by staining methods. *Journal Sedimentary Petrology*, vol. 29, pp. 87–97.

Friedman, G.M. 1961. Distinction between dune, beach, and river sands from their textural characteristics. Journal Sedimentary Petrology, vol. 31, pp. 514-529.

Friedman, G.M. 1962. Comparison of moment measures for sieving and thin-section data in sedimentary petrological studies. *Journal Sedimentary Petrology*, vol. 32, pp. 15–25.

Friedman, G.M. 1965. Terminology of crystallization textures and fabrics in sedimentary rocks. *Journal Sedimentary Petrology*, vol. 35, pp. 643–655.

Friedman, G.M. 1967. Dynamic processes and statistical parameters compared for size frequency distribution of beach and river sands. *Journal Sedimentary Petrology*, vol. 37, pp. 327–354.

Friedman, G.M. 1979. Differences in size distributions of populations of particles among sands of various origins. Pres. Address IAS. *Sedimentology*, vol. 26, pp. 3–32.

Friedman, G.M. and Sanders, J.E. 1974. Positive relief bedforms on modern tidal flats that resemble molds of flutes and grooves: implications for geopetal criteria and for classification of bedforms. *Journal Sedimentary Petrology*, vol. 44, pp. 181–189.

Friedman, G.M. and Sanders, J.E. 1978. *Principles of Sedimentology*. John Wiley & Sons, Inc., New York, 792 pp.

Fritz, W.J. and Moore, J.N. 1988. *Basics of Physical Stratigraphy and Sedimentology*. John Wiley & Sons, Inc., New York, 371 pp.

Fryberger, S.G. 1979. Dune forms and wind regimes. *In* E.D. McKee (ed.), *A Study of Global Sand Seas*. U.S. Geol. Survey Prof. Pap. no. 1052.

Füchtbauer, Hans. 1974. Some problems of diagenesis in sandstones. *Bulletin Centre Rech. Pan-SNPA*, vol. 8, pp. 391–403.

Gaal, Gabor. 1964. Pre-Cambrian flysch and molasse—tectonics and sedimentation around Rakha mines and Jaikan in Singhbhum District, Bihar, India. *Report of XXII International Geological Congress, New Delhi*, pt. IV, pp. 331–356.

Galloway, W.E. 1975. Process framework for describing the morphologic and stratigraphic evolution of deltaic depositional systems. *In* M.L. Broussard (ed.), *Deltas*, pp. 87-98, Houston Geol. Society, Taxas, 555 pp.

Galloway, W.E. 1989. Genetic stratigraphic sequences in basin analysis. I: Architecture and genesis of flooding-surface bounded depositional units. *American Association Petroleum Geologists Bulletin*, vol. 73, pp. 125–142.

Galloway, W.E. and Hobday, D.K. 1983. *Terrigenous Clastic Depositional Systems*. Springer-Verlag, New York, 423 pp.

Garde, R.J. and Ranga Raju, K.G. 1977. *Mechanics of Sediment Transporation and Alluvial Stream Problems*. (A Halsted Press Book). John Wiley & Sons, Inc., New York, 483 pp.

Gardner, T.W. 1983. Paleohydrology and Paleomorphology of a Carboniferous, meandering, fluvial sandstone. *Journal Sedimentary Petrology*, vol. 53, pp. 991-1005.

Gardner, G.H.F., Gardner, L.W. and Gregory, A.R. 1974. Formation velocity and density, the diagnostic basis for stratigraphic traps. *Geophysics*, vol. 39, pp. 770–780.

Garrels, R.M. and Christ, C.L. 1965. *Solutions, Minerals and Equilibria*. Harper & Row, New York, 450 pp.

Garrels, R.M. and Mackenzie, F.T. 1971. *Evolution of Sedimentary Rocks*. W.W. Norton & Company, Inc., New York, 397 pp.

Geological Survey of India (GSI). 1986. *Analysis of Rocks, Ores and Minerals by X-ray Spectrometry*. Bulletin Series C.No. 1. Airborne Mineral Surveys & Exploration Wing. Geological Survey of India, Bangalore, 486 pp.

Ghosh, J.K., Mazumder, B.S. and Sengupta, S. 1981. Methods of computation of suspended load from bed materials and flow parameters. *Sedimentology*, vol. 28, pp. 781-791.

Ghosh, J.K., Mazumder, B.S., Saha, M.R. and Sengupta, Supriya. 1986. Deposition of sand by suspension currents: experimental and theoretical studies. *Journal Sedimentary Petrology*, vol. 56, pp. 57–66.

Ghosh, Santosh Kumar, 1972. Comparative study of the Lower Vindhyan Limestones around Churhat, Sidhi District, Madhya Pradesh, *Quart. Journ. Geological, Mining & Metallurgical Society India*, vol. 44, pp. 41-53.

Ghosh, S.K. and Lahiri, Samhita. 1983. Morphology of penecontemporaneous interpenetrative contortions and their modifications by diastrophic movements in the Ghatsila-Galudih area, Singhbhum, Eastern India. *In* S. Sinha-Roy (ed.), *Structure and Tectonics of the Precambrian Rocks of India*, pp. 144-157. Hindusthan Publishing Corporation, New Delhi, 252 pp.

Ghosh, S.K. and Mukhopadhyay, A. 1985. Tectonic history of the Jharia Basin—an intracratonic Gondwana basin in eastern India. *Quart. Jour. Geological, Mining & Metallurgical Society India*, vol. 57, pp. 33–58.

Gilbert, G.K. 1885. The topographic features of lake shores. *U. S. Geological Survey, Annual Report*, 5, pp. 104–108.

Gilbert, G.K. 1914. Transportation of debris by running water. U.S. Geological Survey, Professional paper, No. 86, 263 pp.

Gilbert, G.K. 1954. *In* H. Williams, F.J. Turner and G.K. Gilbert. *Petrography. An Introduction to the Study of Rocks in Thin Sections*. W.H. Freeman & Co., San Francisco, 406 pp.

Gilluly, J., Waters, A.C. and Woodford, A.O. 1975. *Principles of Geology.* W.H. Freeman & Co., San Francisco, 2nd ed., 527 pp.
Gingerich, P.D. 1969. Markov analysis of cyclic alluvial sediments. *Journal Sedimentary Petrology,* vol. 39, pp. 330–332.
Ginsburg, R.N. (ed.) 1975. *Tidal Deposits*, Springer, New York, 428 pp.
Glennie, K.W. and Evans, G. 1976. A reconnaissance of the Recent sediments of the Rann of Kutch, India. *Sedimentology,* vol. 23, pp. 625–647.
Goldhaber, Martin. 1978. Euxinic facies. *In* R.W. Fairbridge and Joanne Bourgeois (eds.), *Encyclopedia of Sedimentology,* pp. 296–300. Dowden, Hutchinson & Ross, Inc., Stroudsburg, 901 pp.
Goldich, S.S. 1938. A study in rock weathering. *Journal Geology,* vol. 46, pp. 17–58.
Goodwin, A.M. 1982. Distribution and origin of Precambrian banded iron formation (Abs.). *Intl. Symposium on Archean and Early Proterozoic Geological Evolution and Metallogenesis, Salvador, Brazil.* pp. 15–16.
Goodwin, P.W. and Anderson, E.J. 1985. Punctuated aggradational cycles: a general hypothesis of episodic stratigraphic accumulation. *Journal Geology,* vol. 93, pp. 515–533.
Goswami, D.C. 1988. Estimation of bedload transport in the Brahmaputra River, Assam. *Indian Journal Earth Sciences*, vol. 15, pp. 14–26.
Grabau, A.W. 1904. On the classification of sedimentary rocks. *Amer. Geol.*, vol. 33, pp. 228–247. (Cited by Pettijohn, 1975.)
Graton, L.C. and Fraser, H.J. 1935. Systematic packing of spheres—with particular reference to porosity and permeability. *Jour. Geol.*, vol. 43, pp. 785-909.
Griffiths, J.C., 1967. *Scientific Methods in Analysis* of *Sediments.* McGraw-Hill, New York, 508 pp.
Hand, B.M. and Bartberger, C.E. 1988. Leeside sediment fallout patterns and the stability of angular bedforms. *Journal Sedimentary Petrology,* vol. 58, pp. 33–43.
Hansen, E.C., Newton, R.C. and Janardhan, A.S. 1984. Pressures, temperatures and metamorphic fluids across an unbroken amphibolite facies to granulite facies transition in Southern Karnataka, India. *In* A. Kroner et al. (eds.), *Archean Geochemistry,* pp. 161–181. Springer-Verlag, Berlin-Heidelberg, 286 pp.
Haq, B.U., Hardenbol, J. and Vail, P.R. 1987. Chronology of fluctuating sea levels since the Triassic (250 m.y. to present). *Science,* vol. 235, pp. 1156–1167.
Harms, J.C. and Fahnestock, R.K. 1965. *Stratification, bedforms and flow phenomena (with example from the Rio Grande).* Society of Economic Paleontologists and Mineralogists. Spl. Pub. No. 12, pp. 84–115.
Harms, J.C., Southard, J.B., Spearing, D.R. and Walker, R.G. 1975. *Depositional Environments as Interpreted from Primary Sedimentary Structures and Stratification Sequences.* Short course No. 2, Society of Economic Paleontologists and Mineralogists, Tulsa, OK, 161 pp.
Harms, J.C., Southard, J.B. and Walker, R.G. 1982. *Structures and Sequences in Clastic Rocks.* Lecture Notes for Short Course No. 9, Society of Economic Paleontologists and Mineralogists, Tulsa, OK, 8–49 pp.
Harper, C.W. Jr. 1984. Improved methods of facies sequence analysis. *In* R.G. Walker (ed.), *Facies Models*, 2nd ed. pp. 11–13, Geological Association Canada, Toronto, Ont. 317 pp.
Hatch, F.H. and Rastall, R.H. 1913. *The Petrology of the Sedimentary Rocks.* George Allen, London, 425 pp.
High, L.R. Jr. and Picard, M.D. 1974. Reliability of cross-stratification types as paleocurrent indicator. *Journal Sedimentary Petrology,* vol. 44, pp. 158–168.
Hjulström, F. 1935. Studies of the morphological activity of rivers as illustrated by the river Fyris. *Bulletin of Geological Institution University of Uppsala*, vol. 25, pp. 221–346.
Hoffman, Paul, Dewey, J. F. and Burke, Kevin. 1974. Aulacogens and their genetic relation to geosynclines, with a Proterozoic example from Great Slave Lake, Canada. *In* R.H. Dott

Jr. and R.H. Shaver (eds.), *Geosynclinal Sedimentation*, pp. 38-55. Society of Economic Paleontologists and Mineralogists, Spl. Pub. No. 19, Tulsa, OK, 380 pp.

Holmes, Arthur and Holmes, Doris L. 1978. Holmes Principles of Physical Geology (3rd edition). Thomas Nelson and Sons Ltd. Middlesex. 730 pp.

Howell, D.G. and McLean, Hugh. 1976. Middle Miocene paleogeography, Santa Cruz and Santa Rosa Island. *In* D.G. Howell (ed.), *Aspects of the Geologic History of California Continental Borderland.* pp. 266–293. AAPG Pacific Section Misc. Pub. 24.

Howell, D.G. and Normark, W.R. 1982. Sedimentology of submarine fans. *In* P.A. Scholle and D. Spering (eds.), *Sandstone Depositional Environments*, pp. 365–404, American Association of Petroleum Geologists, Tulsa, OK., 410 pp.

Hough, J.L. 1958. Freshwater environment of deposition of Precambrian banded iron formations. *Journal Sedimentary Petrology*, vol. 28, pp. 414–430.

Hubbert, M., King., 1967. Critique of the principle of uniformity, *In* C.C. Albritton (ed.), *Uniformity and Simplicity*, pp. 3–33, Geological Society of America Spl. paper 89.

Hunter, R.E. 1977. Basic types of stratification in eolian dunes, *Sedimentology*, vol. 24, pp. 363–387.

Hunter, R.E. 1981. Stratification styles in eolian sandstones: some Pennsylvanian to Jurassic examples from the Western Interior U.S.A. *in* F.G. Ethridge and R.M. Floves (eds.), *Recent and Ancient Nonmarine Depositional Environments*, pp. 315–329, Society of Economic Paleontologists and Mineralogists, Spl. Pub. No. 31.

Hutchinson, R.W. and Engels, G.G. 1970. Tectonic significance of regional geology and evaporite lithofacies in northeastern Ethiopia. Philosophical Transactions Royal Society, Ser.A, vol. 267, pp. 313–329.

Imam, M.B. and Shaw, H.F. 1984. The diagenesis of Neogene clastic sediments from the Bengal basin, Bangladesh. *Journal Sedimentary Petrology*, vol. 55, pp. 665–671.

James, H.L. 1966. Chemistry of the iron-rich sedimentary rocks. U.S. Geological Survey Professional Paper 440-W, 61 pp.

James, N.P. 1984. Shallowing upward sequences in carbonates. *In* R.G. Walker (ed.), *Facies Models*, 2nd ed. pp. 213-228. The Geological Association of Canada, Toronto, Ont. 317 pp.

Jarvis, G.T. 1984. An extensional model of graben subsidence—the first stage of basin evolution. *Sedimentary Geology*, vol. 40, pp. 13–31.

Jarvis, G.T. and McKenzie, D.P. 1980. Sedimentary basin formation with finite extension rates. *Earth & Planetary Science Letters*, vol. 48, pp. 42-52.

Johansson, C.E. 1976. Structural studies of frictional sediments. *Geografiska Annaler*, 58A, pp. 201-300.

Jopling, A.V. 1961. Origin of regressive ripples explained in terms of fluid mechanic processes. *U.S. Geol. Survey Prof. Paper* 424-D, 408 p.

Jopling, A.V. 1965. Hydraulic factors controlling the shape of laminae in laboratory deltas. *Journal Sedimentary Petrology*, vol. 35, pp. 777-791.

Jopling, A.V. 1967. Origin of laminae deposited by movements of ripples along a streambed: A laboratory study. *Journal Geology*, vol. 75, pp. 287-305.

Jopling, A.V. 1980. Palaeohydraulics of a tabular, cross-stratified sand in the Brampton esker, Ontario. *Sedimentary Geology*, vol. 25, pp. 169-188.

Jopling, A.V. and Walker, R.G. 1968. Morphology and origin of ripple drift cross-lamination, with examples from the Pleistocene of Massachusetts. *Journal Sedimentary Petrology*, vol. 38, pp. 971-984.

Kay, Marshall. 1951. North American Geosynclines. *Geological Society America Memoir*, vol. 48, 143 pp.

Kendall, A.C. 1984. Evaporites. *In* R.G. Walker (ed.), *Facies Models*. 2nd ed., pp. 259-296. Geol. Assocn. of Canada, Toronto, Ont., 317 pp.

King, B.C. 1970. Vulcanicity and rift tectonics in East Africa. *In* T.N. Clifford and I.G. Glass (eds.), *African Magmatism and Tectonics*, pp. 263–283. Oliver & Boyd, Edinburg.

Kolb, C.R. and Van Lopik, J.R. 1966. Depositional environments of the Mississippi River delta plain, southeastern Louisiana. *In* M.L. Shirley and J.A. Regsdale (eds.), *Deltas in Their Geological Framework*, pp. 19–61. Houston Geological Society, Houston, Texas, 251 pp.

Komar, P.D. 1986. 'Breaks' in grain-size distributions and application of suspension criterion to turbidites—Reply to a discussion by J.V. Tassell. *Sedimentology*, vol. 33, pp. 437–440.

Kraig, D.E., Lawrence, M.B., Moore, G.F. and Curray, J.R. 1980. Structural framework of the fore-arc basin, NW Sumatra. *Journal Geological Society London*, vol. 137, pp. 77–91.

Krinsley, D.H. and Donahue, J. 1968. Environmental interpretation of sand grain-size textures by electron microscopy. *Geological Society America Bulletin*, vol. 79, pp. 743–748.

Krinsley, D.H. and Doornkamp, J.C. 1973. *Atlas of Quartz Sand Grain Surface Textures*. Cambridge Earth Science Series, Cambridge Univ. Press, New York, 91 pp.

Krishnan, M.S. 1964. Iron-ores in India. Intl. Geological Congress, XXII Session, New Delhi. 21 pp.

Krumbein, W.C. 1934. Size frequency distribution of sediments. *Journal Sedimentary Petrology*, vol. 4, pp, 65–77.

Krumbein, W.C. and Pettijohn, F.J. 1938. *Manual of Sedimentary Petrography*. Appleton-Century-Crofts, New York, 549 pp.

Krumbein, W.C. and Sloss, L.L. 1951. *Stratigraphy and Sedimentation*. W.H. Freeman & Co., San Francisco, 497 pp; 2nd ed. (1963), 660 pp.

Krumbein, W.C. and Garrels, R.M. 1952. Origin and classification of chemical sediments in terms of pH and oxidation-reduction potentials. *Journal Geology*, vol. 60, pp. 1–33.

Krynine, P.D. 1948. The megascopic study and field classification of sedimentary rocks. *Journal Geology*, vol. 56, pp. 130–165.

Kuenen, Ph.H. 1950. *Marine Geology*. John Wiley, New York, 568 pp.

Kuenen, Ph.H. 1953. Significant features of graded bedding. *American Assoc. Petrol. Geol. Bulletin*, vol. 37, pp. 1044–66.

Kuenen, Ph.H. 1958. Experiments in geology. *Transactions Geological Society Glasgow*, vol. 23, pp. 1-28.

Kuenen, Ph.H. 1960a. Experimental abrasion of sand grains. *21st Session. International Geological Congress, Norden*, pt. 10, pp. 50–53.

Kuenen, Ph.H. 1960b. Experimental abrasion. 4. Eolian action. *Journal Geology*, vol. 68, pp. 427–449.

Kuenen, Ph.H. 1964. Experimental abrasion. 6. Surf action. *Sedimentology*, vol. 3, pp. 29–43.

Kuenen, Ph.H. 1966. Matrix of turbidites: experimental approach. *Sedimentology*, vol. 7, pp. 267-297.

Kuenen, Ph.H. and Migliorini, C.I. 1950. Turbidity currents as a cause of graded bedding. *Journal Geology*, vol. 58, pp. 91-127.

Kuenen, Ph.H. and Sengupta, S. 1970. Experimental marine suspension currents, competency and capacity. *Geologie en Mijnbouw*, vol. 49(2), pp. 89–118.

Kukal, Z. 1970. *Geology of Recent Sediments*. Czech. Academy of Science, Prague, 490 pp.

Kumar, R.K. and Nanda, A.C. 1989. Sedimentology of the Middle Siwalik sub-group of Mohand area, Dehra Dun Valley, India. *Jour. Geol. Soc. India*, vol. 34, pp. 597–616.

Kumar, N. and Sanders, J.E. 1974. Inlet sequence: a vertical succession of sedimentary structures and textures created by the lateral migration of tidal inlets. *Sedimentology*, vol. 21, pp. 491-532.

Lambert, A. and Hsü, K.J. 1979. Non-annual cycles of varve-like sedimentation in Walensee, Switzerland. *Sedimentology*, vol. 26, pp. 453–461.

Lane, E.W. et al. 1947. Report of the Subcommittee on Sediment Terminology. *American Geophysical Union Transactions*, vol. 28, pp. 936–938.

La Pichon, X., Francheteau, J. and Bonnin, J. 1973. *Plate Tectonics*. Elsevier, Amsterdam, 300 pp.

Larsonneur, C. 1975. Tidal deposits, Mont Saint-Michel Bay, France. *In* R.N. Ginsburg (ed.), *Tidal Deposits*, pp. 21-30, Springer, New York.

Laursen, E.M. 1958. The total sediment load of streams. *Journal Hydraulics Division*. American Society of Civil Engineers (ASCE), vol. 54, (HY 1). Proceedings. Paper 1530, pp. 1-36.

Lee, Y.I. and Klein, G.deV. 1986. Diagenesis of sandstones in the bac-arc basins of the Western Pacific Ocean. *Sedimentology*, vol. 33, pp. 651-675.

Leopold, L.B., Wolman, M.G. and Miller, J.P. 1964. *Fluvial Processes in Geomorphology*. Freeman, San Francisco, 522 pp.

Levorsen, A.I. 1958. *Geology of Petroleum*. W.H. Freeman & Co., San Francisco. 703 pp.

Lewis, K.B. 1980. Quaternary sedimentation on the Hikurangi oblique-subduction transform margin, New Zealand. *In* P.F. Ballance and R.G. Reading (eds.), *Sedimentation in Oblique-Slip Mobile Zones*, pp. 171-189. International Assoc. Sedimentologists, Spl. Pub. No. 4.

Ljunggren, Pontus and Sundborg, Åke. 1968. Some aspects of fluvial sediments and fluvial morphology. II. A study of some heavy mineral deposits in the valley of the river Alv. *Geografiska Annaler*, vol. 50, pp. 121-135.

Longman, M.W. 1981. Carbonate diagenesis as a control on stratigraphic traps (with examples from Williston basin). *A.A.P.G. Education Course Note Series* no. 21, 159 pp.

Lowe, D.R. 1975. Water escape structures in coarse-grained sediments. *Sedimentology*, vol. 22, pp. 157-204.

Mader, Detlef. 1985. Aspects of Fluvial Sedimentation in the Lower Triassic Buntsandstein of Europe. *Lecture Notes in Earth Sciences*, vol. 4. Springer, Berlin.

Mahoney, J.J., MacDougall, J.D., Lugmair, G.W. and Copalan, K. 1983. Kerguelen hotspot source for Rajmathal Traps and Ninetyeast Ridge? *Nature*, vol. 303, pp. 385-389.

Mather, K.F. and Mason, S.L. 1939. *A Source Book in Geology 1400-1900*. Harvard University Press. Cambridge, Mass., 2nd printing (1967), 702 pp.

McBridge, E.F. 1974. Significance of colour in red, green, purple, olive, brown and gray beds of DiFunta Group, North-eastern Mexico. *Journal Sedimentary Petrology*, vol. 44, pp. 760-773.

McCave I.N. and Syvitski, J.P.M. 1991. Principles and methods of geological particle size analysis. *In* J.P.M. Syvitski (ed.) *Principles, Methods, and Application of Particle Size Analysis*, pp. 3-21, Cambridge University Press, Cambridge, 368 pp.

McIntyre, D.B. 1963. James Hutton and the philosophy of geology, pp. 1-23. *In* C.C. Albritton (ed.) *The Fabric of Geology*. Addison-Wesley Publishing Co. Reading. 372 pp.

McKee, E.D. 1945. Cambrian history of the Grand Canyon region. Part 1. *Stratigraphy and Ecology of the Grand Canyon Cambrian*. Carnegie Institute Washington Publication 563, pp. 3-168.

McKee, E.D. 1949. Facies changes in the Colorado Plateau. *Geological Society America Memoir*, vol. 39, pp. 35-48.

McKee, E.D. 1954. Stratigraphy and history of the Moenkopi Formation of Triassic age. *Geological Society America Memoir*, vol. 61, 133 pp.

McKee, E.D. 1979. Introduction to a study of global sand seas (pp. 1-19), and sedimentary structures in dunes (pp. 83-134). *In* E.D. McKee (ed.), *A Study of Global Sand Seas*. U.S. Geological Survey Professional Paper 1052, pp. 83-134.

McKee, E.D. and Weir, G.W. 1953. Terminology for stratification and cross-stratification in sedimentary rocks. *Geological Society America Bulletin*, vol. 64, pp. 381-390.

McKee, E.D. and Goldberg, M. 1969. Experiments on formation of contorted structures in mud. *Geological Society America Bulletin*, vol. 80, pp. 231-244.

McKenzie, D.P. 1978. Some remarks on the development of sedimentary basins. *Earth and Planetary Science Letters*, vol. 40, pp. 25-32.

McKenzie, D.P. and Parker, R.L. 1967. The North Pacific: an example of tectonics on a sphere. *Nature*, vol. 216, pp. 1276–1280.

Mehrotra, M.N., Srivastava, R.A.K. and Sinhamahaptra, P.K. 1975. Physico-chemical behaviour of the carbonate rocks of western Son Valley region, India. *Journal Thermal Analysis*, vol. 7, pp. 667–674.

Miall, A.D. 1973. Markov chain analysis applied to ancient alluvial plain succession *Sedimentology*, vol. 20, pp. 347–364.

Miall, A.D. 1977. A review of the braided river depositional environment. *Earth Science Reviews*, vol. 13, pp. 1–62.

Miall, A.D. (ed.) 1978. *Fluvial Sedimentology*. Canadian Society of Petroleum Geologists. Calgary, 859 pp.

Miall, A.D. 1978a. Lithofacies types and vertical profile models in braided river deposits: a summary. *In* A.D. Miall (ed.) *Fluvial Sedimentology*, pp. 597–604. Canadian Society of Petroleum Geologists. Calgary, 859 pp.

Miall, A.D. 1984. *Principles of Sedimentary Basin Analysis*. 2nd ed. (1989) Springer, New York, 490 pp.

Miall, A.D. 1984a. Deltas. *In* R.G. Walker (ed.), *Facies Models*, 2nd ed., pp. 105–118. The Geological Society of Canada, Toronto, Ont., 317 pp.

Middleton, G.V. 1973. Johannes Walther's Law of Correlation of Facies. *Geological Society America Bulletin*, vol. 84, pp. 979–988.

Middleton, G.V. 1978. Sedimentology—History. *In* F.W. Fairbridge and Joanne Bourgeois (eds.), *The Encyclopedia of Sedimentology*, pp. 707-712. Dowden, Hutchinson & Ross. Stroudsburg, 901 pp.

Middleton, G.V. 1978a. Facies. *In* F.W. Fairbridge and Joanne Bourgeois (eds.), *The Encyclopedia of Sedimentology*, pp. 323–325. Dowden, Hutchinson & Ross, Stroudsburg, 901 pp.

Middleton, G.V. and Hampton, M.A. 1976. Subaqueous sediment transport and deposition by sediment gravity flows. *In* D.J. Stanley and D.J.P. Swift (eds.), *Sediment Transport and Environmental Management*, pp. 197–218. John Wiley, New York.

Miller, M.C., McCave, I.N. and Komar, P.D. 1977. Threshold of sediment motion under unidirectional currents. *Sedimentology*, vol. 24, pp. 507–527.

Milner, H.B. 1922. *An Introduction to Sedimentary Petrography*. Murby, London, 125 pp.

Milner, H.B. 1962. *Sedimentary Petrography* (in 2 vols.). Macmillan, New York, 643 pp. and 715 pp.

Mitchum, R.M. 1977. Glossary of terms used in seismic stratigraphy. *In* C.E. Payton (ed.). *Seismic Stratigraphy*, pp. 205–212. *AAPG Mem.* 26, 516 pp.

Mitchum, R.M. Jr., Vail, P.R. and Thompson, III, S. 1977. Seismic stratigraphy and global changes in sea level Part 2: The depositional sequence as a basic unit for stratigraphic analysis. *In* C.E. Payton (ed.), *Seismic Stratigraphy*, pp. 53–62, *AAPG. Mem.* 26, 516 pp.

Moiola, R.J. and Weiser, D. 1968. Textural parameters: an evaluation. *Journal Sedimentary Petrology*, vol. 38, pp. 45–53.

Moore, D.G., Curray, J.R., Raitt, R.W. and Emmel, F.J. 1974. Stratigraphic seismic section correlations and implications to Bengal Fan history. *In* C.G. Von der Borch and J.G. Sclater (eds.), Initial Reports of the Deep Sea Drilling Project, vol. 22, U.S. Govt. Printing Office.

Morrow, D.W. 1990. Dolomite. Part 2—Dolomitization models and ancient dolostones. *In* I.A. McIlreath and D.W. Morrow (*eds*)., *Diagenesis*, pp. 125–139, Geoscience Canada Reprint Ser. 4.

Mukherjee, K.K., Das, S. and Chakrabarti, A. 1987. Common physical sedimentary structures in a beach-related open-sea siliciclastic tropical tidal flat at Chandipur, Orissa, India, and evaluation of the weather conditions through discriminant analysis. *Senckenbergiana Marit.*, vol. 19, pp. 261–293.

Mukhopadhyay, A. and Chanda, S.K. 1972. Silica diagenesis in the banded hematite jasper and bedded chert associated with the Iron Ore Group of Jamad-Koira Valley, Orissa, India, *Sedimentary Geology*, vol. 8, pp. 113–135.

Mutti, E. and Ricci-Lucchi, F. 1975. Examples of turbidite facies and facies associations from selected formations of the northern Apennines. 9th International Sedimentological Congress, Nice, France, Guidebook to Field Trip A-11, 120 pp.

Naha, K. 1961. Precambrian sedimentation around Ghatsila in east Singhbhum, eastern India. *Proceedings National Institute Sciences India*, vol. 27, pp. 361–372.

Naidu, A.S. 1966. Lithological and chemical facies changes in the Recent deltaic sediments of the Godavari River, India. *In* M.L. Shirley and J.A. Ragsdale (eds.), *Deltas in Their Geological Framework*, pp. 126–157. Houston Geological Society, Houston, Texas, 251 pp.

Narain, Hari and Kaila, K.L. 1982. Inferences about the Vindhyan basin from geophysical data. *In* K.S. Valdiya et al. (eds.), *Geology of Vindyanchal*, pp. 179–192, Hindusthan Publishing Corporation (India), Delhi, 231 pp.

Nelson, C.M. 1985. Facies in stratigraphy: from 'terrains' to 'terranes'. *Journal Geological Education*, vol. 33, pp. 175–187.

Nemec, W. 1988. The shape of the rose. *Sedimentary Geology*, vol. 59, pp. 149–152.

Newton, R.C. 1988. Nature and origin of fluids in granulite facies metamorphism. *Journal Geological Society India*, vol. 31, pp. 103–105.

Newton, R.S. 1968. Internal structure of wave-formed ripple marks in the nearshore zone. *Sedimentology*, vol. 11, pp. 275–292.

Normark, W.R. 1970. Growth patterns of deep-sea fans. *American Association Petroleum Geologists Bulletin*, vol. 54, pp. 2170-2195.

Normark, W.R. 1978. Fan valleys, channels and depositional lobes on modern submarine fans: characters for recognition of sandy turbidite environments. *American Association Petroleum Geologists Bulletin*, vol. 62, pp. 912-931.

North American Commission on Stratigraphic Nomenclature (NACSN). 1983. North American Stratigraphic Code. *American Association Petroleum Geologists Bulletin*, vol. 67, pp. 841–875.

Novak, I.D. 1973. Predicting coarse sediment transport: the Hjulström curve revisited. *In* M. Morisawa (ed.), *Fluvial Geomorphology*, pp. 13–25. Allen and Unwin, London. 164 pp.

Okada, H. 1966. Non-Greywacke turbidite sandstones in the Welsh geosyncline. *Sedimentology*, vol. 7, pp. 211-232.

Okada, H. 1971. Classification of sandstone: analysis and proposal. *Journal Geology*, vol. 79, pp. 509-525.

Parkash, B., Awasthi, A.K. and Gohain, K. 1983. Lithofacies of the Markanda terminal fan, Kurukshetra District, Haryana, India. International Association of Sedimentologists, Spl. Pub. vol. 6. pp. 337–344.

Passega, R. 1957. Texture as a characteristic of clastic deposition. *American Association Petroleum Geologists Bulletin*, vol. 41, pp. 1952-1984.

Passega, R. 1964. Grain-size representation by CM patterns as a geological tool. *Journal Sedimentary Petrology*, vol. 34, pp. 830–847.

Paul, D.D. and Lian, H.M. 1975. Offshore Tertiary basins of south-east Asia, Bay of Bengal to South China Sea. *Proceedings IX World Petroleum Congress*, vol. 3, pp. 107-121.

Petters, S.W. 1978. Stratigraphic evolution of the Benue Trough and its implications for the Upper Cretaceous Paleogeography of West Africa. *Journal Geology*, vol. 86, pp. 311–322.

Pettijohn, F.J. 1949. *Sedimentary Rocks*. Harper & Row, New York, 526 pp.; 2nd ed. (1957), 718 pp.; 3rd ed. (1975), 628 pp.

Pettijohn, F.J. 1962. Paleocurrent and paleogeography. *American Association Petroleum Geologists Bulletin*, vol. 46, pp. 1468–1493.

Pettijohn, F.J., Potter, P.E. and Siever, R. 1973. *Sand and Sandstone.* Springer-Verlag, New York, 618 pp.; 2nd ed. (1987), 553 pp.

Picard, M.D. and High, L.R. Jr. 1972. Criteria for recognizing lacustrine rocks. *In* J.K. Rigby and W.K. Hamblin (eds.), *Recognition of Ancient Sedimentary Environment,* pp. 108–145. Society of Economic Paleontologists and Mineralogists, Spl. Pub. no. 16, 340 pp.

Pitman III, W.C. 1978. Relationship between eustacy and stratigraphic sequences of passive margins. *Geological Society America Bulletin,* vol. 89, pp. 1389–1403.

Posamentier, H.W., Jervey, M.T. and Vail, P.R. 1988. Eustatic controls on clastic deposition. *In* C.K. Wilgans, B.S. Hastings and G.G.St.C. Kendall (eds.), *Sea Level Research—an Integrated Approach,* pp. 109–154. Society of Economic Paleontologists and Mineralogists, Spl. Pub. no. 42.

Potter, P.E. 1978. Significance and origin of big rivers. *Journal Geology,* vol. 86, pp. 13–33.

Potter, P.E. 1978a. Petrology and chemistry of modern big river sands. *Journal Geology,* vol. 86, pp. 423–449.

Potter, P.E. and Pettijohn, F.J. 1963. *Paleocurrent and Basin Analysis.* Springer-Verlag, New York, 296 pp., 2nd ed. (1977), 425 pp.

Potter, P.E., Maynard, J.B. and Pryor, W.A. 1980. *Sedimentology of Shale.* Springer-Verlag, New York, 306 pp.

Powers, M.C. 1953. A new roundness scale for sedimentary particles. *Journal Sedimentary Petrology,* vol. 23, pp. 117-119.

Rai, K.L., Sarkar, S.N. and Paul, P.R. 1980. Primary depositional and diagenetic features in the Banded Iron Formation and associated iron-ore deposits of Noamundi, Singhbhum district, Bihar, India. *Mineral. Deposita* (Berl.), vol. 15, pp. 189-200.

Rao, J.S. and Sengupta, Supriya, 1970. An optimum hierarchical sampling procedure for cross-bedding data. *Journal Geology,* vol. 78, pp. 533–544.

Rao, J.S. and Sengupta, Supriya, 1972. Mathematical techniques for palaeocurrent analysis: treatment of directional data. *Mathematical Geology,* vol. 4, pp. 235–248.

Rao, M.B. Ramchandra and Sengupta, S.N. 1964. Hidden ridges in the Indo-Gangetic Plains. *Advancing Frontiers in Geology & Geophysics* (Krishnan Volume). Indian Geophysical Union, Hyderabad, pp. 147–158.

Reading, H.G. (ed.). 1978. *Sedimentary Environments and Facies.* Elsevier, New York, 569 pp.; 2nd ed. (1986), 615 pp.

Reading, H.G. 1982. Sedimentary basins and global tectonics. *Proceedings Geological Association,* 93 (4), pp. 321–350.

Reineck, H.E. and Singh, I.B. 1973. *Depositional Sedimentary Environments.* Springer-Verlag, New York, 439 pp. 2nd ed. (1980), 549 pp.

Reineck, H.E. and Wunderlich, F. 1968. Classification and origin of flaser and lenticular bedding. *Sedimentology,* vol. 11, pp. 99–104.

Reinson, G.E. 1984. Barrier island and associated strand-plain systems. *In* R.G. Walker (ed.), *Facies Models,* 2nd ed., pp. 119-140. The Geological Society of Canada, Toronto, Ont., 317 pp.

Reyer, T.A. 1983. Transgressive-regressive cycles and the occurrence of coal in some Upper Cretaceous strata of Utah. *Geology,* vol. 11, pp. 207-210.

Ricci-Lucchi, F. 1975. Depositional cycles in two turbidite formations of Northern Apennines (Italy), *Journal Sedimentary Petrology,* vol. 45, pp. 3–43.

Rodolfo, K.S. 1969. Bathymmetry and marine geology of the Andaman Basins and tectonic implications for southeast Asia. *Geological Society America Bulletin,* vol. 80, pp. 1203–30.

Rouse, Hunter, 1938. *Fluid Mechanics for Hydraulic Engineers.* McGraw Hill Book Co. (Dover edition printed in 1961). Dover Publications, Inc., New York, 422 pp.

Roybarman, A. 1983. Geology and hydrocarbon prospects of West Bengal. *In* L.L. Bhandari et al. (eds.), Petroliferous Basins of India, Part I, pp. 51-56, *Petroleum Asia Journal,* V-VI (4), 189 pp.

Roychoudhury, S.C. and Deshpande, S.V. 1982. Regional distribution of carbonate facies, Bombay offshore region, India. *American Association Petroleum Geologists Bulletin*, vol. 66, pp. 1483–1496.

Rubin, D.M. and McCulloch, D.S. 1980. Single and superimposed bedforms: a synthesis of San Francisco Bay and flume observations. *Sedimentary Geology*, vol. 26, pp. 207–231.

Rust, B.R. 1978. A classification of alluvial channel system. *In* A.D. Miall (ed.), *Fluvial Sedimentology*, pp. 187–198. *Canadian Society of Petroleum Geologists Memoir 5*, Calgary, Alberta, 859 pp.

Rust, B.R. 1978a. Depositional models for braided alluvium. *In* A.D. Miall (ed.) Fluvial Sedimentology, pp. 605–625. Canadian Society of Petroleum Geologists. Calgary, 859 pp.

Rust, B.R. and Koster, E.M. 1984. Coarse alluvial deposits. *In* R.G. Walker (ed.), *Facies Models*, 2nd. ed. pp. 53–70. Geological Association of Canada, Toronto, Ont., 317 pp.

Saha, Dilip and Chakraborty, Tapan. 1990. TILTVEC: A Fortran-77 program for the tilt correction of paleocurrent data with resolution of incongruities. *Computers and Geosciences*, 16, 8, pp. 1193–1207.

Saha, Dilip and Ghosh, Gautam, 1987. Tectonic setting of Proterozoic sediments around Somanpalli, Godavari Valley. *Journal Indian Association Sedimentologists*, vol. 7, pp. 29–45.

Sahu, B.K. 1964. Transformation of weight frequency and number frequency data in size distribution studies of clastic sediments. *Journal Sedimentary Petrology*, vol. 34, pp. 768–773.

Sahu, B.K. 1964a. Depositional mechanisms from the size analysis of clastic sediments. *Journal Sedimentary Petrology*, vol. 34, pp. 73–83.

Sarkar, Bhagabati, 1984. Microfossils from the Banded Iron Formation from the Noamundi basin, Eastern India. *Quarterly Journal Geological, Mining & Metallurgical Society India*, vol. 56, pp. 41–46.

Scheidegger, A.E. and Potter, P.E. 1968. Textural studies of graded bedding. *Sedimentology*, vol. 11, pp. 163–170.

Schenck, H.G. and Muller, S.W. 1941. Stratigraphic terminology. *Geological Society America Bulletin*, vol. 52, pp. 1419–1426.

Schlumberger Ltd. 1970. *Fundamentals of Dipmeter Interpretation*. Schulmberger, Houston, Texas, 145 pp.

Schreiber, B.C., Friedman, G.M., Decima, A. and Schreiber, E. 1976. Depositional environments of Upper Miocene (Messinian) evaporite deposits of the Sicilian basin. *Sedimentology*, vol. 23, pp. 729–760.

Schumm, S.A. 1972. Fluvial Paleochannels. *In* J.K. Rigby and W.K. Hamblin (eds.), *Recognition of Ancient Sedimentary Environments*, pp. 98–107. Society of Economic Paleontologists and Mineralogists, Spl. Pub. no. 16, 340 pp.

Schumm, S.A. 1973. Geomorphic thresholds and complex response of drainage systems. *In* M. Morisawa (ed.), *Fluvial Geomorphology*, State University of New York, Binghampton, Publications in Geomorphology, vol. 4, pp. 299–310.

Schumm, S.A. 1977. *The Fluvial System*. Wiley Interscience, New York, 335 pp.

Selley, R.C. 1968. A classification of paleocurrent models. *Journal Geology*, vol. 76, pp. 99–110.

Selley, R.C. 1976. *An Introduction to Sedimentology*. Academic Press London, 408 pp.

Sellwood, B.W. 1986. Shallow-marine carbonate environments. *In* R.G. Reading (ed.), *Sedimentary Environments and Facies*, 2nd ed. pp. 283–342. Blackwell Scientific Publications, Oxford, 615 pp.

Srivastava, B.N., Rana, M.S. and Verma, N.K. 1983. Geology and hydrocarbon prospects of the Vindhyan basin. *In* L.L. Bhandari et al. (eds.) Petroliferous Basins of India, vol. I. pp. 179–189. *Petroleum Asia Journal*, V–VI, 189 pp.

Sen, D.P. 1991. Sedimentation patterns of the Talchir Group in the Giridih Gondwana basin, India: A case of multiple glacial advance and retreat. *Palaeogeography, Palaeoclimatology, Palaeoecology*, vol. 86, pp. 339–352.

Sen, D.P., Bhattacharyya, S.C. and Ray, S.K. 1969. Turbidite deposits around the Simla hills. A study from the Simla series, Himachal Pradesh (India). *Sedimentary Geology*, vol. 3, pp. 317–329.

Sengupta, Supriya, 1957. Petrology of para-lavas of eastern part of Jharia coalfield. *Quarterly Journal Geological, Mining & Metallurgical Society India*, vol. 29, pp. 79–101.

Sengupta, Supriya. 1966. Studies on orientation and imbrication of pebbles with respect to cross-stratification. *Journal Sedimentary Petrology*, vol. 36, pp. 363–369.

Sengupta, Supriya. 1966a. Geological and geophysical studies in western part of Bengal basin, India. *American Association Petroleum Geologists Bulletin*, vol. 50, pp. 1001–1017.

Sengupta, Supriya. 1970. Gondwana sedimentation around Bheemaram (Bhimaram), Pranhita-Godavari Valley, India. *Journal Sedimentary Petrology*, vol. 40, pp. 140–170.

Sengupta, Supriya, 1974. Fluviatile cross-stratifications—ancient, recent and experimental. Q.J. Geological, Mining & Metallurgical Society of India, Golden Jubilee Volume (vol. 46), pp. 87–99.

Sengupta, Supriya, 1975. Size-sorting during suspension transportation—log normality and other characteristics. *Sedimentology*, vol. 22, pp. 257–273.

Sengupta, Supriya, 1979. Grain-size distribution of suspended load in relation to bed materials and flow velocity. *Sedimentology*, vol. 26, pp. 63–82.

Sengupta, S., Bose, D., Siva Prasad, K. and Das, S.S. 1988. Karharbari and Barakar Sedimentation in the Giridih basin. *Indian Journal Geology*, vol. 60, pp. 35–56.

Sengupta, Supriya, Ghosh, J.K. and Mazumder, B.S. 1991. Experimental-theoretical approach to interpretation of grain-size frequency distributions. *In* J.P.M. Syvitski (ed.), *Principles, Methods and Application of Particle Size Analysis*, pp. 264–280. Cambridge University Press, Cambridge, 368 pp.

Sengupta, Supriya and Rao, J.S. 1966. Statistical analysis of cross-bedding azimuths from the Kamthi Formation around Bheemaram, Pranhita-Godavari Valley, Sankhyā, *Indian Journal Statistics*, vol. 28-B, pp. 165–174.

Sengupta, Supriya and Veenstra, H.J. 1968. On sieving and settling techniques for sand analysis. *Sedimentology*, vol. 11, pp. 83–98.

Shannon, P.M. and Naylor, D. 1989. *Petroleum Basin Studies*. Graham & Trotman, London, 206 pp.

Shea, J.H. 1982. Uniformitarianism and sedimentology. *Journal Sedimentary Petrology*, vol. 52, pp. 701-702 and Reply to discussions in v. 53, (1983), p. 680.

Shen, H.W. 1978. Sediment transport models. *In* A.D. Miall (ed.), *Fluvial Sedimentology*, pp. 49–60. Canadian Society of Petroleum Geologists, Calgary, 859 pp.

Sheriff, R.E. 1982. *Seismic Stratigraphy* (EBD Indian Student Reprint). The English Book Depot, Dehra Dun, 227 pp.

Shrock, R.R. 1948. *Sequence in Layered Rocks*. McGraw-Hill, New York, 507 pp.

Siever, Raymond. 1983. The dynamic earth. *Scientific American*, vol. 249, pp. 30–39.

Sindowski, F.K.H. 1949. Results and problems of heavy-mineral analysis in Germany; a review of sedimentary-petrological papers, 1936–1948. *Journal Sedimentary Petrology*, vol. 19, pp. 3–25.

Simons, D.B., Richardson, E.V. and Nordin, C.F. Jr. 1965. Sedimentary structures generated by flow in alluvial channels. *In* G.V. Middleton (ed.), *Primary Sedimentary Structures and Their Hydrodynamic Interpretations*, pp. 34–52. Society of Economic Paleontologists and Mineralogists, Tulsa, OK., Spl. Pub. no. 12, 265 pp.

Singh, I.B. 1980. The Bijaigarh Shale, Vindhyan System (Precambrian), India—an example of a lagoonal deposit. *Sedimentary Geology*, vol. 25, pp. 83–103.

Singh, I.B. 1980a. Precambrian sedimentary sequences of India: their peculiarities and comparison with modern sediments. *Precambrian Research*, vol. 12, pp. 411–436.

Sinha, R.C. and Singh, R.M. 1964. Distribution and behaviour of trace elements during the diagenesis of Vindhyan sediments of the Son Valley near Robertsganj, U.P., India. *Report of the XXII International Geological Congress, New Delhi*, Part XV, pp. 205–225.

Sippel, R.F. 1968. Sandstone petrology: evidence from luminescence petrography. *Journal Sedimentary Petrology*, vol. 38, pp. 530–554.

Slingerland, R.L. and Williams, E.G. 1979. Paleocurrent analysis in light of trough cross-stratification geometry. *Journal Geology*, vol. 87, pp. 724–732.

Sloss, L.L. 1988. Forty years of sequence stratigraphy. *Geological Society America Bulletin*, vol. 100, pp. 1661-1665.

Smith, A.J. 1963. Evidence for a Talchir (Lower Gondwana) glaciation: striated pavement and boulder bed at Irai, Central India. *Journal Sedimentary Petrology*, vol. 33, pp. 739–750.

Smith, D.G. and Smith, N.D. 1980. Sedimentology in anastomosing river systems: examples from alluvial valleys near Banff, Alberta, *Journal Sedimentary Petrology*, vol. 50, pp. 157–164.

Sneed, E.D. and Folk, R.L. 1958. Pebbles in the lower Colorado River, Texas—a study in particle morphogenesis. *Journal Geology*, vol. 66, pp. 114–150.

Southard, J.B. and Boguchwal, L.A. 1990. Bed configuration in steady unidirectional water flows. Part 2. Synthesis of flume data. *Journal Sedimentary Petrology*, vol. 60, pp. 658–679.

Srinivasan, R., Naqvi, S.M., Uday Raj, B., Subba Rao, D.V. and Balaram, V. 1989. Geochemistry of the Archaean grawywackes from the northwestern part of the Chitradurga Schist Belt, Dharwar Craton, South India—evidence for granitoid upper crust in the Archaean. *Journal Geological Society India*, vol. 34, pp. 505–516.

Srinivasan, R. and Ojakangas, R.W. 1986. Sedimentology of quartz-pebble conglomerates and quartzites of the Archean Bababudan Group, Dharwar Craton, South India: evidence for early crustal stability. *Journal Geology*, vol. 94, pp. 199–214.

Srivastava, B.N., Rana, M.S. and Verma, N.K. 1983. Geology and hydrocarbon prospects of the Vindhyan basin. *In* L.L. Bhandari et al. (eds.) Petroliferous Basins of India, vol. I, pp. 179–189. *Petroleum Asia Journal*, V-VI, 189 pp.

Stanton, R.L. 1972. *Ore Petrology*. McGraw-Hill Book Co., New York, 713 pp.

Sundborg, Åke. 1956. The River Klarälven. A Study of Fluvial Process. *Geografiska Annaler*, 38, 2: 127–316.

Suttner, L.J. and Dutta, P.K. 1986. Alluvial sandstone composition and paleoclimate, I. Framework mineralogy. *Journal Sedimentary Petrology*, vol. 56, pp. 329–345.

Suttner, L.J., Basu, Abhijit and Mack, G.H. 1981. Climate and the origin of quartz arenites. *Journal Sedimentary Petrology*, vol. 51, pp. 1235–1246.

Swift, D.J.P., Figueiredo, A.G. (Jr.), Freeland, G.L. and Oertel, G.F. 1983. Hummocky, cross-stratification and megaripples: a geological double standard? *Journal Sedimentary Petrology*, vol. 53, pp. 1295–1317.

Syvitski, J.P.M. (ed.) 1991. *Principles, Methods and Application of Particle Size Analysis*. Cambridge University Press, Cambridge, 368 pp.

Tada, Ryuji and Siever, Raymond. 1989. Pressure solution during diagenesis. *Annual Review Earth and Planetary Sciences*, vol. 17, pp. 89–118.

Tandon, S.K. 1976. Siwalik sedimentation in a part of Kumaun Himalalya, India. *Sedimentary Geology*, vol. 16, pp. 131–154.

Tandon, S.K. 1990. The Himalayan foreland: focus on Siwalik basin. *In* S.K. Tandon et al. (eds.), *Sedimentary Basins of India: Tectonic Context*, pp. 171-201, Gyanodaya Prakashan, Naini Tal.

Tandon, S.K., Kumar R. and Singh, P. 1985. Syntectonic controls of palaeoflow reversals and variability: sediment-vector sequences in the late orogenic fluvial Siwalik basin, Punjab sub-Himalaya, India, *Sedimentary Geology*, vol. 41, pp. 97-112.

Terry, R.D. and Chilingar, G.V. 1955. Summary of 'Concerning some additional aids in studying sedimentary formations' by M.S. Shvetsov (in Russian). *Journal Sedimentary Petrology*, vol. 25, pp. 229–234.

Thakur, V.C. and Virdi, N.S. 1979. Lithostratigraphy, structural framework, deformation and metamorphism of the SE region of Ladakh, Kashmir Himalaya, India. *Himalayan Geology*, vol. 9, pp. 63–78.

Thiel, G.A. 1945. Mechanical effects of stream transportation on mineral grains of sand size (Abs.), *Geological Society America Bulletin*, vol. 56, p. 1207.

Trevena, A.S. and Nash, W.P. 1981. An electron microprobe study of detrital feldspar. *Journal Sedimentary Petrology*, vol. 51, pp. 137-150.

Twenhofel, W.H. 1928. *Treatise on Sedimentation.* Williams and Wilkens, Baltimore, 2nd ed. (1932), 926 pp.

Twenhofel, W.H. 1950. *Principles of Sedimentation.* McGraw-Hill, New York, 673 pp.

Udden, J.A. 1914. The mechanical composition of clastic sediments. *Geological Society of America Bulletin*, vol. 25, pp. 655–744.

Vail, P.R., Mitchum, R.M. and Thompson, III S. 1977. Seismic stratigraphy and global changes of sea level. *In* C.E. Payton (ed.), *Seismic Stratigraphy—Applications to Hydrocarbon Exploration*, pp. 83–97. American Association of Petroleum Geologists, Memoir no. 26, Tulsa, OK. 516 pp.

Vail, P.R., Todd, R.G. and Sangree, J.B. 1977. Chronostratigraphic significance of seismic reflections. *In* C.E. Payton (ed.), *Seismic Stratigraphy—Applications to Hydrocarbon Exploration*, pp. 99–116. American Association of Petroleum Geologists, Memoir no. 26, Tulsa, OK. 516 pp.

Vail, P.R. 1987. Sesmic stratigraphy interpretation using sequence stratigraphy. Part I. Seismic stratigraphy interpretation. *In* A.W. Bally (ed.), *Atlas of Seismic Stratigraphy*, vol. I, *AAPG Studies in Geology*, vol. 27, pp. 1–10.

Valdiya, K.S. 1970. Simla Slates. The Precambrian flysch of the Lesser Himalaya, its turbidites, sedimentary structures and paleocurrents. *Geological Society America Bulletin*, vol. 81, pp. 451–468.

van Houten, F.B. 1974. Northern Alpine molasse and similar Cenozoic sequences of Southern Europe. *In* R.H. Dott and R.H. Shaver (eds.), *Modern and Ancient Geosynclinal Sedimentation*, pp. 260–273. Society of Economic Paleontologists and Mineralogists, Spl. Pub. no. 19, Tulsa, OK., 380 pp.

Vanoni, V.A., Brooks, N.A. and Kennedy, J.F. 1961. Lecture notes on sediment transportation and channel stability. Report No. KH-R-1, W.M. Keck Laboratory of Hydraulics and Water Resources, California Institute of Technology, Pasadena, California.

van Straaten, L.M.J.U. 1954. Composition and structure of Recent marine sediments in the Netherlands. *Leidse Geol. Medel.*, vol. 19, pp. 1–110.

van Straaten, L.M.J.U. 1954a. Sedimentology of Recent tidal flat deposits and the psammites du Condroz (Devonian). *Geologie Mijnbouw*, vol. 16, pp. 25–47.

Van Wagoner, J.C., Mitchum, R.M., Jr., Campion, K.M. and Rahmanian, V.D. 1990. Siliciclastic sequence stratigraphy in well logs, cones and concepts for high resolution correlation of time and facies. *AAPG Methods in Exploration Series* no. 7, 55 pp.

Vine, J.D. and Tourtelot, E.B. 1970. Geochemistry of black shale deposits—a summary report. *Economic Geology*, vol. 65, pp. 253–272.

Virdi, N.S. 1986. Indus-Tsangpo suture in the Himalaya: Crustal expression of a palaeo-subduction zone. *Annales Soc. Geol. Poloniae*, vol. 56, pp. 3–31.

Visher, G.S. 1969. Grain-size distributions and depositional processes. *Journal of Sedimentary Petrology*, vol. 39, pp. 1074–1106.

von Engelhardt, Wolf. 1977. *The Origin of Sediments and Sedimentary Rocks.* (trans. W.D. Johns). A Halsted Press Book. John Wiley & Sons, New York, 2nd rev. ed., 359 pp.

Wadell, H. 1932. Volume, shape and roundness of rock particles. *Journal Geology*, vol. 40, pp. 443–451.

Wadell, H. 1935. Volume, shape and roundness of quartz particles. *Journal Geology*, vol. 43, pp. 250–280.

Walcott, R.I. 1978. Geodetic strains and large earthquakes in the axial tectonic belt of North Island, New Zealand. *Journal Geophysical Research*, vol. 83, pp. 4419–4429.

Walker, R.G. (ed.) 1978. *Facies Models.* Geoscience Canada Reprint Series 1. 2nd ed. (1984) Geological Association of Canada, Toronto, Ont., 317 pp.

Walker, R.G. 1984a. General introduction: facies, facies sequences and facies models. In R.G. Walker (ed.) *Facies Models* (2nd ed.), pp.1-10. Geological Association of Canada, Toronto Ont., 317 pp.

Walker, R.G. 1984b. Shelf and shallow marine sands. *In* R.G. Walker (ed.) *Facies Models* (2nd ed.), pp. 141-170. Geological Association of Canada, Toronto, Ont., 317 pp.

Walker, R.G. 1984c. Turbidites and associated coarse clastic deposits. *In* R.G. Walker (ed.) *Facies Models* (2nd ed.), pp. 171-188. Geological Association of Canada, Toronto, Ont., 317 pp.

Walker, R.G. and Mutti, E. 1973. Turbidite facies and facies associations. *In* G.V. Middleton and A.H. Bouma (eds.), *Turbidites and Deep Water Sedimentation.* Pacific Section, SEPM, Short course (Anaheim, 1973), pp. 119–157.

Walker, R.G. and Cant, D.J. 1984. Sandy fluvial systems. *In* R.G. Walker (ed.), *Facies Models*, 2nd ed., pp. 71–89. Geological Association of Canada, Tontonto, Ont., 317 pp.

Wasson, R.J. and Hyde, R. 1983. Factors determining desert dune type. *Nature*, vol. 304, pp. 337–339.

Wilcox, R.E., Harding, T.P. and Seely, D.R. 1973. Basic wrench tectonics. *American Association Petroleum Geologists Bulletin*, vol. 57, pp. 74-96.

Watts, A.B., Karner, G.D. and Steckler, M.S. 1982. Lithospheric flexure and the evolution of sedimentary basins. *Philosophical Transactions Royal Society, London*, Ser. A. Vol. 305, pp. 249–281.

Weeks, L.G. 1952. Factors of sedimentary basin development that control oil occurrence. *American Association Petroleum Geologists Bulletin*, vol. 36, pp. 2071-2124.

Weimer, R.J. 1976. Deltaic and shallow marine sandstones: Sedimentation, tectonics, and petroleum occurrences. *AAPG Education Course Note Series* no. 2.

Wentworth, C.K. 1922. A scale of grade and class terms for clastic sediments. *Journal Geology*, vol. 30, pp. 377–392.

Wernicke, B.P. 1981. Low angle normal faults in the Basin and Range Province; nappe tectonics in an extending orogen, *Nature*, vol. 291, pp. 645–648.

Wilson, I.G. 1972. Aeolian bedforms—their development and origins. *Sedimentology*, vol. 19, pp. 193–210.

Winkelmolen, A.M. 1971. Rollability as a function of shape property of sand grains. *Journal Sedimentary Petrology*, vol. 41, pp. 703–714.

Winkelmolen, A.M. 1972. Shape sorting in Lower Oligocene, North Belgium. *Sedimentary Geology*, vol. 7, pp. 183–227.

Zenger, D.H. 1986. Lyell and Episodicity, *Journal Geological Education*, vol. 34, pp. 10–13.

Zingg, Th. 1935. Beitrage zur Schotteranalyse. *Min. Petrog. Mitt. Schweiz.*, vol. 15, pp. 39–140 (cited by Pettijohn, 1975).

SUBJECT INDEX

D

E